REMINGTON 1100 22 INCH BARREL

W9-DFZ-697

ALLOWABLE STRESS FOR BEAM 287

SEW SHAPES 287

10.00

EQUASIONS 4-32

# Statics and Strength of Materials

# Statics and Strength of Materials
## Third Edition

**MILTON G. BASSIN, B.M.E., M.M.E., P.E.**
York College

**STANLEY M. BRODSKY, B.M.E., M.M.E., Ph.D., P.E.**
N.Y.C. Community College

**HAROLD WOLKOFF, B.M.E., M.B.A., P.E.**
N.Y.C. Community College

Gregg Division
McGRAW-HILL BOOK COMPANY

New York · St. Louis · Dallas · San Francisco · Auckland · Bogotá · Düsseldorf ·
Johannesburg · London · Madrid · Mexico · Montreal · New Delhi · Panama · Paris ·
São Paulo · Singapore · Sydney · Tokyo · Toronto

**Library of Congress Cataloging in Publication Data**

Bassin, Milton G
   Statics and strength of materials.
   Includes index.
   1. Mechanics, Applied. 2. Strength of materials. I. Brodsky, Stanley M., joint
author. II. Wolkoff, Harold, joint author. III. Title.
TA350.B32   1979   620.1'12   78-11813
ISBN 0-07-004030-3

   4 5 6 7 8 9 0 DODO  8 6 5 4 3 2 1

The editors for this book were Cary Baker and Mary Levai, the designer was Eileen Thaxton,
the cover designer was Barbara Soll, the art supervisor was George T. Resch, and the pro-
duction supervisor was Regina R. Malone. It was set in Baskerville by Monotype Composition
Company.
Printed and bound by R. R. Donnelley and Sons Company.

# Contents

**STATICS**

# List of Tables

# Preface to the Third Edition

As the previous editions of *Statics and Strength of Materials* have done, this third edition focuses on courses in statics and strength of materials which do not require the use of calculus. The practical orientation of the text has been retained and emphasized to ensure comprehension of the underlying principles and their applications. The text can therefore also serve as a reference for courses in materials testing, materials selection, structural design, and machine design.

Although changes have been introduced in this revision, the flavor and general content of much of the text have been retained. One of the more significant additions is the introduction of SI metric units in illustrative examples as well as in student problems, with given metric data rather than conversions. However, U.S. customary units are still employed throughout the text. The addition of metrics will permit instructors to selectively include proper SI units in their present courses so that students may begin to learn to think in terms of these units in preparation for the anticipated increasing usage in United States practice. Also, the latest changes in pertinent codes have been incorporated so that the text keeps pace with design procedures which students are likely to encounter in industry. In addition to the foregoing, many of the comments and suggestions of those who have used the second edition have been incorporated.

As in the previous editions, the problems are graded for difficulty. Chapter 1, on fundamental terms, has been revised to include an introduction to metrics. Chapter 9 has been revised extensively in order to update the design of connections. In keeping with current industrial practice, the sections on rivets have been essentially eliminated and the treatment of bolted connections has been greatly expanded. Up-to-date codes pertaining to bolted and welded connections are employed. Chapter 12 has been revised so that beam design is based on the latest American Institute of Steel

Construction (AISC) code. In addition, tables and beam specifications have been revised to show the new designations for rolled sections. To reflect the trend in connections, the section on eccentrically loaded riveted joints has been replaced with a new section on eccentrically loaded bolted connections, in Chapter 14. Chapter 15 has been updated through the introduction of the latest code provisions for columns. The section on continuous beams by the three-moments theorem method has been eliminated from Chapter 16. Where necessary throughout the text, figures have been revised to show more clearly pinned joints and beam supports.

As before, the text consists of two main sections: Statics (chapters 1 through 6) and Strength of Materials (chapters 7 through 16). For those who have previously completed a course in statics, it is recommended that they be referred to the review of statics in Appendix A.

The most frequently used tables have been updated and are shown in Appendix B. A list of important formulas is again placed at the end of the appendix for convenient reference. The symbols used throughout the book conform to the latest recommendations of the American National Standards Institution (ANSI).

The authors invite suggestions, comments, and corrections by persons using the text.

We wish to acknowledge with thanks the many useful suggestions received from Professors G. J. Cavaliere; S. B. Foreman, P.E.; A. Peters, P.E.; and other members of both the construction technology and the mechanical technology departments of New York City Community College (N.Y.C.C.C.) of The City University of New York. In addition, we wish to offer our special thanks to Professors P. Granek, P.E., and S. P. Stern, P.E., of the construction technology department at N.Y.C.C.C.

MILTON G. BASSIN, P.E.
STANLEY M. BRODSKY, P.E.
HAROLD WOLKOFF, P.E.

# CHAPTER
# 1
# Fundamental Terms

## 1-1   INTRODUCTION TO MECHANICS

The importance of a thorough knowledge of fundamentals in any field cannot be overemphasized. Fundamentals have always been stressed in the learning of new skills, whether it be football or physics.

Similarly, the science of mechanics is founded on basic concepts and forms the groundwork for further study in the design and analysis of machines and structures.

Mechanics can be divided into two parts: (*a*) statics, which relates to bodies at rest, and (*b*) dynamics, which deals with bodies in motion. This text is concerned with statics.

In mechanics, the term *strength of materials* refers to the ability of the individual parts of a machine or structure to resist loads. It also permits the selection of materials and the determination of dimensions to ensure sufficient strength of the various parts.

## 1-2   BASIC TERMS

It is essential that the following 13 basic terms be understood, since they continually recur in all phases of this technical study.

*Length.*   This term is applied to the linear dimension of a straight or curved line. For example, the diameter of a circle is the length of a straight line which divides the circle into two equal parts; the circumference is the length of its curved perimeter.

*Area.*   The two-dimensional size of a shape or a surface is its area. The shape may be flat (lie in a plane) or curved; for example, the size of a plot of land, the surface of a fluorescent bulb, or the cross-sectional size of a shaft.

*Volume.*   The three-dimensional or cubic measure of the space occupied by a substance is known as its volume.

1

*Force.*   This term is applied to any action on a body which tends to make it move, change its motion, or change its size and shape. A force is usually thought of as a push or a pull, such as a hand pushing against a wall or the pull of a rope fastened to a body.

*Pressure.*   The external force per unit area, or the total force divided by the total area on which it acts, is known as pressure. Water pressure against the face of a dam, steam pressure in a boiler, and earth pressure against a retaining wall are some examples.

*Mass.*   The amount of matter in a body is called its mass, and for most problems in mechanics, mass may be considered constant.

*Weight.*   The force with which a body is attracted toward the center of the Earth by the gravitational pull is called its weight.

*Density.*   This term may refer either to weight density or mass density. Weight density is the weight per unit volume of a body or substance; mass density is the mass per unit volume of a body or substance.

*Load.*   This term is used to indicate that a body of some weight is applying a force against some supporting structure or part of a structure. For example, a desk weighing 100 lb is applying a load (and thus a force) of 100 lb against the floor on which it sits.

*Moment.*   The *tendency* of a force to cause rotation about an axis through some point is known as a moment.

*Torque.*   The action of a force which *causes* rotation to take place is known as torque. The action of a belt on a pulley causes the pulley to rotate because of torque. Also, if you grasp a piece of chalk near each end and twist your hands in opposite directions, it is the developed torque that causes the chalk to twist and, perhaps, snap.

*Work.*   The energy developed by a force acting through a distance against a resistance is known as work. The distance may be along a straight line or along a curved path. When the distance is *linear*, the work can be found from *work = force × distance*. When the distance is along a *circular path*, the work can be found from *work = torque × angle*. Common forms of work include a weight lifted through a height, a pressure pushing a volume of a substance, and torque causing a shaft to rotate.

*Power.*   The rate of doing work, or the work done per unit time, is called power. For example, a certain amount of work is required to raise an elevator to the top of its shaft. A 5-kW motor can raise the elevator, but a 20-kW motor can do the same job *four times faster*.

## 1-3   INTRODUCTION TO METRICS

Because industry has taken major steps toward a changeover to metrics, the technically trained individual should be familiar with its usage.

Historically, a large number of different metric units have been in use throughout the world (countries have not necessarily used the same units as their neighbors or trading partners). In order to overcome the obvious disadvantages of dealing with a variety of systems, an international agreement was reached which limits and standardizes the metric units to be used. The common system adopted is called the International System of Units, usually referred to as SI.* The United States has agreed to adopt this system of metrics. Thus, all reference and use of metric units in this text will be in the SI system.

The metric system offers major advantages relative to the U.S. customary system (our historical system of units). For example, the metric system uses only one basic unit for length—the meter; whereas, the U.S. customary system uses many basic units for length—inch, foot, yard, mile, league, fathom, etc. Also, because the metric system is based on multiples of 10, it is easier to use and learn. For example, in the metric system there are 100 centimeters in 1 meter, 1000 meters in 1 kilometer, etc; in the U.S. customary system there are 12 inches in 1 foot, 3 feet in 1 yard, 5280 feet in 1 mile, etc.

A major distinction exists between the SI system and the U.S. customary system. Whereas the U.S. system is a *gravitational* system with *force as a basic unit*—from which mass is derived—the SI system is an *absolute* system with *mass as a basic unit*—from which force is derived. Simply put, this means that in the U.S. customary system an object is referred to as having a *weight* of some number of pounds (lb), while in the SI system the same object would be referred to as having a *mass* of some number of kilograms (kg).

The following example will illustrate the difference in procedure between U.S. customary units and SI in order to determine the force exerted by an object resting on a table. In dealing with the U.S. customary system, the problem would state that the object has a weight of 100 lb, and we would then indicate that the force exerted by the object on the table is 100 lb, and, if needed, the mass would be calculated. In the SI system, the problem would state that the object has a mass of 45 kg. In order to establish the force the object exerts on the table, we must first calculate the weight of the object by using the formula $W = m \cdot g$ (weight = mass·gravitational constant). Thus, for $g = 9.81$ m/s$^2$ (gravitational constant for sea level on Earth), $W = 45$ kg·9.81 m/s$^2$ = 441 kg·m/s$^2$. Since 1 kg·m/s$^2$ is defined as a newton (N), $W = 441$ N. Since the force is equal to the weight, we can now indicate that $F = 441$ N.

The above example introduced you to SI units and symbols that you may not be familiar with—kg for kilograms, m for meters, s for seconds. As you continue into the text you will be introduced to additional SI units and symbols. You will find that once you become familiar with the SI system, you will enjoy working with it, since the system employs powers

* SI comes from the official name in French: *Le Système International d'Unités.*

**TABLE 1-1**   COMMON SI METRIC PREFIXES

| Factor | Prefix | SI Symbol |
|---|---|---|
| $1\ 000\ 000\ 000\ =\ 10^9$ | giga | G |
| $1\ 000\ 000\ =\ 10^6$ | mega | M |
| $1\ 000\ =\ 10^3$ | kilo | k |
| $0.001\ =\ 10^{-3}$ | milli | m |
| $0.000\ 001\ =\ 10^{-6}$ | micro | $\mu$ |
| $0.000\ 000\ 001\ =\ 10^{-9}$ | nano | n |

of 10. Those problems dealing with the SI system of metrics will be preceded by an asterisk. For example, the basic unit for length is the meter (m). A kilometer (km) is equal to $10^3$ m. A millimeter (mm) is equal to $10^{-3}$ m. As we just discovered, the basic unit for weight and force is the newton (N), which is equal to 1 kg·m/s². A meganewton (MN) is equal to $10^6$ N. A micronewton ($\mu$N) is equal to $10^{-6}$ N. Thus, we see that the SI system employs prefixes (kilo, mega, milli, etc.) in front of a basic unit. Table 1-1 lists the most common prefixes used in engineering. Table 13, App. B, lists the SI units used in this text.

There are a number of specific rules that apply to the use of SI units. For example, a period is never used after a symbol, except when required to indicate the end of a sentence (e.g., kg not kg.); commas are not used to separate groupings of three digits as is done in the U.S. customary system—instead a space is used (e.g., 40,096 is written as 40 096).

## 1-4  VECTORS AND SCALARS

Any quantity that is specified by a *magnitude only* is a *scalar* term. That is, a scalar is not related to any definite direction in space. Examples of scalars are $10, 5 ft, 18°C, 16 hp, 6 volts, etc.

A *vector* quantity has *magnitude* and is *related to a definite direction in space*. Force is such a quantity. To properly specify a force, its magnitude and direction must be known.

A vector quantity is represented by a line carrying an arrowhead at one end. The length of the line (to a convenient scale) equals the magnitude of the vector. The line, together with its arrowhead, defines the direction of the vector.

Suppose a man is using a rope to pull a box along the floor and in order to exert as little effort as possible he tilts the box onto one edge, as shown in Fig. 1-1. He must exert a pull of 30 lb, at an angle of 30° with the horizontal, in order to keep the box moving. This force can be represented as a vector, as shown in Fig. 1-2, by drawing it to scale (each unit of length equals 10 lb of force) at the same angle and sense as the force. Sense refers to the placement of the arrow on the vector.

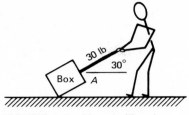

FIGURE 1-1   Man pulling box with rope.

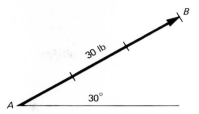

FIGURE 1-2   Force represented as a vector.

## 1-5   NUMERICAL ACCURACY

Because the data used in engineering calculations vary in accuracy, it is important that we develop some means of indicating the accuracy of the data. This is necessary so that after performing a series of calculations, we arrive at an answer no more accurate than the original data.

We can establish the accuracy of a number by the use of *significant figures*. The larger the number of significant figures, the more accurate the data. For example, the number 2300.65 has six significant figures and is more accurate than the number 2301, which has four significant figures.

To establish the significant figures, we must keep in mind that the decimal point has nothing to do with the number of significant figures. Thus, each of the following has four significant figures: 27.61; 276.1; 0.2761; 0.002761; 2761; 276 100; and $2.761 \times 10^6$.

One of the ways you can identify well-trained technicians or engineers is that they carefully avoid giving numerical answers which have too many significant figures. They recognize that the *least* accurate number in the original data establishes the accuracy of the result. The following sample problem will illustrate the above principle.

**Sample Problem 1**   Determine the amount of work done (ft·lb) by a 55-lb force acting on a body for a distance of 1075 ft.

**Solution:**   Work equals force times the distance through which the force acts.

$$\text{Work} = F \times d$$
$$\text{Work} = 55 \times 1075 = 59\ 125 \text{ ft·lb}$$

This is the result which would be displayed by a calculator. But this answer has five significant figures. It is not proper to give such an accurate result from original data in which the least accurate number had only two significant figures (55 lb). Thus, the answer should be rounded off to two significant figures:

$$\text{Work} = 59\ 000 \text{ ft·lb}$$

This means that with the information available in this Sample Problem we can only determine the work done to the nearest 1000 ft·lb. It should be noted that most engineering calculations do not require an accuracy greater than three significant figures.

## 1-6   ROUNDING OFF NUMBERS

If a number is given to the nearest hundredth and we wish to express it to the nearest tenth, what procedures do we follow? The following will apply to the rounding off of all numbers.

1.   If the digit to be dropped is 5 or greater, increase the digit to the left by 1. Example: 36.48 becomes 36.5.

2.   If the digit to be dropped is less than 5, simply drop it without changing the value of the digit to the left. Example: 36.42 becomes 36.4.

## 1-7   DIMENSIONAL ANALYSIS

In dealing with data, not only are we concerned with magnitude, but we must also be concerned with the unit. For example, if we measure the length of a room and express it as 30, this would be meaningless. Do we mean 30 in, or 30 ft, or perhaps 30 yd? Thus, we see that the *dimension* of length consists of both *magnitude* and *unit*.

Very frequently we must change a dimension expressed in one unit to the same dimension expressed in another unit. For example, a length of 30 ft is equal to how many inches? To solve this, we have to use a conversion factor which will enable us to convert from feet to inches. The conversion factor will have to be used so that the value of the dimension will not be changed. This can be accomplished if we *multiply* the original form of the dimension by the conversion factor set up as a *fraction*. Thus,

$$30 \text{ ft} \times \frac{12 \text{ in}}{1 \text{ ft}} = 360 \text{ in}$$

Take note that the conversion factor is equal to 1, since the numerator is equal to the denominator. Take note also that units can be canceled. In fact, units may be divided, multiplied, squared, etc., just as numbers.

Example: express 30 mi/hr as ft/s.

$$30 \frac{\text{mi}}{\text{hr}} \times \frac{5280 \text{ ft}}{1 \text{ mi}} \times \frac{1 \text{ hr}}{3600 \text{ s}} = 44 \frac{\text{ft}}{\text{s}}$$

Example: express 60 km/h as m/s.

$$60 \frac{\text{km}}{\text{h}} \times \frac{1000 \text{ m}}{\text{km}} \times \frac{1 \text{ h}}{3600 \text{ s}} = 16.7 \text{ m/s}$$

# CHAPTER
# 2
# Resultant and Equilibrant of Forces

## 2-1 DEFINITIONS

*External Force.* When a force is applied to a body, it is called an external force.

*Internal Force.* The resistance to deformation, or change of shape, exerted by the material of a body is called an internal force.

Suppose a load $F$ is suspended by a cable as in Fig. 2-1. The load tends to stretch or break the cable. The fibers of the cable resist this tendency. Thus, the load applies an *external force* to the cable, while the cable fibers exert an *internal force* to prevent being pulled apart.

*Collinear Forces.* These are forces whose vectors lie along the same straight line.

*Concurrent Forces.* Forces whose lines of action pass through a common point are called concurrent forces. In Fig. 2-2 forces $F_1$, $F_2$, and $F_3$ pass through the common point $O$ and are concurrent.

*Coplanar Forces.* Forces whose lines of action lie in the same plane are called coplanar forces. Note that collinear forces must also be coplanar.

*Rigid Body.* When a body is acted upon by forces and the size and shape of the body are not changed, the body is said to be *rigid*. All real bodies are more or less elastic, and the application of a force would produce some change. In the study of mechanics, the bodies on which the forces act are often considered rigid.

*Transmissibility of a Force.* A force may be considered as acting at any point on its line of action so long as the direction and magnitude are unchanged. Suppose a body (Fig. 2-3) is to be moved by a horizontal force $F$ applied by hooking a rope to some point on the body. The force $F$ will have the same effect if it is applied at points $A$, $B$, $C$ (Fig. 2-4), or

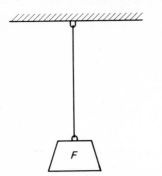

FIGURE 2-1   Load suspended
from cable.

FIGURE 2-2   Concurrent forces.

any point on its line of action. This property of a force is called
*transmissibility.*

The principle of transmissibility may be used freely where overall
effects on a body are being determined. It should be noted, however, that
the local effects on individual members of the body will change when the
force is moved. This can be seen in Fig. 2-4 by visualizing three eyebolts
located at points *A*, *B*, and *C*. Only that eyebolt on which the rope is
hooked will feel the effects of force *F*; the two unused eyebolts are
unaffected by the force.

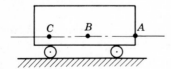

FIGURE 2-3   Transmissibility of a force.

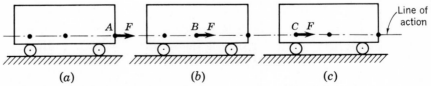

FIGURE 2-4   Transmissibility of a force.

## 2-2   TYPES OF FORCE SYSTEMS

Most practical problems in statics involve several forces which act simultaneously. Such force systems may be classified as follows.

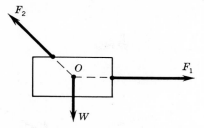

FIGURE 2-5    Concurrent-coplanar forces.

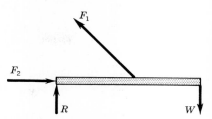

FIGURE 2-6    Nonconcurrent-coplanar forces.

*Concurrent-Coplanar.*    This system occurs where the lines of action of all forces lie in the same plane and pass through the common point. In Fig. 2-5, forces $F_1$, $F_2$, and $W$ all lie in the same plane (the plane of the paper) and all their lines of action have point $O$ in common. Collinear forces are the simplest type and a special case of concurrent-coplanar forces. This chapter deals with concurrent-coplanar force systems.

*Nonconcurrent-Coplanar.*    Such a system exists where the lines of action of *all* forces lie in the same plane but *do not* pass through a common point. Figure 2-6 shows a situation of this type. These systems will be treated in Chaps. 3 and 4. It should be noted that parallel force systems, such as Fig. 2-7, are a special case of nonconcurrent-coplanar systems.

*Concurrent-Noncoplanar.*    This system is evident where the lines of action of all forces *do not* lie in the same plane but *do* pass through a common point. An example of this force system is the forces in the legs of a tripod support for a camera (Fig. 2-8) or a transit. Chapter 5 is devoted to concurrent-noncoplanar systems.

*Nonconcurrent-Noncoplanar.*    Where the lines of action of all forces *do not* lie in the same plane and *do not* pass through a common point, a nonconcurrent-noncoplanar force system is present. This type will not be discussed in the text.*

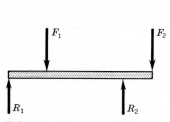

FIGURE 2-7    Parallel forces.

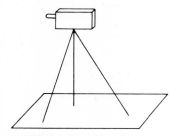

FIGURE 2-8    Tripod support for camera.

* For a discussion of nonconcurrent-noncoplanar force systems see A. Jensen and H. H. Chenoweth, *Applied Engineering Mechanics*, 3d ed., McGraw-Hill Book Company, New York, 1972.

## 2-3    RESULTANT OF CONCURRENT FORCE SYSTEMS

A *resultant* is a single force which can replace two or more concurrent forces and produce the same effect on the body as the concurrent forces.

It is a fundamental principle of mechanisms, demonstrated by experiment, that when a force acts on a body which is free to move, the motion of the body is in the direction of the force, and the distance traveled in a unit time depends on the magnitude of the force. Then, for a system of concurrent forces acting on a body, the body will move in the direction of the *resultant* of that system, and the distance traveled in a unit time will depend on the magnitude of the *resultant*.

## 2-4    RESULTANT OF COLLINEAR FORCES

The simplest type of a concurrent force system involves collinear forces. When two or more forces act on a body along the same straight line, their resultant will also act along the same line.

If a body is acted upon by forces of 20 and 30 lb, as shown in Fig. 2-9, it is evident that a single force of 20 + 30 = 50 lb would be the resultant.

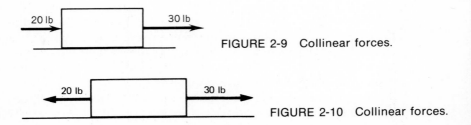

20 lb          30 lb

FIGURE 2-9    Collinear forces.

20 lb          30 lb

FIGURE 2-10    Collinear forces.

If the forces act in opposite directions (Fig. 2-10), their resultant would be 30 − 20 = 10 lb acting to the right.

As a matter of convenience, when a force acts in a certain direction, say to the right, it is considered positive. If it acts to the left, it is negative. The force is read or expressed as so many pounds, with a plus or minus sign attached, depending on the direction in which the force acts. Thus, the forces of Fig. 2-10 are +30 and −20 lb. The purpose of the sign is to indicate direction of action only.

The rule for determining the resultant of collinear forces may be stated as follows: *The resultant of any number of forces acting along the same straight line is their algebraic sum.*

Although the choice of signs is a matter of convenience, the custom in mechanics is to observe the signs as in mathematics. Thus, for forces acting horizontally, a force acting to the right is positive, while a force acting to the left is negative. For forces acting on a vertical line, up forces are positive and down forces are negative.

**Sample Problem 1** Find the resultant of the collinear force system in Fig. 2-11.

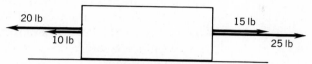

20 lb                                    15 lb

10 lb                                         25 lb

FIGURE 2-11   Diagram for Sample Problem 1.

**Solution:**   $R = + 15 + 25 - 20 - 10 = +10$ lb
     Thus, the four given forces may be replaced by a single force of 10 lb acting to the right.

*\*Sample Problem 2* Find the resultant of the collinear forces in Fig. 2-12.

**Solution:**   $R = + 8 + 12 - 10 = +10$ N

12 N

8N

10N

FIGURE 2-12   Diagram for Sample Problem 2.

Therefore, the three given forces may be replaced by a single up force of 10 N.

**Sample Problem 3**   Find the resultant of the forces in Fig. 2-13.

**Solution:**   There are two sets of collinear forces which will be analyzed separately.

In the vertical direction, $R_y = + 10 - 10 = 0$.

In the horizontal direction, $R_x = - 6 - 4 + 8 = -2$ lb.

Then the five forces may be replaced by a single force of 2 lb acting horizontally to the left.

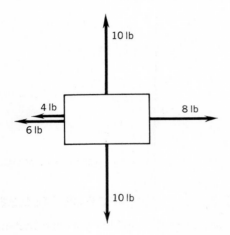

FIGURE 2-13   Diagram for Sample Problem 3.

## 2-5   EQUILIBRIUM OF CONCURRENT FORCE SYSTEMS

When the resultant of any number of concurrent forces acting on a body is zero, the forces are said to be in *equilibrium*. The student should notice that when forces are in equilibrium, the body on which they act may be at rest or in uniform motion.

*Statics* is a study of forces in equilibrium.

## 2-6   EQUILIBRIUM OF COLLINEAR FORCES

If the resultant of forces acting on a body in the same straight line is zero, the forces are in *equilibrium*. The body on which these forces act will not move or, if in motion, will continue to move with the same speed. This condition of equilibrium might be stated in another way by saying that the sum of the forces in one direction must be equal to the sum of the forces in the opposite direction. Direction is not limited to either the horizontal or the vertical, but may make any angle with the horizontal.

## 2-7  ACTION AND REACTION

To every *action* there is an equal and opposite *reaction* (Newton's third law).

When one pushes against a wall, one exerts an action. The wall resists the push with an equal and opposite force called a reaction.

When a load is suspended from a hook, the load exerts an action downward. The hook exerts an equal, opposite force called the reaction.

A body weighing 10 lb rests on a table. The action force is the downward pull due to the Earth's attraction. The table exerts an upward reaction $N$ (Fig. 2-14). Since the body does not move, the forces are in equilibrium. Then $R = 0$; but $R = N - 10$. Thus $N - 10 = 0$, and $N = 10$ lb.

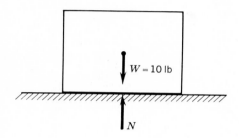

$W = 10$ lb

$N$                FIGURE 2-14  Action and reaction.

## 2-8  TENSION AND COMPRESSION: TWO-FORCE MEMBERS

When two equal and opposite forces $F$ act on a body as the bar in Fig. 2-15, the body is in *tension*, because the tendency is to stretch or pull the bar apart. A hoisting cable is a good illustration of this action.

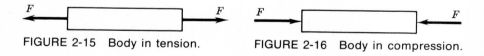

FIGURE 2-15  Body in tension.          FIGURE 2-16  Body in compression.

If the forces act in the reversed directions, as in Fig. 2-16, the body is in *compression*, because the tendency of the forces $F$ is to shorten, compress, or crush the body. This action occurs in a prop, brace, or column.

When a bar is acted upon by two opposite forces that are in equilibrium, the member is always either in tension or compression, and the direction of the external forces is along the axis of the bar. These forces are termed *axial* forces.

It is important to distinguish between axial tension and axial compression forces. A long member subjected to axial tension force is not normally affected by its length. On the other hand, a long member which is subjected to axial compression force *is* affected by its length. If a member subjected to compression force is relatively *long* compared to its cross-sectional area, it will be subjected to both buckling and compression stresses. This type of member is discussed in Chap. 15. If a member subjected to compression force is relatively *short* compared to its cross-sectional area, it will be subjected to compression stress only. Unless otherwise indicated, it will be assumed in this text that a member subjected to compression force is a short compression member.

For a simple experiment in tension, a rope of any convenient length is chosen and a force of 50 lb applied at each end, as in Fig. 2-17.

50 lb        *C*              50 lb          FIGURE 2-17   Rope in tension.

The rope is cut at any point $C$, and two spring balances are fastened to the ends and hooked together (Fig. 2-18).

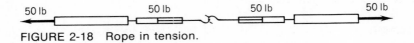

50 lb              50 lb              50 lb                50 lb

FIGURE 2-18   Rope in tension.

Each balance then reads 50 lb, an indication that a force of 50 lb was required to hold the parts of the rope together. This is just the force that was exerted by the fibers of the rope before it was cut.

Members subjected solely to axial forces or to forces whose resultants are axial forces are called *two-force members*. The ability to recognize two-force members will be useful in analyzing structures and mechanisms.

## 2-9  RESULTANT OF TWO CONCURRENT FORCES

Two forces that are concurrent can be added vectorially by means of the parallelogram law or the triangle law.

Figure 2-19 shows a body acted on by two concurrent forces, $P$ and $F$. By constructing a parallelogram whose sides are equal to $P$ and $F$, respectively, the resultant of these forces can be determined. The diagonal of the parallelogram will be the resultant $R$ (Fig. 2-20a).

The resultant $R$ can also be found by constructing one-half of the parallelogram. This figure will be a triangle. From any convenient point $A$, using an appropriate scale, draw a vector equal in magnitude and direction to one of the forces (Fig. 2-20b). From the tip of the arrowhead of the first vector, lay out the second vector to the same scale so that the vectors are connected head-to-tail. The closing side of the triangle, drawn from the starting point $A$ of the first vector to the tip of the arrowhead

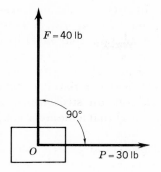

FIGURE 2-19   Diagram for Sample Problem 4.

of the second vector, is the resultant of the two forces (Fig. 2-20$b$). Note that the arrowhead of the resultant $R$ is at the end away from the starting point. The magnitude of the resultant is determined by measuring its length with the same scale used for drawing the other vectors.

Most cases of concurrent force systems are best solved by algebraic methods. Such a solution is demonstrated in the following examples, along with the graphical method described above.

**Sample Problem 4** Two concurrent forces of 30 and 40 lb act on a body, making an angle of 90° with each other, as in Fig. 2-19. Find the magnitude and direction of the resultant by ($a$) graphical method, ($b$) algebraic method.

**Solution a (Graphical):**    From any point $A$, draw **AB** equal to 30 lb to any convenient scale and parallel to force $P$, as in Fig. 2-20$b$. From $B$, draw a line **BC** in the direction of force $F$ and in length to represent 40 lb. Then the line **AC** is the resultant $R$ in magnitude and direction. By

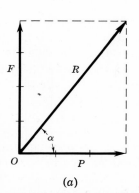

(*a*)

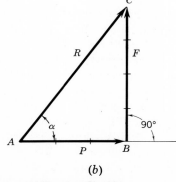

(*b*)

FIGURE 2-20   (*a*) Parallelogram solution for Sample Problem 4. (*b*) Force triangle.

measurement, $R$ is found to be 50 lb. Angle $\alpha$ can be measured by protractor.

Figure 2-20$a$ indicates the graphical solution by means of the parallelogram law.

**Solution b (Algebraic):**   Notice that since the resultant forms a triangle with the two given forces, an algebraic solution of the triangle will give $R$. Since $ABC$ is a right triangle,

$$R = \sqrt{P^2 + F^2} = \sqrt{900 + 1600} = 50 \text{ lb}$$

The angle $\alpha$ that $R$ makes with $P$ may be found from

$$\tan \alpha = \frac{\mathbf{BC}}{\mathbf{AB}} = \frac{F}{P} = \frac{40}{30} = 1.333$$
$$\alpha = 53.13° \text{ or } 53°8'$$

**\*Sample Problem 5** Find the magnitude and direction of the resultant of two concurrent forces of 50 and 75 N acting on a body at an angle of 50° with each other, as in Fig. 2-21. Use ($a$) graphical method, ($b$) algebraic method.

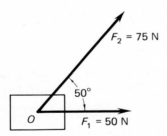

FIGURE 2-21   Diagram for Sample Problem 5.

**Solution a (Graphical):**   From any point $A$, draw $\mathbf{AB}$ equal and parallel to $F_1$, as in Fig. 2-22$a$. From $B$, draw line $\mathbf{BC}$ equal and parallel to $F_2$. Line $\mathbf{AC}$ is the resultant $R$. By measurement, $R$ is found to be 114 N.

Figure 2-22$b$ indicates the graphical solution by means of the parallelogram law.

**Solution b (Algebraic):**   Triangle $ABC$ (Fig. 2-22$a$) may be solved for $R$ by using the law of cosines.

$$R^2 = F_1^2 + F_2^2 - 2F_1F_2 \cos \theta$$

But $\theta = 180 - 50 = 130°$.

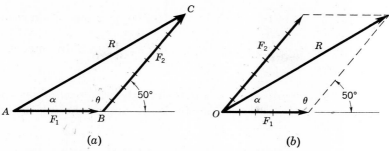

FIGURE 2-22   (a) Force triangle for Sample Problem 5.
(b) Parallelogram solution.

$$\cos 130° = -\cos 50° = -0.6428$$
$$R^2 = 50^2 + 75^2 - 2(50)(75)(-0.6428)$$
$$= 2500 + 5625 + 4820 = 12\,945$$
$$R = 114 \text{ N}$$

Angle $\alpha$ may be calculated by the law of sines.

$$\frac{F_2}{\sin \alpha} = \frac{R}{\sin \theta} \qquad \sin \alpha = \frac{F_2}{R} \sin \theta$$
$$\sin \theta = \sin 130° = \sin 50° = 0.7660$$
$$\sin \alpha = \frac{75}{114}(0.7660) = 0.505$$
$$\alpha = 30.3° \text{ or } 30°18'$$

The student should appreciate that the degree of accuracy obtainable with the graphical method depends upon the care with which the drawing is made and the scale that is used.

## 2-10 EQUILIBRANT AND THE FORCE TRIANGLE

If two forces are acting on a body, the third force that will hold them in equilibrium is called the *equilibrant*, or the *balancing force*.

The relations among three forces in equilibrium are best illustrated by an experiment. In Fig. 2-23, let *OA*, *OB*, and *OC* be fastened to a ring at *O*. Spring balances are fastened at *A*, *B*, and *C*. The springs are then stretched and the rings fastened to pins *E*, *F*, and *G*. Since the forces acting on *OA*, *OB*, and *OC* are axial, these are two-force members.

Balances *E*, *F*, and *G* show forces of 12, 15, and 20 lb, respectively. These forces are in equilibrium because the ring *O* is at rest.

Now draw **MN** parallel to *OA* and to scale equal to 12 lb, as in Fig. 2-24. From *N*, draw **NP** equal to 15 lb and parallel to *OB*. Join *P* and *M*. **PM** by careful measurement will be found to scale 20 lb. If the angles of

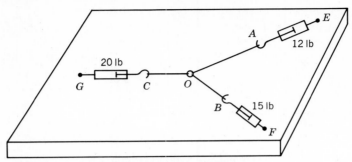

FIGURE 2-23   Three forces in equilibrium.

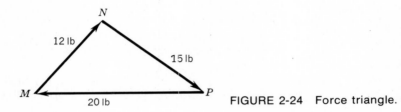

FIGURE 2-24   Force triangle.

the triangle *MNP* are found by a protractor and compared with angles *AOC* and *BOC*, **PM** will be found parallel to *OC*.

The equilibrant, or balancing force, of any two forces is the closing line of the triangle, with the arrowhead pointing in such a direction that the arrowheads of all the forces appear to follow each other around the triangle.

Triangle *MNP* is called the *force triangle*. Thus, three concurrent forces in equilibrium may be represented as sides of a force triangle.

## 2-11 PRINCIPLE OF CONCURRENCE

*If three nonparallel forces acting on a body are in equilibrium, their lines of action must meet in a common point.* As already shown, the resultant of any two concurrent forces is a single force acting through their point of intersection. When equilibrium exists, their resultant is balanced by a third force (equilibrant). However, in order to be in equilibrium, the resultant and the equilibrant must act along the same straight line. To meet this condition, the lines of action of the original forces must meet at a point. This principle of concurrence will be found useful in later problems in which the direction of some of the unknown forces is not at first evident.

## 2-12 METHODS OF SOLUTION

Since three concurrent forces in equilibrium may be represented as a triangle of forces, the solution of such a problem, in mechanics, is the solution of a triangle.

If the triangle of forces is constructed accurately, the unknown parts of the triangle may be found by measurement. This is called the *graphical solution*.

When the triangle is solved by means of algebra or trigonometry, the method is called *algebraic*.

The three theorems used in the algebraic method are as follows.

1. *Pythagorean theorem. In any right triangle, the square of the hypotenuse is equal to the sum of the squares of the two legs:*

$$c^2 = a^2 + b^2 \tag{2-1}$$

2. *Law of sines. In any triangle, the sides are to each other as the sines of the opposite angles:*

$$\frac{a}{\sin A} = \frac{b}{\sin B} = \frac{c}{\sin C} \tag{2-2}$$

3. *Law of cosines. In any triangle, the square of any side is equal to the sum of the squares of the other two sides minus twice the product of the sides and cosine of their included angle:*

$$c^2 = a^2 + b^2 - 2ab \cos C \tag{2-3}$$

**Sample Problem 6**  Two forces of 75 and 100 lb, respectively, make an angle of 110° with each other as shown in Fig. 2-25. Find their resultant and the angle it makes with the 100-lb force.

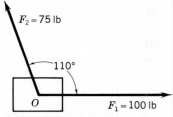

FIGURE 2-25  Diagram for Sample Problem 6.

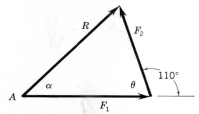

FIGURE 2-26  Force triangle for Sample Problem 6.

**Solution:**  Sketch a force-vector diagram as in Fig. 2-26.

$$\theta = 180 - 110 = 70°$$

By Eq. (2-3),

$$\begin{aligned}
R^2 &= F_1{}^2 + F_2{}^2 - 2F_1F_2 \cos \theta \\
&= 100^2 + 75^2 - 2(100)(75)(0.3420) \\
&= 10\,000 + 5625 - 5130 = 10\,495 \\
R &= 102 \text{ lb}
\end{aligned}$$

By Eq. (2-2),

$$\frac{F_2}{\sin \alpha} = \frac{R}{\sin \theta}$$

$$\sin \alpha = \frac{F_2}{R} \sin \theta = \frac{75}{102}(0.9397) = 0.691$$

$$\alpha = 43.7° \text{ or } 43°42'$$

**\*Sample Problem 7**  Three concurrent forces of 45, 60, and 90 N, respectively, are in equilibrium (Fig. 2-27). Find the angles that these forces must make with each other.

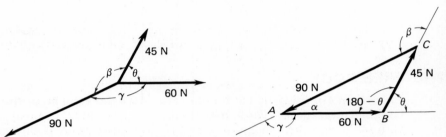

FIGURE 2-27   Diagram for Sample Problem 7.

FIGURE 2-28   Force triangle for Sample Problem 7.

**Solution:**   Since these forces are in equilibrium they may be represented as the three sides of a triangle (Fig. 2-28).

With $A$ as a starting point, draw **AB** parallel and equal to the 60 N force. With $B$ as a center and a line equal to 45 N as a radius, strike an arc. With $A$ as a center and a radius equal to 90 N, draw an arc intersecting the first arc at $C$. Draw **AC** and **BC**. Then $ABC$ is the required force triangle (Fig. 2-28) with sides parallel to the given forces. The required angles can be found by means of a protractor. A more accurate method is to apply the law of cosines.

By Eq. (2-3),

$$45^2 = 60^2 + 90^2 - 2(60)(90)\cos \alpha$$

$$\cos \alpha = \frac{60^2 + 90^2 - 45^2}{2(60)(90)} = \frac{9675}{10\ 800} = 0.896$$

$$\alpha = 26.4° \text{ or } 26°24'$$

By Eq. (2-2),

$$\frac{90}{\sin (180 - \theta)} = \frac{45}{\sin \alpha} \qquad \text{but} \qquad \sin (180 - \theta) = \sin \theta$$

$$\sin \theta = \frac{90}{45} \sin \alpha = \frac{90}{45} (0.4446) = 0.8892$$
$$\theta = 62.8° \text{ or } 62°48'$$

Referring to Fig. 2-28,

$$\gamma = 180 - \alpha = 180 - 26.4$$
$$= 153.6° \text{ or } 153°36'$$

Referring to Fig. 2-27,

$$\beta = 360 - \theta - \gamma = 360 - 62.8 - 153.6$$
$$= 143.6° \text{ or } 143°36'$$

The student should note that the angles between the forces (Fig. 2-27) are supplements of the angles of the force triangle (Fig. 2-28).

## 2-13 FREE BODY

In solving problems of mechanics, it is often convenient to disregard bodies acting upon a single body and to replace these actions by force vectors, having their magnitude and direction the same as those of the actions produced by each of the acting bodies. A body thus represented is said to be a *free* body.

It is essential to designate the free body clearly and to account for all the forces that act on it. A sketch showing the body with the forces acting is an important aid in the investigation and solution of problems and should always be drawn.

Take, for example, a 10-ft ladder resting against a smooth wall. The ladder weighs 25 lb. A 155-lb man stands on it, as shown in Fig. 2-29, while his helper holds his foot against the bottom to prevent the ladder from slipping on the smooth floor.

In order to analyze this problem, consider the ladder as a free body. To sketch a free-body diagram, draw a line to represent the ladder, so that the direction and length are proper. Force vectors are placed at the points on the line to represent the external forces which act on the ladder. The free-body diagram for the ladder is shown in Fig. 2-30.

In this free-body diagram, the weight of the ladder and the man are shown acting downward. $R_1$ and $R_2$ are the reactions of the floor and wall, respectively. $F$ is the force applied by the helper's foot. This particular problem involves nonconcurrent-coplanar forces and will be solved in Chap. 4 (Sample Problem 1). Free-body diagrams are useful for all types of equilibrium force systems.

*Sample Problem 8** A body having a mass of 300 kg is suspended from a beam by two ropes, making angles of 20° and 30° with the vertical, respectively. Find the forces in the ropes.

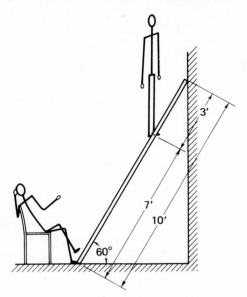

FIGURE 2-29   Man on a ladder.

**Solution:**   Let $AB$ and $BC$ be the ropes that support the body (Fig. 2-31). $AB$ and $BC$ are both two-force members and are in tension.

Set point $B$ out as a free body. There are three forces: the weight of the suspended body acting down ($W = m \cdot g = 300$ kg$\cdot$9.81 m/s$^2$ = 2940 N), and $F_1$ and $F_2$ up, at angles 20° and 30°, as shown in Fig. 2-32.

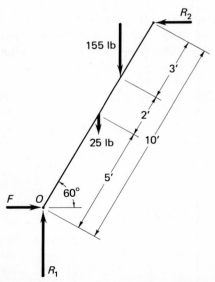

FIGURE 2-30   Free-body diagram of ladder.

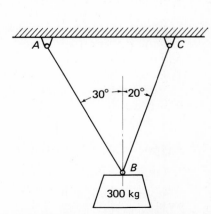

FIGURE 2-31   Diagram for Sample Problem 8.

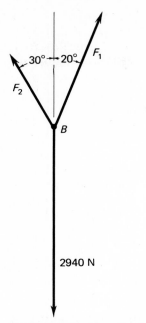

FIGURE 2-32   Free-body diagram
for Sample Problem 8.

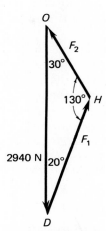

FIGURE 2-33   Force triangle
for Sample Problem 8.

From any point $O$ in Fig. 2-33 draw **OD** to represent 2940 N. From $D$, draw a line parallel to $F_1$, but indefinite in length. $F_1$ is known in direction but not in magnitude. Since the three forces must form a triangle, force $F_2$ must end at the point $O$. Therefore, draw a line through $O$, parallel to $F_2$, cutting the second line at $H$. Triangle $ODH$ is the force triangle, and **DH** and **HO** to scale represent $F_1$ and $F_2$. These values may be obtained by direct measurement. By the law of sines,

$$\frac{F_1}{\sin 30°} = \frac{2940}{\sin 130°} \quad \text{and} \quad \frac{F_2}{\sin 20°} = \frac{2940}{\sin 130°}$$

$$F_1 = \frac{0.50(2940)}{0.7660} = 1919 \text{ N} \quad \text{and} \quad F_2 = \frac{0.3420(2940)}{0.7660} = 1313 \text{ N}$$

Say, $F_1 = 1920$ N $= 1.92$ kN (tension)   and   $F_2 = 1310$ N $=$
1.31 kN (tension)

## 2-14 ANALYSIS OF A SIMPLE STRUCTURE

A simple structure is a body formed of tension and compression members (two-force members) hinged or pinned together. The hinges or pins are assumed frictionless, and the weight of the parts is neglected. Most roof trusses, many bridge trusses, and certain mechanical devices are treated as simple structures.

Since the parts are in equilibrium, free-body diagrams may be drawn for any of the parts in a simple structure. However, it is usually more practical to use the pins at the joints as free bodies, since these pins are generally subjected to concurrent forces which are in equilibrium.

Consider a structure made up of a boom $BC$ connected to a wall by pin $C$ and a tie rod $AB$ connected to the boom at pin $B$ and to the wall at pin $A$. A load of $W$ lb is suspended from the end of the boom, as shown in Fig. 2-34.

Now, the tie rod $AB$ and the boom $BC$ are two-force members, since tie rod $AB$ is a tension member and boom $BC$ is a compression member. This is so because the boom acts as a strut, supporting pin $B$ while the tie rod prevents the boom from rotating about pin $C$. The student should recall that internal tensile or compressive forces act in the direction of the axis of their respective members. The weight of the boom $BC$ is assumed to be small enough, compared to load $W$, to be neglected.

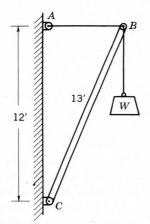

FIGURE 2-34   A simple structure.

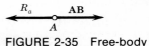

FIGURE 2-35   Free-body diagram of pin A.

This structure is in equilibrium; hence, the pins $A$, $B$, and $C$ are in equilibrium. Free-body diagrams of the three pins may be drawn. For pin $A$, Fig. 2-35 shows the free-body diagram. This diagram shows the forces acting *on the pin*. **AB** is the tension which tends to tear the pin from the wall, while $R_a$ is the wall reaction which is necessary to maintain pin $A$ in equilibrium.

Similarly Fig. 2-36 is a free-body diagram for pin $B$. **AB** is again the tension of member $AB$ which prevents pin $B$ from swinging clockwise, **BC** is the compression, or thrust, of the boom, and $W$ is the suspended load.

Figure 2-37 is the free-body diagram for pin $C$. The boom force **BC** tends to drive pin $C$ into the wall, while $R_c$ is the wall reaction to maintain pin $C$ in equilibrium. Note that pins $A$ and $C$ are each subjected to two

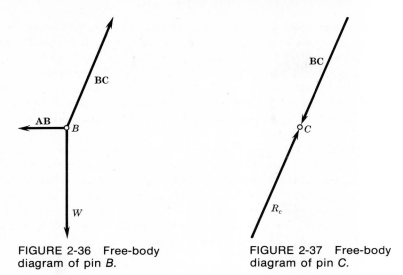

FIGURE 2-36  Free-body
diagram of pin *B*.

FIGURE 2-37  Free-body
diagram of pin *C*.

collinear forces, while pin *B* is acted upon by three concurrent (and coplanar) forces.

**Sample Problem 9**  For the structure of Fig. 2-34, the distance between pins *A* and *C* is 12 ft, the length of member *BC* = 13 ft, and *W* = 4000 lb. Find the forces in members *AB* and *BC*.

**Solution a:**  *ABC* is a right triangle; hence, by Eq. (2-1) the length of member *AB* = 5 ft. Take pin *B* as a free body, as in Fig. 2-36, where *W* = 4000 lb. From *M*, draw a line parallel to *BC* and indefinite in length. Since the figure must close, **AB** terminates at *O*. But **AB** must be horizontal, since member *AB* is a two-force member and is horizontal. Then from *O*, sketch a line parallel to *AB* to intersect the second line drawn at point *N*. Thus, *MNO* is a force triangle (Fig. 2-38). Since *MNO* is a right triangle, **AB** and **BC** may be found trigonometrically.

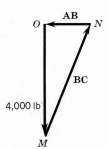

FIGURE 2-38  Force triangle for Sample Problem 9.

$$\cos M = \frac{4000}{\mathbf{BC}}$$

But angle $M$ equals angle $C$ in Fig. 2-34. Thus:

$$\cos M = \cos\ C = \frac{12}{13}$$

$$\frac{12}{13} = \frac{4000}{\mathbf{BC}}$$

$$\mathbf{BC} = \frac{13(4000)}{12} = 4330 \text{ lb (compression)}$$

Also, $\quad \sin M = \dfrac{\mathbf{AB}}{\mathbf{BC}} \quad$ and $\quad \sin M = \sin C = \dfrac{\text{length } AB}{\text{length } BC} = \dfrac{5}{13}$

$$\frac{5}{13} = \frac{\mathbf{AB}}{4330}$$

$$\mathbf{AB} = \frac{5(4330)}{13} = 1670 \text{ lb (tension)}$$

**Solution b (Similar Triangles):** Since force triangle $MNO$ is similar to triangle $ABC$, or the original structure, a direct solution is possible. Since corresponding sides of similar triangles are proportional, then

$$\frac{\mathbf{BC}}{13} = \frac{4000}{12}$$

$$\mathbf{BC} = 4330 \text{ lb (compression)}$$

and $\qquad \dfrac{\mathbf{AB}}{5} = \dfrac{4000}{12}$

$$\mathbf{AB} = 1670 \text{ lb (tension)}$$

*Whenever the principle of similar triangles is applicable, it should be used.*

## 2-15 COMPONENTS OF A FORCE

The resultant of two forces was previously defined as the single force that will produce the same effect as the two forces. Also, it was stated that the two forces with their resultant form a triangle.

The converse of this statement is also true, that a force may be replaced by any two forces which, with the given force, form a triangle. In Fig. 2-39, **AB** and **BC** are two forces that, if applied to a body at $A$, produce the same effect as **AC**. **AB** and **BC** are called the *components* of **AC**. Note the direction of the arrows.

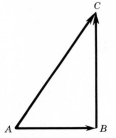

FIGURE 2-39    Components
of a force.

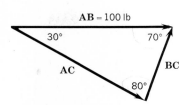

FIGURE 2-40    Diagram for
Sample Problem 10.

**Sample Problem 10** Resolve a force of 100 lb into two components making angles of 30° and 70°, respectively, with the force.

**Solution:**   Let **AB** be the given force (Fig. 2-40). Draw **AC** and **BC** making angles of 30° and 70°, respectively. Then **AC** and **BC** are the components. By the law of sines,

$$\frac{\mathbf{AC}}{\sin 70°} = \frac{100}{\sin 80°} \qquad \mathbf{AC} = \frac{0.9397(100)}{0.9848} = 95.4 \text{ lb}$$

$$\frac{\mathbf{BC}}{\sin 30°} = \frac{100}{\sin 80°} \qquad \mathbf{BC} = \frac{0.50(100)}{0.9848} = 50.8 \text{ lb}$$

Say, **AC** = 95 lb and **BC** = 51 lb.

Therefore, forces of 95 and 51 lb acting at $A$ (as in Fig. 2-41) are the components of **AB**.

*__Sample Problem 11__ A force of 93 N has forces of 60 and 75 N for components. Find their directions with the given force (Fig. 2-42).

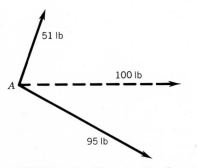

FIGURE 2-41    Diagram for Sample
Problem 10.

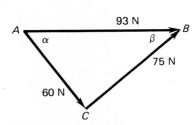

FIGURE 2-42    Diagram for Sample
Problem 11.

**Solution:** Triangle $ABC$ may be solved for angles $\alpha$ and $\beta$ by Eqs. (2-2) and (2-3). Using the law of cosines,

$$(\mathbf{BC})^2 = (\mathbf{AB})^2 + (\mathbf{AC})^2 - 2(\mathbf{AB})(\mathbf{AC})\cos\alpha$$

$$\cos\alpha = \frac{(\mathbf{AB})^2 + (\mathbf{AC})^2 - (\mathbf{BC})^2}{2(\mathbf{AB})(\mathbf{AC})}$$

$$= \frac{93^2 + 60^2 - 75^2}{2(93)(60)} = \frac{8650 + 3600 - 5625}{11\,160} = \frac{6625}{11\,160}$$

$$= 0.594$$

$$\alpha = 53.6° \text{ or } 0.935 \text{ rad}$$

(Refer to App. B, Table 14, for conversion from degree to radians.) By the law of sines,

$$\frac{60}{\sin\beta} = \frac{75}{\sin 53.56°}$$

$$\sin\beta = \frac{60(0.8045)}{75} = 0.6436$$

$$\beta = 40.1° \text{ or } 0.699 \text{ rad}$$

# 2-16 RECTANGULAR COMPONENTS OF A FORCE

When a force is resolved into two components that are at right angles to each other, they are called *rectangular components*. These components with the given force as hypotenuse form a right triangle.

In the solution of many problems in mechanics, it is customary to choose an origin and $x$ and $y$ axes at right angles to each other. The force or forces acting are then resolved into their rectangular components along these axes. If the $x$ axis is horizontal and the $y$ axis vertical, the components of a force along these axes are called *horizontal* and *vertical* *components*.

The axes are not always horizontal and vertical (such as in Fig. 2-45), but the components will be referred to as $x$ and $y$ components and represented by $F_x$ and $F_y$, when the given force is $F$.

Let $F$ be a force making an angle $\theta$ with the horizontal, as shown in Fig. 2-43. From the point $B$, drop a perpendicular to the $x$ axis. Then $F_x = \mathbf{OA}$, and $F_y = \mathbf{AB}$.

From trigonometry,

$$\cos\theta = F_x/F$$
$$\sin\theta = F_y/F$$
$$F_x = F\cos\theta$$
$$F_y = F\sin\theta$$

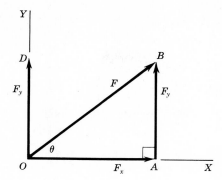

FIGURE 2-43   Rectangular components of a force.

But $F_x$ is the projection of $F$ on the $x$ axis, and since **AB = OD**, $F_y$ is the projection of $F$ on the $y$ axis.

The following definitions may then be given.

The $x$ component of a force is the force times the cosine of the angle that the force makes with the $x$ axis.

The $y$ component of a force is the force times the cosine of the angle that the force makes with the $y$ axis, or the force times the sine of the angle that the force makes with the $x$ axis.

When a force is at right angles with a line, its component is zero along the line; that is, the force has no effect on a body in a direction at right angles to the force.

**Sample Problem 12** A force of 50 lb makes an angle of 30° with the horizontal. Find $F_x$ and $F_y$ (Fig. 2-44).

**Solution:**

$$F_x = F \cos \theta = 50 \cos 30° = 43.3 \text{ lb (say, 43 lb)}$$
$$F_y = F \sin \theta = 50 \sin 30° = 25 \text{ lb}$$

**\*Sample Problem 13** A body having a mass of 50 kg rests on an inclined plane, making an angle of 20° with the horizontal. Find $F_x$ and $F_y$ of the body's weight (Fig. 2-45).

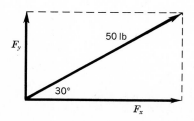

FIGURE 2-44   Diagram for Sample Problem 12.

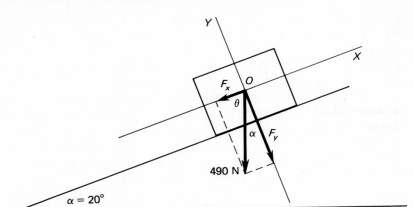

FIGURE 2-45 Diagram for Sample Problem 13.

**Solution:** Take the $x$ axis parallel and the $y$ axis perpendicular to the plane. The weight of the body is a force of $(50 \text{ kg})(9.81 \text{ m/s}^2) = 490 \text{ N}$ acting down. But $\theta = 70°$ and $\alpha = 20°$. Then

$$F_x = 490 \cos 70° = 490(0.3420) = 167.6 \text{ N}$$
$$F_y = 490 \cos 20° = 490(0.9397) = 460.5 \text{ N}$$

Say, $F_x = 170 \text{ N}$ and $F_y = 460 \text{ N}$.

## 2-17 INCLINED PLANE

Let a body of weight $W$ rest on an inclined plane making an angle $\alpha$ with the horizontal (Fig. 2-46).

Choose the axes along and perpendicular to the face of the plane. Then

$$F_x = W \sin \alpha$$
$$F_y = W \cos \alpha$$

That is, the component of the weight of a body parallel to the plane is equal to the weight times the sine of the angle that the plane makes with the horizontal.

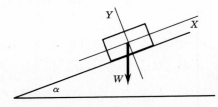

FIGURE 2-46 Body on an inclined plane.

The component of the weight perpendicular to the plane is the weight times the cosine of the angle that the plane makes with the horizontal.

Since the $y$ component is the one perpendicular to the plane, it produces no motion of the body.

The component of the weight parallel to the plane $(F_x)$ is the force that acts to drag the body down the plane. If the body does not move, it is because of the reaction of some external force, such as friction, or of an applied force. (Friction is the force that acts between two surfaces in contact and tends to resist motion of one surface over the other. A more complete discussion of friction is found in Chap. 6.)

If there is no friction, then it can be seen that the applied force necessary to prevent the body from sliding down the plane will be equal to $W \sin \alpha$. A force parallel to the plane and slightly greater than $W \sin \alpha$ by the amount of the friction encountered will be needed to move the body up the plane. Frequent use is made of the inclined plane in loading heavy machinery on trucks, in jackscrews, and in ordinary screw threads. The general principle involved is that a relatively small force exerted parallel to the plane is capable of displacing the body along the plane and thereby raising the body to the desired height.

**Sample Problem 14**   The floor of a truck is 3 ft above the ground. An inclined plane is formed by using a 12-ft plank. It is desired to load a box of machine parts weighing 300 lb by pushing it up the 12-ft plane. How much force must be exerted parallel to the plane, neglecting friction?

**Solution:**

$$\sin \alpha = \frac{3}{12} = \frac{1}{4}$$

$$F_x = W \sin \alpha = 300\left(\frac{1}{4}\right) = 75 \text{ lb}$$

A force slightly in excess of 75 lb and applied to the box in a direction parallel to the plane will slide the box up the plane. It is seen that a vertical lift without using the inclined plane would require 300 lb.

## 2-18  RESULTANT OF MORE THAN TWO FORCES IN A PLANE

Suppose that three or more forces act through a common point $O$, as in Fig. 2-47. Find the resultant.

*Graphical Solution:*   From any convenient point $A$, in Fig. 2-48, draw **AB** parallel and equal to $F_1$. From $B$, draw **BC** equal and parallel to $F_2$. **AC** is the resultant of $F_1$ and $F_2$. From $C$, draw **CD** equal and parallel to $F_3$. **AD** is the resultant of **AC** and **CD**. It is therefore the resultant of $F_1$, $F_2$,

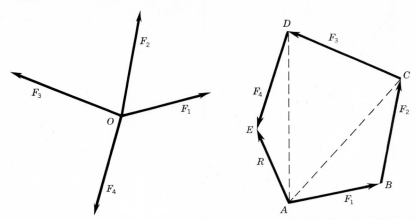

FIGURE 2-47   Four concurrent forces.     FIGURE 2-48   Force polygon.

and $F_3$. From $D$, draw **DE** equal and parallel to $F_4$. **AE**, the resultant of **AD** and **DE**, is then the resultant of the four forces $F_1$, $F_2$, $F_3$, and $F_4$. *ABCDEA* is known as the *polygon of forces*.

The method for determining the resultant of any number of concurrent forces is as follows. From any point, construct the force polygon by adding the vectors taken in order, either clockwise or counterclockwise around the point. The closing line of the polygon drawn from the starting point of the first vector to the terminal point of the last vector is the resultant of the concurrent forces. (Notice the direction of the arrowheads.)

**Sample Problem 15**  Given concurrent forces of 20, 30, 10, and 40 lb making angles of 10°, 30°, 120°, and 125°, respectively, with a horizontal line through the common point, find the resultant.

**Solution:**   Figure 2-49 shows the given concurrent force vectors. A force polygon may now be constructed. From any point $A$, lay out the 20-lb vector **AB**. From point $B$, draw **BC** to represent the 30-lb force. Similarly, draw **CD** and **DE** for the 10- and 40-lb forces, respectively. Now, the resultant is the vector **AE** which closes the polygon, as in Fig. 2-50. By careful measurement, $R$ is found to be 63 lb acting at an angle of 74° with the horizontal. These results may be verified by progressively solving triangles *ABC*, *ACD*, and *ADE* (Fig. 2-51) for unknown sides and angles by Eqs. (2-2) and (2-3). Note that $\theta$ is the sum of $\angle EAD$, $\angle DAC$, $\angle CAB$, and 10°.

## 2-19 EQUILIBRIUM OF MORE THAN TWO FORCES

When three or more forces are in equilibrium, their resultant is zero. From Fig. 2-51, the resultant of the four forces shown is **AE**. The single

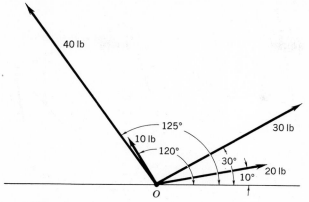

FIGURE 2-49   Diagram for Sample Problem 15.

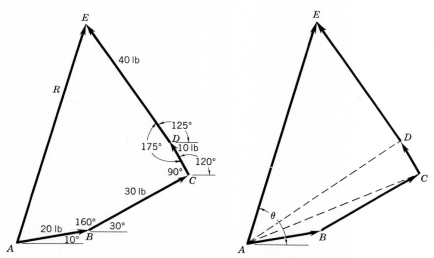

FIGURE 2-50   Force polygon for
Sample Problem 15.

FIGURE 2-51   Force polygon.

force to be in equilibrium with **AE** is **EA**, a force equal in magnitude and opposite in direction. But **EA** closes the polygon. *Then when concurrent forces are in equilibrium, the force diagram must be a closed polygon.*

**Sample Problem 16**  A body weighing 50 lb rests on a rough horizontal plane. A force of 10 lb acts horizontally to the right. The body does not move. Find the force resisting motion and the reaction of the plane (Fig. 2-52).

**Solution:**   Let the body be considered as a particle (point $O$), and draw a free-body diagram (Fig. 2-53). At $O$, the weight of 50 lb acts down. The

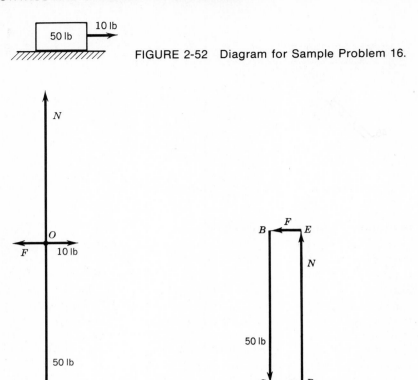

FIGURE 2-52   Diagram for Sample Problem 16.

FIGURE 2-53   Free-body diagram
for Sample Problem 16.

FIGURE 2-54   Force polygon
for Sample Problem 16.

reaction of the plane is up and is represented by $N$. The force of 10 lb acts to the right. Since the body does not move, there must be a force acting between the two surfaces in contact to oppose the motion. Call it $F$. Construct the force polygon $BCDEB$ (Fig. 2-54), taking the forces in order, counterclockwise. Since $N$ and $F$ are known in direction only, first draw the forces completely known. $\mathbf{BC}$ is equal to 50 lb and $\mathbf{CD}$ to 10 lb. At $D$, erect a perpendicular indefinite in length. Since the four forces are in equilibrium, they must form a closed polygon. Then force $F$ must end at $B$. From $B$, draw a line perpendicular to $\mathbf{BC}$ to meet the perpendicular from $D$ at the point $E$. Then $\mathbf{DE}$ is $N$, and $\mathbf{EB}$ is $F$. Since $BCDE$ is a rectangle, lines $\mathbf{BC}$ and $\mathbf{DE}$ are equal in magnitude, or $\mathbf{DE} = 50$ lb $= N$, and $\mathbf{EB} = 10$ lb $= F$.

\*Sample Problem 17  A bar $AC$ (Fig. 2-55), pinned to the wall at $A$, supporting a mass of 1000 kg at $C$, is held in a horizontal position by the strut, or compression member $DB$ pinned at $D$ and $B$. Find the force in $DB$ and the pin reaction at $A$.

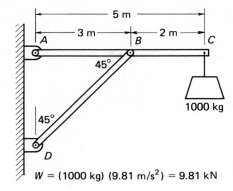

$W = (1000 \text{ kg}) (9.81 \text{ m/s}^2) = 9.81 \text{ kN}$

FIGURE 2-55   Diagram for Sample Problem 17.

**Solution:**   The bar $AC$ is in equilibrium under the action of three forces at $A$, $B$, and $C$, respectively. By the *principle of concurrence* (Sec. 2-11), *the lines of action of three (nonparallel) forces in equilibrium must pass through a common point.* The direction of the 9.81-kN load is vertical at point $C$. Since $DB$ is a two-force member in compression, the direction of the thrust at point $B$ must be along the axis of strut $DB$. Sketch $AC$ as a free body and show the 9.81-kN load and the line of action of the force at $B$. Prolong these lines of action to meet at point $E$ (Fig. 2-56). Then the reaction at pin $A$ is a force that must also pass through point $E$, owing to concurrence. The force triangle may now be drawn, as in Fig. 2-57. The angle $\theta$ which $R_a$ makes with the horizontal is found from Fig. 2-56. Since $BCE$ is a right triangle and $\angle B = 45°$, then length $EC =$ length $BC =$ 2 m. Now from triangle $ACE$,

$$\tan \theta = \tfrac{2}{5} = 0.40$$
$$\theta = 21.8°$$

Thus, in Fig. 2-57,          $\alpha = 90 + \theta = 111.8°$

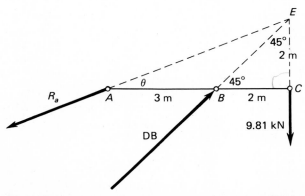

FIGURE 2-56   Free-body diagram of member $AC$ for Sample Problem 17.

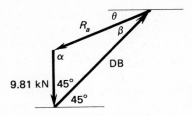

FIGURE 2-57.   Force triangle at point E for Sample Problem 17.

Hence,        $\beta = 180 - 111.8 - 45 = 23.2°$

Now, by the law of sines,

$$\frac{R_a}{\sin 45°} = \frac{9.81}{\sin 23.2°} \qquad R_a = \frac{0.7071(9.81)}{0.3939} = 17.61 \text{ kN (say, 17.6 kN)}$$

$$\frac{DB}{\sin 111.8°} = \frac{9.81}{\sin 23.2°}$$

But          $\sin 111.8° = \sin (90 + \theta)$
And          $\sin (90 + \theta) = \sin (90 - \theta)$
Then         $\sin 111.8° = \sin 68.2°$
             $= 0.9285$

$$DB = \frac{0.9285(9.81)}{0.3939}$$

$$DB = 23.12 \text{ kN (say, 23.1 kN) (compression)}$$

**Sample Problem 18**  The beam $AD$ is pinned to the wall at $A$ and held in a horizontal position by the tension member $BC$. There is a load of 1000 lb at $D$. What are the force in $BC$ and the reaction at $A$ (Fig. 2-58)?

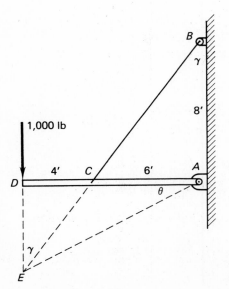

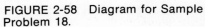

FIGURE 2-58   Diagram for Sample Problem 18.

**Solution:**    The beam $AD$ is in equilibrium under the action of three forces at $A$, $C$, and $D$, respectively. Since the member $BC$ is acted on by forces at its ends only, this member is a two-force member. The force at $C$ is in the direction of $BC$. The load is vertical. Prolong line $BC$ to intersect the vertical through $D$. Then $E$ will be the point of concurrence of the forces. The reaction $A$ is a force whose line of action must pass through $E$. The directions of all the forces are now known, and the force triangle can be constructed. Angles $\theta$ and $\gamma$ may be determined from Fig. 2-58. From right triangle $CAB$,

$$\tan \gamma = \tfrac{6}{8} = 0.75$$
$$\gamma = 36.9° \text{ or } 36°54'$$

Note that angle $CED$ is also $\gamma$.
    Now triangle $ABC$ is similar to triangle $CED$. Thus,

$$\frac{\text{Length } DE}{\text{Length } AB} = \frac{\text{Length } DC}{\text{Length } AC}$$
$$DE = \frac{8(4)}{6} = 5.33 \text{ ft}$$

In triangle $AED$,

$$\tan \theta = \frac{5.33}{10} = 0.533$$
$$\theta = 28.1° \text{ or } 28°6'$$

    Figure 2-59 is the force triangle for the three concurrent forces acting on member $AD$

$$\alpha = 90 + \theta = 118.1° \text{ or } 118°6'$$
$$\beta = 180 - \alpha - \gamma = 180 - 118.1 - 36.9 = 25°$$

By the law of sines,

$$\frac{R_a}{\sin \gamma} = \frac{1000}{\sin \beta} \qquad R_a = \frac{0.6004(1000)}{0.4226} = 1420 \text{ lb}$$
$$\frac{\textbf{BC}}{\sin \alpha} = \frac{1000}{\sin \beta}$$
$$\sin \alpha = \sin (90 + \theta) = \sin (90 - \theta)$$
$$\sin (90 - \theta) = \sin 61.9° = 0.8821$$
$$\textbf{BC} = \frac{0.8821(1000)}{0.4226} = 2087 \text{ lb (say, 2100 lb) (tension)}$$

    Note that the point of concurrence need not lie on the body in equilibrium.

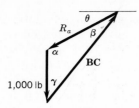

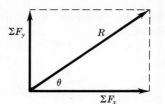

FIGURE 2-59  Force triangle for Sample Problem 18.

FIGURE 2-60  Resultant by summation of components.

## 2-20 RESULTANT OF CONCURRENT FORCES BY SUMMATION

Any force may be resolved into two components at right angles to each other. If $x$ and $y$ axes are taken through the point common to all the forces and each force is resolved into its $x$ and $y$ components the given forces are then replaced by two sets of forces acting along the axes of reference. The forces or component forces along the $x$ axis may be combined by algebraic addition into a single force. This force is represented by $\Sigma F_x$. Similarly, there is found a force acting along the $y$ axis and represented by $\Sigma F_y$ (Fig. 2-60). The hypotenuse of the right triangle with legs equal to $\Sigma F_x$ and $\Sigma F_y$ is the resultant force in magnitude and direction of the given system of forces. That is,

$$F_x = F \cos \alpha \tag{2-4}$$
$$F_y = F \sin \alpha \tag{2-5}$$
$$R = \sqrt{(\Sigma F_x)^2 + (\Sigma F_y)^2} \tag{2-6}$$
$$\tan \theta = \frac{\Sigma F_y}{\Sigma F_x} \tag{2-7}$$

**Sample Problem 19**  Two 120-lb forces, one which is horizontal and the other at an angle of 60° with the horizontal, act on a body (point $O$). Find their resultant (Fig. 2-61).

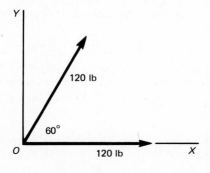

FIGURE 2-61  Diagram for Sample Problem 19.

**Solution:**    Establish $x$ and $y$ axes. The components may be tabulated as follows:

| Force | Angle with x axis | $F_x$ | $F_y$ |
|---|---|---|---|
| 120 | 0 | 120 | 0 |
| 120 | 60 | 60 | 104 |
| | | 180 | 104 |

Then

$$\Sigma F_x = 180 \text{ lb}$$
$$\Sigma F_y = 104 \text{ lb}$$

In Fig. 2-62, $\Sigma F_x$ and $\Sigma F_y$ are shown at 90° with each other. Their resultant is the hypotenuse $R$. By Eq. (2-6),

$$R = \sqrt{180^2 + 104^2} = \sqrt{43\ 200} = 208 \text{ lb}$$

By Eq. (2-7), $\tan \theta = \dfrac{\Sigma F_y}{\Sigma F_x} = \dfrac{104}{180} = 0.5778$

$$\theta = 30°$$

Then a force of 208 lb acting at an angle of 30° with the horizontal is the resultant of the given forces.

That the angle of the resultant is 30° could have been seen by constructing the parallelogram of forces as in Fig. 2-63. Since the figure is equilateral, the diagonal bisects the 60° angle.

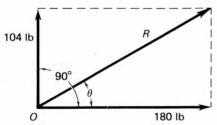

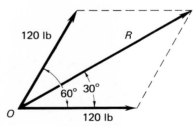

FIGURE 2-62   Summation of components for Sample Problem 19.

FIGURE 2-63   Parallelogram solution for Sample Problem 19.

If the body at $O$ were a pin fastened to a vertical wall and held rigid, the pin would exert a force equal to $R$ in the opposite direction, by the law that action equals reaction. Therefore, the pin reaction is a force of 208 lb acting from $O$, opposite $R$, the angle being 30° below the negative side of the $x$ axis.

**Sample Problem 20**  Find the resultant of the forces with magnitude and direction, as shown in Fig. 2-64.

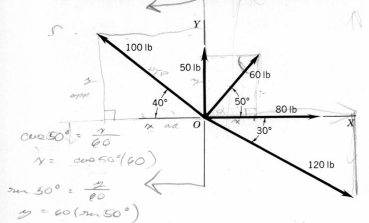

FIGURE 2-64   Diagram for Sample Problem 20.

**Solution:**   Resolve each force into its $x$ and $y$ components and arrange in a table.

| Force | Angle with $x$ axis | $F_x$ | $F_y$ |
|---|---|---|---|
| 80 | 0 | + 80 | 0 |
| 60 | 50 | + 38.6 | + 46 |
| 50 | 90 | 0 | + 50 |
| 100 | 40 | − 76.6 | + 64.3 |
| 120 | 30 | + 104 | − 60 |
| | | + 146 | + 100.3 |

Notice that the forces in the left-hand column cannot be added algebraically because they do not act in the same straight line, whereas those in the $F_x$ and $F_y$ columns are added algebraically because they do act in the same line.

$$\Sigma F_x = 80 + 38.6 + 0 - 76.6 + 104 = +146 \text{ lb}$$
$$\Sigma F_y = 0 + 46 + 50 + 64.3 - 60 = +100.3 \text{ lb}$$

Then, from Fig. 2-65,

$$R = \sqrt{146^2 + 100.3^2} = \sqrt{31\,380} = 177 \text{ lb}$$
$$\tan \theta = \frac{\Sigma F_y}{\Sigma F_x} = \frac{100.3}{146} = 0.687$$
$$\theta = 34.5° \text{ or } 34°30'$$

Therefore, the resultant is a force of 177 lb acting at an angle of 34.5° with the horizontal.

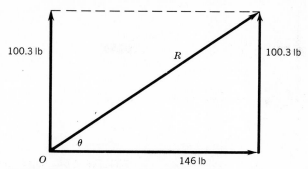

100.3 lb

$R$

100.3 lb

$\theta$

$O$                          146 lb

FIGURE 2-65    Summation of components for
Sample Problem 20.

## 2-21 EQUILIBRIUM OF CONCURRENT FORCES

In Sec. 2-20, it was found that, when concurrent forces act on a body, the
resultant

$$R = \sqrt{(\Sigma F_x)^2 + (\Sigma F_y)^2}$$

Now, when forces are in equilibrium, their resultant is zero. Then

$$(\Sigma F_x)^2 + (\Sigma F_y)^2 = 0$$

But both expressions are plus, being squares of numbers. The sum
of two positive numbers cannot be zero unless each one is zero. That is,

$$\Sigma F_x = 0 \quad \text{and} \quad \Sigma F_y = 0 \quad\quad (2\text{-}8)$$

*Then, when concurrent forces are in equilibrium, the sum of the components
along each of two rectangular axes must be zero.*

*Conversely, if $\Sigma F_x = 0$ and $\Sigma F_y = 0$, the concurrent forces are in equilibrium
and their resultant is zero.*

The foregoing law can be illustrated in another way. A fundamental
principle of mechanics is that a body cannot be in equilibrium when acted
upon by a single force. Now $\Sigma F_x$ is the sum of all the components of
acting forces along any arbitrarily chosen axis. If a body is in equilibrium,
there can be no change in motion. Therefore, there cannot be a force
acting in any direction, or the body would move or change its motion.
Consequently,

$$\Sigma F_x = 0 \quad \text{and} \quad \Sigma F_y = 0$$

$$\sin \alpha = \frac{opp}{hyp}$$

$$\cos \alpha = \frac{adj}{hyp}$$

$$\tan \alpha = \frac{opp}{adj}$$

$$\cot \alpha = \frac{adj}{opp}$$

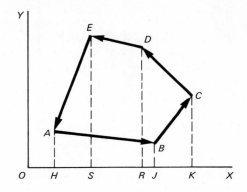

x projection of $AB = HJ$

x projection of $BC = JK$

x projection of $CD = KR$

x projection of $DE = RS$

x projection of $EA = SH$

FIGURE 2-66   Force polygon.

The meaning of Eq. (2-8) can also be illustrated by the force polygon shown in Fig. 2-66. Suppose $ABCDEA$ is the force polygon of five forces in equilibrium. Now the x component of a force is its projection on the x axis as in Fig. 2-66.

But $\Sigma F_x = HJ + JK + KR + RS + SH = 0$. That is, the sum of the positive projections is equal and opposite to the sum of the negative projections.

*A corollary to the preceding law is that, when the resultant effect of concurrent forces along any direction or axis is desired, simply project each of the forces onto the axis and find the algebraic sum of these projections.*

## PROBLEMS

**\*2-1.** A body is acted on by an upward force of 12 N and a horizontal force to the right of 20 N. What are the amount and the direction of the resultant?

**2-2.** Two forces of 100 lb each act on a body at an angle of 120° with each other. Find the resultant in magnitude and direction by constructing the triangle and also by trigonometry.

**2-3.** Two equal forces $P$ act on a body at an angle of 60° with each other. Find the resultant by trigonometry.

X **2-4.** Two concurrent forces of 40 and 50 lb make an angle of 30° with each other. Find the direction and magnitude of their resultant.

**2-5.** Resolve a force of 60 lb into two components, one of which is 38 lb and makes an angle of 45° with the given force. *Hint:* Construct the triangle and apply cosine and sine laws.

X ***2-6.** Three forces of 40, 50, and 75 N, respectively, make angles of 10°, 30°, and 120° with the $x$ axis. What are the amount and the direction of the resultant?

O ***2-7.** A body is supported by two cords, each of which makes an angle of 50° with the vertical. The tension in each cord is 65 N. What is the mass of the body?

O **2-8.** A weight of 6000 lb is suspended from the point $C$ in Fig. Prob. 2-8. Find the forces in $AC$ and $BC$ if $\alpha = 30°$ and $\beta = 60°$. Which member is a tie rod (tension), and which member is a brace (compression)?

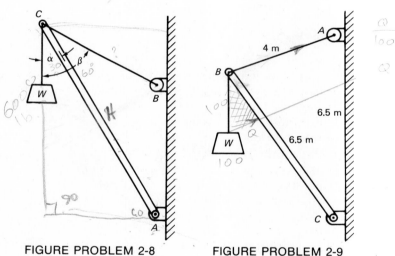

FIGURE PROBLEM 2-8          FIGURE PROBLEM 2-9

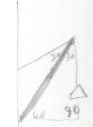

O ***2-9.** A mass of 10 metric tons (see Table 14, Appendix B) is carried by the simple derrick shown in Fig. Prob. 2-9. By making use of similar triangles, find the forces in $AB$ and $BC$.

**2-10.** A body weighing 160 lb is suspended from a hook by a rope 12 ft long. If a horizontal force is applied to the body so as to swing it to one side and cause the rope to make an angle of 30° with the vertical, what will be the tension in the cord, and what is the amount of the horizontal force?

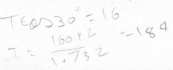

**2-11.** Find the rectangular components of a force of 200 lb making (a) an angle of 10° with the x axis; (b) an angle of 35°; (c) an angle of 55°; (d) an angle of 80°; (e) an angle of 90°.

**2-12.** A simple truss supports a load of 8 kip (1 kip = 1000 lb) (Fig. Prob. 2-12). Find the forces (in kips) in members $AB$, $BC$, and $AC$, and the reactions at $A$ and $C$.

FIGURE PROBLEM 2-12          FIGURE PROBLEM 2-18

**2-13.** A wire 24 in long will stand a straight pull of 100 lb. The ends are fastened to two points 21 in apart on the same level. What weight suspended from the middle of the wire will break it?

**\*2-14.** The lifting force of a balloon is 600 N. The anchor rope makes an angle of 65° with the vertical. Find the tension in the anchor rope and the horizontal force of the wind against the balloon.

**2-15.** A truck weighing 2000 lb rests on a slope of inclination of 30°. Find what force will be needed to start the stalled truck up the grade if the resistance to traction is 50 lb.

**\*2-16.** A tractor is attached to a dump car by a cable which makes a horizontal angle of 10° with the track. It requires 1000 N to move the car. Find the force that the tractor exerts. What is the lateral force on the rails?

**2-17.** A boiler weighing 2 tons is supported by cables making angles of 30° and 40°, respectively, with the vertical, while being hoisted into place. Find the tension in each cable.

**2-18.** Find the forces in $AC$ and $BC$ of Fig. Prob. 2-18.

**2-19.** Find the forces in $BC$ and $AB$ of Fig. Prob. 2-19.

**2-20.** A 300-lb body rests on a plane. Find the components parallel and perpendicular to the plane for the following angles of inclination: (a) 20°, (b) 40°, (c) 70°, and (d) 85°.

**\*2-21.** If a force of 200 N is sufficient to move a body on a horizontal plane, find the force necessary to move it uniformly up a plane of slope 1 in 10, assuming the resistance is the same in both cases.

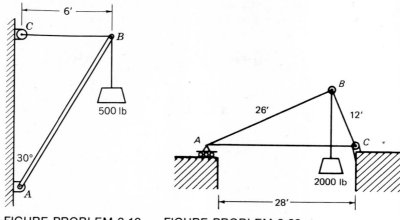

FIGURE PROBLEM 2-19      FIGURE PROBLEM 2-22

**2-22.**   In Fig. Prob. 2-22, find the thrusts in *AB* and *BC*. Also, find the vertical and horizontal components of the reactions at *A* and *C*.

**2-23.**   Figure Problem 2-23 shows a frame made of four pieces of equal length with pivots at points *M, N, S,* and *R*. With the forces acting as shown, find the forces in each of the arms and the horizontal reactions at *S* and *N*.

**2-24.**   Given three concurrent forces of 50, 75, and 90 lb making angles of 0°, 70°, and 120° respectively with the *x* axis, find the resultant.

**\*2-25.**   In a toggle joint as shown in Fig. Prob. 2-25, what vertical force is exerted against the plunger due to the force of 4.0 kN? *AB* = 0.4 m. *BC* = 0.4 m. *BD* = 0.3 m. Assume joint *C* to be stationary.

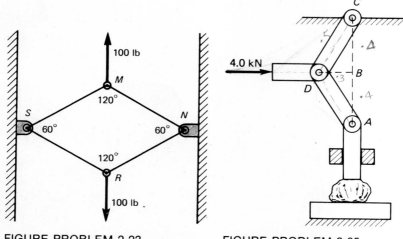

FIGURE PROBLEM 2-23                FIGURE PROBLEM 2-25

# CHAPTER
# 3
# Moments

## 3-1   MOMENTS

A complete study of force systems and their applications requires the understanding of the concept of moments.

Figure 3-1 shows a person balancing a 100-lb load by means of a lever. A downward force of 40 lb applied at $A$ is just sufficient to balance the load of 100 lb. The distance from the pivot to the end of the lever at which the force is applied is 5 ft. Then, 40 lb $\times$ 5 ft = 200 ft·lb is necessary to balance the turning effort of the load, which is also 200 ft·lb; that is, 100 lb $\times$ 2 ft = 200 ft·lb (Fig. 3-2). If the person were to try balancing the load by applying a force at $B$, which is 4 ft from the pivot, 50 lb would be required. That is, 50 lb $\times$ 4 ft = 200 ft·lb. At the 1-ft point, $C$, it would take 200 lb to do the job; 200 lb $\times$ 1 ft = 200 ft·lb. Thus, 200 ft·lb is a measure of the turning effort required to balance the load and is called the *moment* of the force about the pivot.

It should be noted that the moment of a force varies directly with its distance from the pivot. The *moment is measured by the product of the force and the perpendicular distance from the pivot to the line of action of the force*.

For example, it is much easier to turn a revolving door by pushing at the outer edge of the door, as in Fig. 3-3, than by pushing the center, as in Fig. 3-4.

The *moment of a force about an axis* is defined as the force multiplied by the perpendicular distance from the axis to the force. For simplicity, the moment of a force about a point in a plane is to be understood as a moment about an axis perpendicular to the plane. A moment is expressed in foot pounds, or inch pounds, newton-meters, or other combinations of force and distance units.

As a further illustration, if a force is applied to the rim of a pulley mounted on a shaft, the pulley will rotate. Experience shows that it is easier to turn the pulley when the force is applied to the rim than when it is applied closer to the shaft.

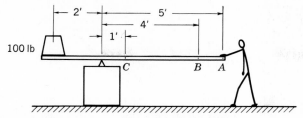

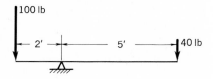

FIGURE 3-1   Use of lever to balance weight.

FIGURE 3-2   Lever.

FIGURE 3-3   Revolving door easily rotated.

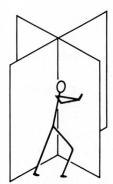

FIGURE 3-4   Revolving door rotated the "hard way."

It is quite apparent that any effort to turn a wheel will not be applied parallel to a spoke, but rather perpendicular to a spoke. Likewise, a force applied parallel to the axle will not rotate the wheel. From such observations, it is seen that no rotation or moment results when the line of action of the force either intersects the axis or is parallel to it. These several conditions will be found useful in deciding about what axis moments may be taken.

## 3-2   SIGN OF MOMENTS

In Fig. 3-5, it is evident that the 100-lb force tends to rotate about $AA$ in one direction, and the 50-lb force tends to rotate about $AA$ in the opposite direction. Some rule for the direction of rotation is then necessary. In this book, the following rotation rules will be used unless stated otherwise.

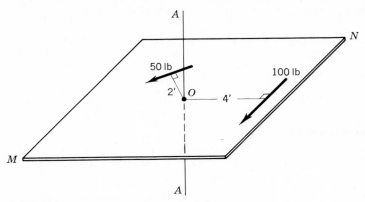

FIGURE 3-5    Rotation about an axis.

*When a force tends to produce clockwise rotation, the moment is positive. When the rotation is counterclockwise, the moment is negative.* For example:

$$\text{Moment of 100-lb force} = +100 \times 4 = +400 \text{ ft·lb}$$
$$\text{Moment of 50-lb force} = -50 \times 2 = -100 \text{ ft·lb}$$
$$\text{Total turning effect} = +400 - 100 = +300 \text{ ft·lb (clockwise rotation)}$$

## 3-3    EQUILIBRIUM OF PARALLEL FORCES

Forces whose lines of action are parallel are called *parallel forces*.

The principles involved in the study and application of parallel forces will be best understood by a simple experiment. Suspend a very light rod $AC$ (Fig. 3-6) by means of two spring balances at $J$ and $K$. Now suspend 120 lb from point $B$. The scale at $J$ will indicate 40 lb, while the one at $K$ will show 80 lb. The forces at $A$, $B$, and $C$ are parallel and in equilibrium (as shown in the free-body diagram, Fig. 3-7). Their sum, $40 + 80 - 120 = 0$. Take moments of all forces about point $A$ as an axis. The algebraic sum of moments about $A$ is represented by $\Sigma M_a$:

$$\Sigma M_a = (8 \times 120) - (12 \times 80) = 960 - 960 = 0$$

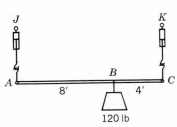

FIGURE 3-6    Parallel forces in equilibrium.

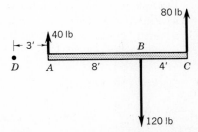

FIGURE 3-7    Free-body diagram.

The algebraic sum of moments about $B$ is

$$\Sigma M_b = (8 \times 40) - (4 \times 80) = 320 - 320 = 0$$

Similarly, about $C$,

$$\Sigma M_c = (12 \times 40) - (4 \times 120) = 480 - 480 = 0$$

About point $D$,

$$\Sigma M_d = -(3 \times 40) + (11 \times 120) - (15 \times 80)$$
$$= -120 + 1320 - 1200 = 0$$

These results illustrate the principle that *when parallel forces are in equilibrium, the sum of their moments about any axis through any point is zero.*

Parallel forces tend to translate and rotate the body on which they act. If such a body is in equilibrium, it must neither translate nor rotate. Then the *conditions for static equilibrium* are as follows:

1. The algebraic sum of the forces must be zero.

$$\Sigma F = 0 \tag{3-1}$$

2. The algebraic sum of the moments about any axis through any point must equal zero.

$$\Sigma M = 0 \tag{3-2}$$

The resultant of the 40 lb at $A$ and 80 lb at $C$ must be their sum, $80 + 40 = 120$, and must act upward at the point $B$. Then

$$\frac{80}{40} = \frac{\text{length } AB}{\text{length } BC} = \frac{8}{4}$$

The resultant is applied at a point dividing the distance between the forces into two parts that are in an inverse ratio to the forces themselves.

In Fig. 3-8, the resultant of $P_1$ and $P_2$ is their sum, $P_1 + P_2$, and acts at $B$ such that

$$\frac{P_1}{P_2} = \frac{l_2}{l_1} \tag{3-3}$$

Clearing of fractions produces

$$l_1 P_1 = l_2 P_2 \tag{3-3a}$$

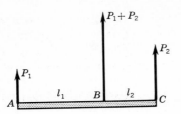

FIGURE 3-8   Resultant of parallel forces.

That is, the moments of the two forces about a point on the resultant are equal to each other.

Taking moments of the forces about $A$,

$$l_1(P_1 + P_2) = (l_1 + l_2)P_2$$

or    $$l_1P_1 + l_1P_2 = l_1P_2 + l_2P_2$$

The last equation could have been obtained by adding $l_1P_2$ to both sides of Eq. (3-3a). The result shows that the moment of the resultant of two parallel forces about any point is equal to the sum of moments of the forces about the same point.

The previous statement can be extended to include any number of coplanar parallel forces. Thus, the resultant of any number of parallel forces is parallel to these forces and equal to their sum, and its moment about any point is equal to the sum of the moments of all the forces about the same point.

**Sample Problem 1** A bar 10 ft long carries a load of 20 lb, 6 ft from the end. What force must be applied at each end to support the rod in equilibrium (Fig. 3-9)?

**Solution a:**   The resultant of $P_1$ and $P_2$ must be 20 lb applied at $A$. By the inverse ratio,

$$\frac{P_1}{P_2} = \frac{4}{6}$$

That is, $\frac{4}{10}$ of the load is at $B$, or $(\frac{4}{10})$ (20) = 8 lb. Also, $\frac{6}{10}$ is at $C$, or $(\frac{6}{10})$ (20) = 12 lb.

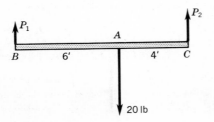

FIGURE 3-9   Diagram for Sample Problem 1.

**Solution b:** By the conditions of equilibrium,

$$\Sigma M_b = (6)(20) - 10P_2 = 0$$
$$P_2 = 12 \text{ lb}$$

Since $\Sigma F_y = 0$
$$P_1 + P_2 - 20 = 0$$
$$P_1 = 20 - 12 = 8 \text{ lb}$$

**Sample Problem 2** A beam resting on two end supports carries concentrated loads, as shown in Fig. 3-10. Find the reactions of the supports.

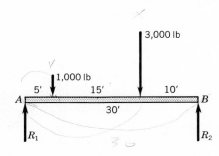

FIGURE 3-10   Diagram for Sample Problem 2.

**Solution a:** Let $R_1$ and $R_2$ be the reactions. By the conditions of equilibrium, the algebraic sum of the moments about any point is zero. A convenient point to choose would be either the right or left support, since one of the unknowns would be eliminated. If the point $A$ is chosen,

$$\Sigma M_a = +(1000)(5) + (3000)(20) - 30R_2 = 0$$
$$R_2 = 2170 \text{ lb}$$
But
$$R_1 + R_2 = 1000 + 3000 = 4000 \text{ lb}$$
$$R_1 = 4000 - 2170 = 1830 \text{ lb}$$

**Solution b:** The problem may also be solved by the inverse ratio. The two components of 1000 lb, acting down at $A$ and $B$, are found as follows:

$$\frac{\text{Component at } A}{\text{Component at } B} = \frac{25}{5} = \frac{5}{1}.$$

Then
$$\text{Component at } A = \tfrac{5}{6}(1000) = 833 \text{ lb}$$
$$\text{Component at } B = \tfrac{1}{6}(1000) = 167 \text{ lb}$$

Similarly,
$$\frac{\text{Component of 3000 at } A}{\text{Component of 3000 at } B} = \frac{10}{20} = \frac{1}{2}$$

Then    Component of 3000 at $A = \frac{1}{3}(3000) = 1000$ lb
        Component of 3000 at $B = \frac{2}{3}(3000) = 2000$ lb
                Total force at $A = 833 + 1000 = 1833$ lb (say, 1830 lb)
                Total force at $B = 167 + 2000 = 2167$ lb (say, 2170 lb)

But $R_1$ and $R_2$ are the equilibrants, or balancing forces:

$$R_1 = 1830 \text{ lb} \qquad R_2 = 2170 \text{ lb}$$

## 3-4 UNIFORMLY DISTRIBUTED LOADS

Thus far, we have considered how to calculate the moment produced by a *concentrated* load. Since a concentrated load is assumed to be acting at a point, its moment arm is simply the perpendicular distance from the line of action of the load to the point about which the moment is assumed to be acting.

Another type of load frequently encountered is a *uniformly distributed load*, which is assumed to be acting over an area rather than at a point. Whereas a concentrated load is designated in terms of a force unit, such as 4000 lb, the uniformly distributed load is usually designated in terms of a force unit per unit length, such as 750 lb/ft.

As an illustration of a uniformly distributed load, consider a multistory structural steel building whose partial typical floor is indicated in Fig. 3-11a. Assuming that the 21- $\times$ 20-ft bay will be subjected to a uniformly distributed load of 100 lb/ft², the beam indicated will be subjected to a uniformly distributed load of 700 lb/ft. This was arrived at as follows.

The floor beam in question is responsible for carrying the load halfway between itself and the adjacent beams on either side. Thus, this beam will be subjected to a 100-lb/ft² uniformly distributed load over an area measuring 20 $\times$ 7 ft (area shown shaded). Each *linear foot* of beam (typical 1-ft strip) will then be subjected to 100 lb/ft² $\times$ 7 ft = 700 lb/ft of uniformly distributed load. Figure 3-11b shows a view of the beam supporting the load and in turn being supported by the girders ($R_1$ and $R_2$ indicate the reactions of the girders at each end of the beam).

In order to calculate the moment developed by a uniformly distributed load, the entire weight of the load is assumed to be concentrated at its centroid, which is the center of gravity of an area (see Sec. 10-2). Thus, the 700-lb/ft load produces a clockwise moment about $R_1$ equal to (700 lb/ft $\times$ 20 ft) (10 ft). The quantity (700 lb/ft $\times$ 20 ft) represents the total weight of the load and the 10 ft represents the moment arm (the distance from the centroid of the area representing the load, to $R_1$).

*Sample Problem 3*   When a load is uniformly distributed over a beam, the total uniform load may be considered as acting at the center of the length of beam over which it is distributed, when the moment of such a

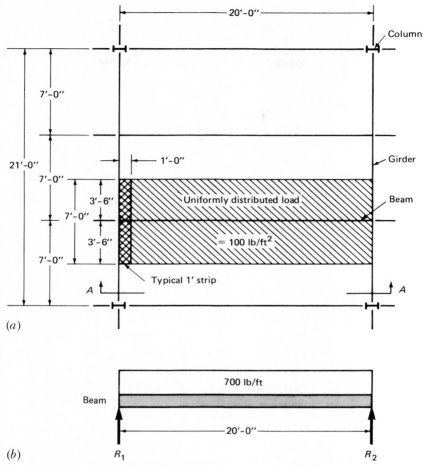

(a)

(b) R₁ R₂

FIGURE 3-11 (a) Partial plan view of structural steel floor framing. (b) Section A-A of Figure 3-11a.

load is desired. The weight of a beam is a uniformly distributed load. Find the reactions of the beam loaded as shown in Fig. 3-12. Ignore the load due to the beam itself.

**Solution:** The beam is in equilibrium; thus, $\Sigma F = 0$.

$$R_1 + R_2 = 1000(3) + 4000 + 1000(1.5) = 3000 + 4000 + 1500 = 8500 \text{ N}$$

For the purpose of taking moments only, the uniform load $AB$ may be assumed to be concentrated at point $C$, and load $EF$ is assumed to be concentrated at point $D$. Taking moments about point $A$,

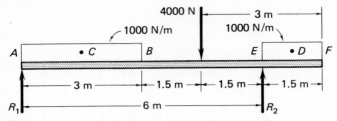

FIGURE 3-12   Diagram for Sample Problem 3.

$$\Sigma M_a = 3000(1.5) + 4000(4.5) + 1500(6.75) - R_2(6)$$
$$R_2 = 5440 \text{ N} = 5.44 \text{ kN}$$
$$R_1 = 8500 - 5440 = 3060 \text{ N} = 3.06 \text{ kN}$$

## 3-5   COUPLES

Figure 3-13 shows a sports car steering wheel. In order to turn the wheel clockwise, the driver must exert equal and opposite forces (20 lb) as shown. These two forces form a couple, because they are *parallel, equal* in magnitude, and are *opposite* in direction. Obviously, a couple can cause rotation only, since $\Sigma F_x = 0$ and $\Sigma F_y = 0$. The perpendicular distance between the two forces is called the *arm* of the couple.

Since the couple causes rotation, the moment, or torque, developed by the couple can be determined by taking the summation of the moments of the forces about some point. First we shall consider moments about the center of the wheel, point $B$,

$$\Sigma M_b = +20(\tfrac{1}{2}) + 20(\tfrac{1}{2}) = +20 \text{ ft·lb}$$

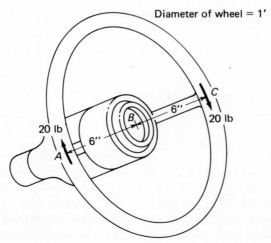

FIGURE 3-13   Steering wheel.

Then, moments about point $A$,

$$\Sigma M_a = +20(1) = +20 \text{ ft} \cdot \text{lb}$$

Choose a point $D$, 1 ft to the left of $A$, as a moment center. Then

$$\Sigma M_d = -20(1) + 20(2) = +20 \text{ ft} \cdot \text{lb}$$

The example shows that the moment of this couple is a constant and is equal to one of the forces times the arm, or distance between the forces. The statement is true, in general, and if the forces are $F$ and $F$ with an arm $a$, the moment is $Fa$.

Since a single force produces translation and a couple produces rotation, a force cannot balance a couple.

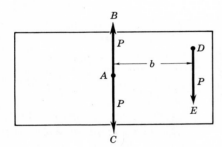

FIGURE 3-14   Eccentric force = force + couple.

In Fig. 3-14, a single force $P$ acts at a point $D$. At $A$, $b$ ft from $D$, draw two equal and opposite forces $P$. They neutralize each other, and the resultant effect remains the same. Now, forces along $AB$ and $DE$ form a couple with moment $bP$, tending to rotate the body clockwise. There remains the force $P$ along $AC$, tending to translate the body.

A single force acting at a point, not the center of the body, causes rotation around the center and a translation of the center.

A single force acting at one point is equivalent to the same force, acting at a different point, plus a couple.

Common illustrations of the couple are seen in the forces that are applied to the steering wheel of the automobile or to the arms of a lug wrench.

## PROBLEMS

**3-1.**   A beam 12 ft long is supported at the ends. A load of 600 lb is placed 5 ft from the left end. What are the reactions?

**\*3-2.**   What should be the values of the forces $R$ and $F$ for the member shown in Fig. Prob. 3-2 to keep the member in equilibrium?

LOVIS
SALOUMEH
9-25-84

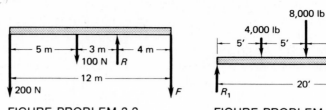

FIGURE PROBLEM 3-2                    FIGURE PROBLEM 3-3

**3-3.** With the loads as shown in Fig. Prob. 3-3, determine the values of $R_1$ and $R_2$ by moments and also by the inverse-ratio principle.

**3-4.** Figure Problem 3-4 shows a beam weighing 40 lb per linear foot. This beam carries a load of 2000 lb, 4 ft from the right end, and an additional load, uniformly distributed, of 300 lb/ft on the 6 ft adjacent to the left end. Find $R_1$ and $R_2$.

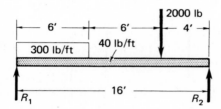

FIGURE PROBLEM 3-4

**3-5.** Find the position of the resultant load for a truck with a wheel base of 140 in, if the distribution of the load is such that the reaction on the rear axle is 3.5 tons and that on the front axle is 2.5 tons.

**\*3-6.** In Fig. Prob. 3-6, the beam overhangs the left support and is supported at the right end. The beam weighs 1000 N/m. With the other loads as shown, what are the reactions?

**3-7.** What are the reactions of the beam shown in Fig. Prob. 3-7? The beam weighs 30 lb per linear foot.

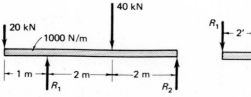

FIGURE PROBLEM 3-6

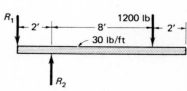

FIGURE PROBLEM 3-7

**\*3-8.** A beam supported at the ends is 6 m long and weighs 700 N/m. It carries a concentrated load of 8 kN, 2.7 m from the right end, and a total uniformly distributed load of 14 kN, which extends the full length of the beam. What are the reactions?

**3-9.** A timber of uniform weight and cross section is shown in position on the rollers while being moved (Fig. Prob. 3-9). What are the roller reactions? With the rollers still 6 ft apart, what should be the position of the timber so that one roller will support twice as much as the other? The timber is 18 ft long.

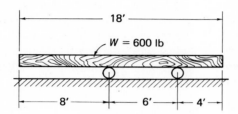

FIGURE PROBLEM 3-9

**3-10.** What are the end reactions for simply supported floor beams 16 ft long and spaced 4 ft from center to center? The total load on the floor including its own weight is 140 psf.

**3-11.** A wheel 3 ft in diameter weighs 2500 lb with its load (Fig. Prob. 3-11). Find the horizontal force necessary to start the wheel over an obstruction 6 in high. (Forces act through the center.)

**\*3-12.** Find the horizontal force $F$ necessary to rotate the block about the point $A$ shown in Fig. Prob. 3-12.

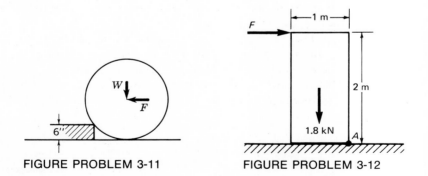

FIGURE PROBLEM 3-11          FIGURE PROBLEM 3-12

**3-13.** What force $P$ is necessary to maintain equilibrium in the arrangement of the beams shown in Fig. Prob. 3-13? $CD$ has a mass of 15.3 kg; $AB$ has a mass of 10.2 kg and is pivoted at $B$.

**3-14.** A safety valve shown in Fig. Prob. 3-14 is 3 in in diameter and is

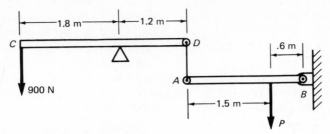

FIGURE PROBLEM 3-13

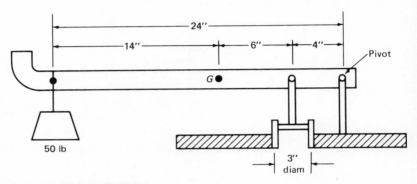

FIGURE PROBLEM 3-14

just on the point of blowing off steam. The lever weighs 7 lb and its center of weight is at $G$. The weight of the valve is 3 lb. If the weight on the end of the lever is 50 lb, find the pressure of the steam in the boiler. If the steam is to blow off at 80 psi, find the amount of the weight on the end of the lever.

*3-15.   Figure Problem 3-15 shows an air cylinder and a system of levers. If the air pressure in the cylinder is 350 kPa, what is the force exerted at $F$? If a force of 40 kN is desired, what should be the air pressure?

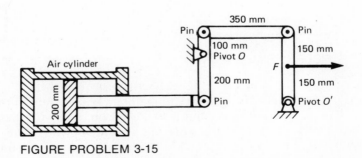

FIGURE PROBLEM 3-15

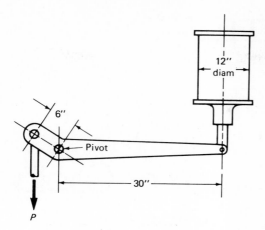

FIGURE PROBLEM 3-16

**3-16.** Figure Problem 3-16 represents a brake cylinder. The lever is bent at an angle of 120°. What force is produced in the brake rod by a pressure of 60 psi in the cylinder?

**3-17.** Forces of 10 lb each are applied on opposite sides of a steering wheel 18 in in diameter to form a couple. What is the moment of the couple? If a single force of 20 lb is applied to the rim of the steering wheel, how is a couple produced, and what is its moment?

**3-18.** An iron pipe 1 in in diameter is embedded in a concrete base (Fig. Prob. 3-18). A horizontal force of 20 lb is applied perpendicular to *BC*. Describe the effect that this force has on each of the lengths of pipe and the elbow.

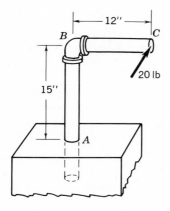

FIGURE PROBLEM 3-18

# CHAPTER
# 4
# Nonconcurrent-Coplanar Forces; Trusses

## 4-1 RESULTANT OF NONCONCURRENT-COPLANAR FORCES

In Chap. 2, on concurrent forces, it was shown that the resultant of a set of concurrent forces was a single force determined by $\Sigma F_x$ and $\Sigma F_y$. Also, in Chap. 3, on parallel forces, the resultant was found to be a single force and a couple, determined by $\Sigma F_x$, $\Sigma F_y$, and $\Sigma M$. In this chapter, it will be shown that the resultant of nonconcurrent nonparallel forces is also a single force and a couple. The principle will be illustrated by an example before taking up the general proof.

Figure 4-1 shows three forces in action. Find the resultant in magnitude and position with reference to point $O$. At $O$, Fig. 4-2a, introduce two forces of 100 lb each, opposite in direction and parallel to the original 100-lb force. This does not change conditions, for the forces introduced are in equilibrium. The two 100-lb forces, drawn as solid lines, form a couple with a moment arm of 2 ft, producing rotation around $O$ in a counterclockwise direction. The other 100-lb force at $O$, shown by a dotted line, causes translation exactly as if the original 100-lb force had been applied at that point. The other forces are treated in the same way. The 200-lb force, Fig. 4-2b, is equivalent to a couple of moment $+4(200) = +800$ ft·lb and a single force of 200 lb at $O$. Also, the 300-lb force, Fig. 4-2c, is equivalent to a couple of moment $-5(300) = -1500$ ft·lb, and a single 300-lb force at $O$. If Figs. 4-2a, 4-2b, and 4-2c are superimposed, the combination is equivalent to the original system. The moment of the entire system is

$$\Sigma M = -200 + 800 - 1500 = -900 \text{ ft·lb}$$

The resultant of the forces (dotted lines) at $O$ is found as in Chap. 2.

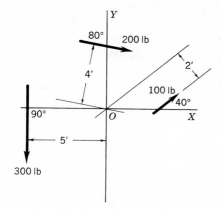

FIGURE 4-1   Three nonconcurrent-coplanar forces.

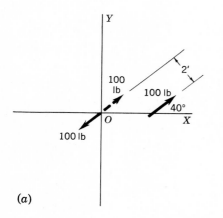

(a)

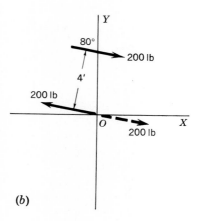

(b)

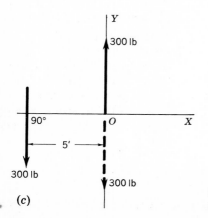

(c)

FIGURE 4-2   Transformation of nonconcurrent forces into equivalent concurrent forces and couples.

$$\Sigma F_x = 100 \cos 40° + 200 \cos 10° + 300 \cos 90° = +274 \text{ lb}$$
$$\Sigma F_y = 100 \sin 40° - 200 \sin 10° - 300 \sin 90° = -270 \text{ lb}$$
$$R = \sqrt{274^2 + 270^2} = 385 \text{ lb}$$

$$\tan \theta = -\frac{270}{274} = -0.985$$
$$\theta = 315.42° \text{ or } 315°25'$$

The resultant is then a force of 385 lb, making an angle of 315.42° with the x axis. It is directed to the right and down. It must have a moment of −900 ft·lb. Then

$$385r = 900$$
$$r = 2.34 \text{ ft}$$

Then $R$ must be drawn to the right and down. Since the moment is negative, $R$ must cause counterclockwise rotation. Its location is shown in Fig. 4-3. The resultant is then a force and a couple.

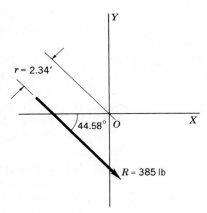

FIGURE 4-3  Resultant of noncon-current force system.

## 4-2  GENERAL METHOD

Now, take a set of forces $F_1$, $F_2$, $F_3$ at the distances $r_1$, $r_2$, $r_3$, respectively, from the origin $O$ in Fig. 4-4. At $O$, insert two forces $F_1$ parallel to $F_1$ and opposite in direction. Two of the forces form a couple of moment $F_1r_1$, and a single force $F_1$ acts at $O$. Doing the same with $F_2$ and $F_3$, we obtain a set of couples tending to rotate around $O$, the resultant moment of which is

$$\Sigma M = F_1r_1 + F_2r_2 + F_3r_3 = \Sigma(Fr) \qquad (4\text{-}1)$$

There is also a set of concurrent forces at $O$, the resultant of which is

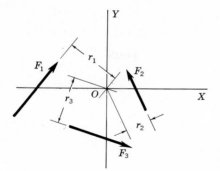

FIGURE 4-4 Nonconcurrent-coplanar forces—general method.

$$R = \sqrt{(\Sigma F_x)^2 + (\Sigma F_y)^2} \qquad (4\text{-}2)$$

$$\tan \theta = \frac{\Sigma F_y}{\Sigma F_x} \qquad (4\text{-}3)$$

$$r = \frac{\Sigma M}{R} \qquad (4\text{-}4)$$

The resultant $R$ is located by drawing a vector, making an angle $\theta$ with the $x$ axis at a distance $r$ from the origin. The resultant, therefore, is equivalent to a force and a couple.

If the system of forces is in equilibrium, $R = 0$. But this result can be true only when $\Sigma F_x = 0$ and $\Sigma F_y = 0$. This condition shows that a body at 0 would not be translated by the system of forces. But, for equilibrium, there must be no rotation, a state that can exist when and only when $\Sigma M = 0$. There are then three equations of condition for the solution of forces in equilibrium. If the original forces $F_1$, $F_2$, and $F_3$ had been resolved into their $F_x$ and $F_y$ components in the position in which they are located, there would then have been two system of forces, one parallel to the $x$ axis and the other parallel to the $y$ axis. But, from Chap. 3, the resultant of a set of parallel forces is their algebraic sum. These sums are $\Sigma F_x$ and $\Sigma F_y$, just as if they had been resolved into their components at point $O$ and their sum taken.

If $\Sigma F_x = 0$ and $\Sigma F_y = 0$ while $\Sigma M$ does not equal 0, the resultant of the system of forces is a couple; also, if $\Sigma M = 0$ while $\Sigma F_x$ and $\Sigma F_y$ are not equal to 0, the resultant is a single force. When all three of the foregoing expressions equal zero, the system of forces is in equilibrium.

## 4-3 GRAPHICAL METHOD

Let $F_1$, $F_2$, and $F_3$ of Fig. 4-5 be three nonconcurrent-coplanar forces. To find their resultant graphically, extend the lines of $F_1$ and $F_2$ to intersect at $A$ (Fig. 4-6a). Lay off $F_1'$ and $F_2'$ equal, respectively, to $F_1$ and $F_2$. Complete the parallelogram. The diagonal $R_1$ is the resultant of $F_1$ and

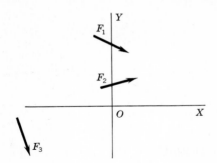

FIGURE 4-5  Nonconcurrent-coplanar forces—graphical method.

$F_2$. Now extend $R_1$ to intersect the line of action of $F_3$ at $B$ (Fig. 4-6$b$). Lay off $R_1'$ and $F_3'$ equal, respectively, to $R_1$ and $F_3$. Complete the parallelogram. The diagonal $R_2$ is the resultant of $R_1$ and $F_3$, or of $F_1$, $F_2$, and $F_3$. With reference to a body at point $O$, not on $R_2$, the resultant is equivalent to a single force $R_2$ at point $O$, and a couple of moment $R_2 r$, causing rotation about point $O$.

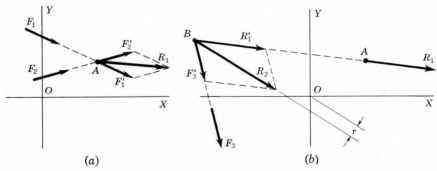

(a)                                   (b)

FIGURE 4-6  (a) Resultant of $F_1$ and $F_2 = R_1$. (b) Resultant of $F_3$ and $R_1 = R_2$.

If point $O$ is on the line of action of $R_2$ (Fig. 4-7$a$), the resultant is a single force $R_2$. The moment of $R_2$ about point $O$ is zero, since the moment arm is zero.

If $F_3$ should be equal and parallel to $R_1$, but opposite in direction (Fig. 4-7$b$), the resultant force is zero. However, there is a resultant couple of moment $R_1 r$, where $r$ is the distance between the parallel forces.

If $F_3$ is equal and opposite to $R_1$ and coincides with it, then the resultant force and couple are each equal to zero. The system of forces is then in equilibrium.

The preceding method of determining a resultant is general and may be extended to include any number of forces.

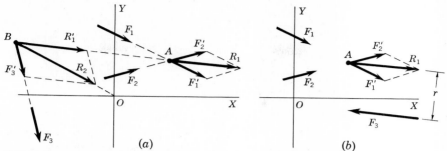

FIGURE 4-7   (a) Resultant $R_2$ has no moment if its line of action passes through point $O$. (b) A moment $R_1 r$ exists and resultant $R_2$ is zero when $F_3$ is equal, parallel, and opposite to $R_1$.

## 4-4 APPLICATIONS

Many common nonconcurrent-coplanar force systems occur in pinned or hinged structures, machine linkages, and in beams. The application of the foregoing principles will be demonstrated by a series of illustrative examples covering a variety of situations. Certain groups of structures will be subjected to more intensive study. For example, beams are discussed in Chaps. 11, 12, and 16, columns in Chap. 15, and trusses in this chapter.

**Sample Problem 1** A 10-ft ladder rests against a smooth wall. The ladder weighs 25 lb. A 155-lb man stands on it, as shown in Fig. 4-8, while his helper holds his foot against the bottom to prevent the ladder from slipping on the smooth floor. Find the unknown force and reactions.

**Solution:**  Use the ladder as a free body and indicate the various known and unknown forces, as in Fig. 4-9.

Since the system is in equilibrium, the sum of vertical forces must be zero, the sum of horizontal forces must be zero, and there can be no unbalanced moment.

$$\Sigma F_y = 0$$
$$R_1 - 25 - 155 = 0$$
$$R_1 = 25 + 155 = 180 \text{ lb}$$
$$\Sigma F_x = 0$$
$$F - R_2 = 0$$
$$F = R_2$$
$$\Sigma M_0 = 0$$
$$25(5 \cos 60°) + 155(7 \cos 60°) - R_2(10 \sin 60°) = 0$$
$$R_2 = 70 \text{ lb}$$
$$F = 70 \text{ lb}$$

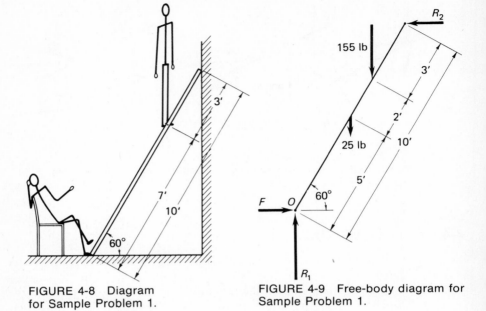

FIGURE 4-8   Diagram
for Sample Problem 1.

FIGURE 4-9   Free-body diagram for
Sample Problem 1.

*Sample Problem 2   A beam 6 m long having a mass of 45.9 kg is pinned
to the floor at $A$ and rests against a smooth wall at $D$. It carries a load of
900 N at $C$. Find the reaction at $A$ and $D$ (Fig. 4-10).

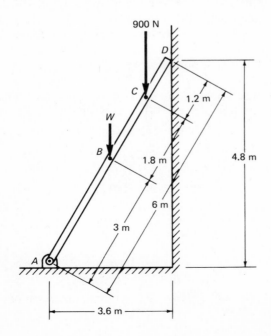

FIGURE 4-10   Diagram for
Sample Problem 2.

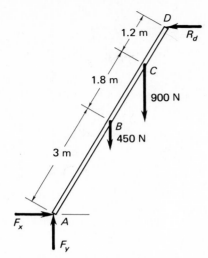

FIGURE 4-11   Free-body diagram
for Sample Problem 2.

**Solution:**   Set $AD$ out as a free body. Since the wall is perfectly smooth,
the wall reaction at $D$ will be normal to the wall. The reaction at $A$, being
unknown in magnitude and direction, will be replaced by its two components $F_x$ and $F_y$. The weight of the beam $W$ is at $B$ and is equal to
$W = m \cdot g = (45.9 \text{ kg})(9.81 \text{ m/s}^2) = 450$ N; the 900-N load is at $C$ (see Fig.
4-11).

$\Sigma F_y = 0$

$$F_y - 450 - 900 = 0$$
$$F_y = 450 + 900 = 1.35 \text{ kN}$$

$\Sigma F_x = 0$

$$F_x - R_d = 0$$
$$F_x = R_d$$

$\Sigma M_a = 0$

$$450 \left(3.6 \times \frac{3}{6}\right) + 900 \left(3.6 \times \frac{4.8}{6}\right) - R_d(4.8) = 0$$

(moment arms are calculated based on similar triangles)
$$R_d = 709 \text{ N} = 0.709 \text{ kN}$$
$$F_x = 709 \text{ N} = 0.709 \text{ kN}$$

By Fig. 4-12,

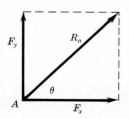

FIGURE 4-12   Summation of force components
at point $A$.

$$R_a = \sqrt{F_x^2 + F_y^2}$$
$$= \sqrt{(0.709)^2 + (1.35)^2} = \sqrt{2.33}$$
$$= 1.53 \text{ kN}$$

$$\tan \theta = \frac{F_y}{F_x}.$$
$$= \frac{1.35}{0.709}$$
$$= 1.90$$
$$\theta = 62.3°$$

**Sample Problem 3**  An A-frame carries a load of 1000 lb, as shown in Fig. 4-13. Find the floor reactions at $A$ and $E$ and the pin reactions at $B$, $C$, and $D$.

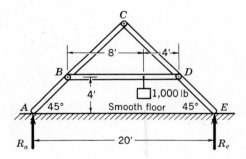

FIGURE 4-13   Diagram for Sample Problem 3.

**Solution:**  Set the entire A-frame out as a free body (Fig. 4-13). Since the load, or action, is vertical, the reactions at $A$ and $E$ will also be vertical, there being no tendency for the frame to move sidewise.

$\Sigma F_y = 0$

$$R_a + R_e - 1000 = 0$$
$$R_a + R_e = 1000$$

$\Sigma M_a = 0$

$$1000(12) - R_e(20) = 0$$
$$R_e = 600 \text{ lb}$$
$$R_a = 400 \text{ lb}$$

Now set all three members out as free bodies (Fig. 4-14). Since all the pin reactions are unknown, they must be represented by their horizontal and vertical components. The student should note very carefully the directions of these components. Since the member $BD$ pulls downward on $AC$ and also prevents the motion of $AC$ to the left, $F_{bx}$ acts to the right on $AC$, and $F_{by}$ acts downward. Since $AC$ prevents $BD$ from falling and also exerts a pull to the left on $BD$ in the free-body diagram of $BD$,

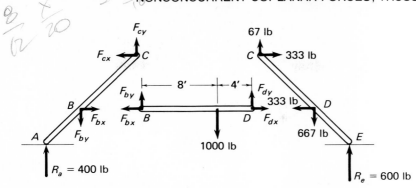

FIGURE 4-14   Free-body diagrams for Sample Problem 3.

$F_{bx}$ acts to the left and $F_{by}$ acts upward. A similar condition is shown at $C$ and $D$. This result verifies the principle that action and reaction are equal. When members of the A-frame are joined together, there must be equilibrium at each of the points $B$, $C$, and $D$. That is, for each of the pins, $\Sigma F_x = 0$ and $\Sigma F_y = 0$. Referring again to the free-body drawing of $AC$, the reader will observe that there are four unknowns. Since only three equations are possible, these unknowns cannot be found without further information.

Now, consider the free-body drawing of $BD$. From this sketch it is possible to determine the two vertical components by applying $\Sigma M = 0$ and $\Sigma F_y = 0$.

$\Sigma F_y = 0$

$$F_{by} + F_{dy} - 1000 = 0$$
$$F_{by} + F_{dy} = 1000$$

$\Sigma M_b = 0$

$$1000(8) - F_{dy}(12) = 0$$
$$F_{dy} = 667 \text{ lb}$$
$$F_{by} = 333 \text{ lb}$$

Inserting this value of $F_{by}$ on the free-body drawing of $AC$, we have but three unknowns remaining, and the solution may be completed. By taking moments around $C$, two of the three remaining unknowns are eliminated.

$\Sigma M_c = 0$

$$+400(10) - 333(6) - F_{bx}(6) = 0$$
$$F_{bx} = 333 \text{ lb}$$

$\Sigma F_x = 0$

$$333 - F_{cx} = 0$$
$$F_{cx} = 333 \text{ lb}$$

$\Sigma F_y = 0$

$$400 - 333 + F_{cy} = 0$$
$$F_{cy} = -67 \text{ lb}$$

The minus sign indicates that the wrong direction was assigned to $F_{cy}$. It acts downward. From the free-body diagram of $BD$, it is evident that $F_{dx} = F_{bx} = 333$ lb. As a check on the solution, insert the known values in the free-body drawing of $CE$ (Fig. 4-14).

$$\Sigma F_x = 0$$
$$+333 - 333 = 0$$
$$\Sigma F_y = 0$$
$$+600 - 667 + 67 = 0$$
$$\Sigma M_e = 0$$
$$-4(667) - 4(333) + 333(10) + 10(67) = 0$$

The resultant pin reaction at $B$ is found as follows:

$$F_b = \sqrt{F_{bx}^2 + F_{by}^2} = \sqrt{(333)^2 + (333)^2} = 471 \text{ lb}$$
$$\tan \theta = \frac{333}{333} = 1.0$$
$$\theta = 45°$$

In the same way, the pin reactions at $C$ and $D$ may be found to be

$$F_c = 340 \text{ lb at } 11.3° \text{ from the horizontal}$$
$$F_d = 745 \text{ lb at } 63.5° \text{ from the horizontal}$$

**Sample Problem 4**  Figure 4-15 shows a derrick carrying a load of 2000 lb. The boom $DG$ weighs 500 lb, and the mast weighs 400 lb. Find the reactions at $E$, $D$, $C$, $B$, and $F$. Neglect the weight of member $CF$.

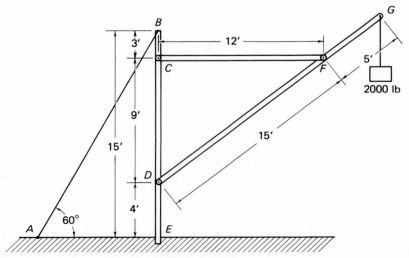

FIGURE 4-15  Diagram for Sample Problem 4.

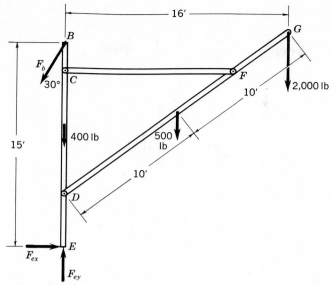

FIGURE 4-16   Free-body diagram for Sample Problem 4.

**Solution:**   First, set the entire derrick out as a free body (Fig. 4-16). The derrick is acted upon by the load, the weights of the members, the cable $AB$, and the ground at $E$. Since $AB$ is a two-force member, the tension $F_b$ in cable $AB$ must act along the cable. The reaction at $E$ is unknown, and therefore we represent it by its components. Taking moments about point $B$,

$$\Sigma M_b = 0$$
$$-15F_{ex} + 8(500) + 16(2000) = 0$$
$$F_{ex} = 2400 \text{ lb}$$
$$\Sigma F_x = 0$$
$$2400 - F_b \sin 30° = 0$$
$$F_b = 4800 \text{ lb}$$
$$\Sigma F_y = 0$$
$$F_{ey} - 400 - 500 - 2000 - 4800 \cos 30° = 0$$
$$F_{ey} = 7060 \text{ lb}$$

Next, classify the members of the derrick. $AB$ and $CF$ are two force members, whereas $EB$ and $DG$ are not. Since there is no point at which only two-force members meet, the next step in the solution will be to set out $DG$ and $EB$ as free bodies (Figs. 4-17 and 4-18). Applying the conditions of equilibrium to Fig. 4-17,

$$\Sigma M_d = 0$$
$$-9F_f + 8(500) + 16(2000) = 0$$
$$F_f = 4000 \text{ lb} = F_c$$

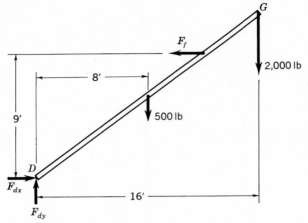

FIGURE 4-17  Free-body diagram of member *DG*.

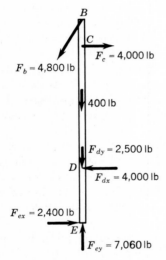

FIGURE 4-18  Free-body diagram of member *EB*.

$$\Sigma F_x = 0$$

$$F_{dx} - 4000 = 0$$
$$F_{dx} = 4000 \text{ lb}$$

$$\Sigma F_y = 0$$

$$F_{dy} - 500 - 2000 = 0$$
$$F_{dy} = 2500 \text{ lb}$$

The reader can now find the resultant reactions by the method shown in the preceding problem. These results are

$F_e = 7460$ lb at $71.2°$ from horizontal
$F_d = 4720$ lb at $32°$ from horizontal
$F_b = 4800$ lb at $60°$ from horizontal
$F_c = F_f = 4000$ lb horizontal

*Check:* Since the forces are all known, the free-body drawing of *BD* may now be used as a check on the solution (Fig. 4-18).

$$\Sigma F_x = 0 \quad 2400 - 4000 + 4000 - 4800 \sin 30° = 0$$
$$\Sigma F_y = 0 \quad 7060 - 2500 - 400 - 4800 \cos 30° = 0$$
$$\Sigma M_b = 0 \quad -15(2400) + 11(4000) - 2(4000) = 0$$

## 4-5  TRUSSES

Trusses are structures whose members are connected to form triangles. The triangle is significant because it is stable and cannot collapse as long as a member does not break or deform. Thus, when designing a truss, it is necessary to find the force in each member of the truss and then select structural members which are adequate with regard to strength and resistance to buckling.

Such structures may exist as roof trusses, bridges, airplane-wing trusses, booms of industrial cranes, and the like, as illustrated in Fig. 4-19. The method of connection of the members may be either with pins, welds, or bolts as shown in Fig. 4-20. At one time, rivets were widely used but are seldom used today.

Roof trusses are generally made of wood or steel depending upon the nature of the structure. Most other trusses would be steel, except for wing trusses, which are usually made of aluminum or magnesium alloys. Wood trusses are generally joined by bolts. Metal trusses are joined either by bolts or welds, with pin connections being used only at the reaction points.

A structure consisting mainly of bolted or welded joints is complex to analyze with theoretical exactness. It is much simpler to assume that all the joints are connected by means of frictionless pins (which, of course, they are not). This assumption yields results which are suitable for many important types of structures. Even for structures which are unusually large or stiff it is possible to first assume the joints are pin-connected and then apply a correction factor.

There are two common methods for analysis of trusses. One is called the *method of joints*; the other, the *method of sections*. Both are based on principles of static equilibrium which were established in Chap. 2, and in both, the free-body method of analysis is important. The method of joints is most convenient in cases where it is desired to determine the forces on all the members of the truss or those near the reactions only. The method of sections is most convenient in cases where it is desired to find the

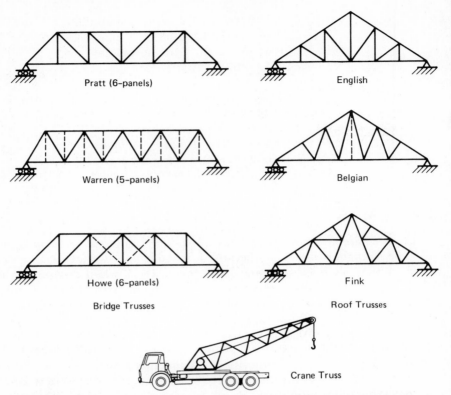

FIGURE 4-19   Several common types of trusses (broken lines represent members occasionally added).

forces on only one or a few members, particularly those which are not at or near the supports. Because each method has its advantages and area of application, both methods will be described here. In each method, the members of the truss are assumed to be connected by frictionless pins, and the members of the truss are assumed to be two-force members.

The reader should recall that two-force compression members may be subjected to both buckling *and* compression, due to column action, as mentioned in Sec. 2-8 and discussed in Chap. 15. In our treatment of trusses, we will only be finding the forces which the members must sustain and transmit. The next design step of selecting the actual sizes of compression members would require the techniques shown in Chap. 15.

## 4-6   METHOD OF JOINTS

Since the members of a truss are two-force members connected by frictionless pins, each force acting at the end of the member is assumed to be acting axially along the member. Furthermore, since the conditions

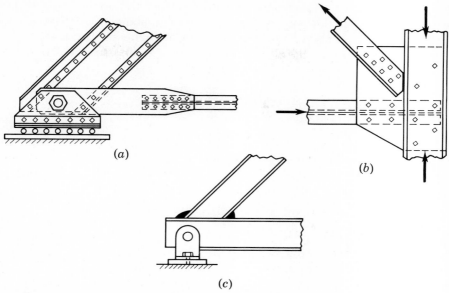

(a)

(b)

(c)

FIGURE 4-20   Methods of connecting truss members.

of static equilibrium apply, the end forces on a member must be equal and opposite.

The method of joints will be demonstrated in the following problem, where it is desired to find the internal forces in each one of the member of the truss (Fig. 4-21).

The reactions at the supports are found by setting out the entire truss as a free body and applying the equations for static equilibrium. First, we should note that because support $A$ is a roller-type support, no horizontal reaction can occur at $A$. However, because support $E$ is a pinned-type support, it is possible for a horizontal reaction to occur at $E$. Since the truss is not subjected to any horizontal forces, the horizontal

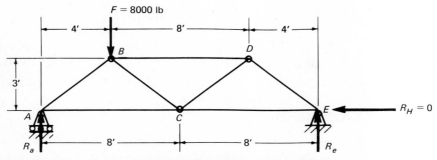

FIGURE 4-21   Truss diagram to illustrate method of joints.

reaction at $E$ $(R_H)$ must be equal to zero and will not be considered in our calculations. Suppose we first take moments about an axis through support $A$ and set the algebraic sum of these moments to zero $(\Sigma M_a = 0)$. This gives us $F(4) - R_e(16) = 0$, from which $R_e$ may be calculated since the load $F$ is known.

$$R_e(16) = 8000(4)$$
$$R_e = 2000 \text{ lb}$$

Now sum the vertical forces algebraically and set this sum to zero $(\Sigma F_y = 0)$. We obtain

$$R_a + R_e - F = 0$$
$$R_a = 8000 - 2000 = 6000 \text{ lb}$$

Now it is necessary to find a joint where there are only two unknown forces. Then, using $\Sigma F_x = 0$, $\Sigma F_y = 0$, and information obtained from the geometry of the situation, it is possible to solve for the two unknown forces.

By referring to Fig. 4-21, it can be seen that joints $A$ and $E$ have two unknown forces each, whereas all the other joints have more than two unknown forces acting (originally, joints $A$ and $E$ had three unknowns before the reactions were computed). Therefore, it is necessary to start either with joint $A$ or $E$.

We shall start with $A$ and draw a free-body diagram of the joint (Fig. 4-22). In the free-body diagram, all the forces acting on the joint $A$ are shown. Included are the reaction $R_a$ and the forces exerted by the members $AB$ and $AC$. Since $AB$ and $AC$ are two-force members, their forces are shown acting axially along the directions of the members. For simplicity, the symbols **AB** and **AC** are used to represent the forces exerted upon the pin by these members.

Although the lines along which the forces act are known, it is necessary to determine in which direction the arrows are pointing. The choice is easily made in this case. $R_a$ is known to be acting upward. Since equilibrium is known to exist (joint $A$ is at rest), $R_a$ must be balanced by a downward force. Since **AC** cannot provide a downward component, such a component must come from **AB**, which then must be acting as shown in Fig. 4-22 in order that $\mathbf{AB}_y$ is downward. This selection of direction for **AB** then automatically makes $\mathbf{AB}_x$ act to the left, so that **AC**

FIGURE 4-22   Free-body diagram of joint $A$.

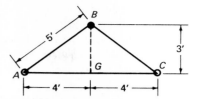

FIGURE 4-23 Free-body diagram of joint $A$ showing force **AB** resolved into its $x$ and $y$ components.

FIGURE 4-24 Part of the truss showing the geometric relationship of the members.

must act to the right in order to balance $\mathbf{AB}_x$. This method of selecting the direction of forces will be found to work in most cases. In the remaining cases, it is necessary only to assume a direction for the force and solve the problem accordingly. Should the answer come out to be negative, then it is only necessary to change the direction of the force.

We can now set up equations for the forces on this joint. It is first desirable to resolve **AB** into its $x$ and $y$ components (Fig. 4-23).

$$\Sigma F_x = 0 \quad \mathbf{AC} - \mathbf{AB}_x = 0 \quad \mathbf{AB}_x = \mathbf{AC}$$
$$\Sigma F_y = 0 \quad 6000 - \mathbf{AB}_y = 0 \quad \mathbf{AB}_y = 6000 \text{ lb}$$

The third equation comes from the geometry of the situation, from which **AB** was resolved into $\mathbf{AB}_x$ and $\mathbf{AB}_y$. The basic information comes from the direction (line of action) in which **AB** is known to act. Member $AB$ is the hypotenuse of a 3-4-5 triangle, as seen in Fig. 4-24. Since force **AB** acts in the same direction as member $AB$, it must make the same angle with the horizontal and the vertical, and the two triangles $ABG$ and $A'B'G'$ (Fig. 4-25) can be seen to be similar. Corresponding sides of similar triangles are proportional, so that

$$\frac{\mathbf{B'G'}}{BG} = \frac{\mathbf{A'G'}}{AG} = \frac{\mathbf{A'B'}}{AB}$$

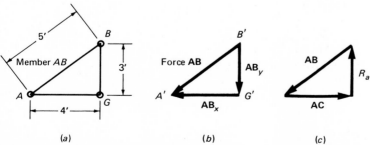

(a)  (b)  (c)

FIGURE 4-25 (a) Part of the truss which forms a 3-4-5 right triangle. (b) Force **AB** and its components. (c) Force triangle for joint $A$.

But     $\mathbf{B'G'} = \mathbf{AB}_y$
$\mathbf{A'G'} = \mathbf{AB}_x$
$\mathbf{A'B'} = \mathbf{AB}$      from the vector triangle Fig. 4-25b

and     $AB = 5$ ft
$BG = 3$ ft
$AG = 4$ ft      from the truss-member triangle Fig. 4-25a

Therefore,

$$\frac{\mathbf{AB}_y}{3} = \frac{\mathbf{AB}_x}{4} = \frac{\mathbf{AB}}{5}$$

Three equations result, since they mean

$$\frac{\mathbf{AB}_y}{3} = \frac{\mathbf{AB}_x}{4}$$

$$\frac{\mathbf{AB}_x}{4} = \frac{\mathbf{AB}}{5}$$

$$\frac{\mathbf{AB}_y}{3} = \frac{\mathbf{AB}}{5}$$

From these equations, we obtain

$$\frac{\mathbf{AB}_y}{3} = \frac{\mathbf{AB}}{5}$$

$$\mathbf{AB}_y = \frac{3\,\mathbf{AB}}{5}$$

and

$$\frac{\mathbf{AB}_x}{4} = \frac{\mathbf{AB}}{5}$$

$$\mathbf{AB}_x = \frac{4\,\mathbf{AB}}{5}$$

We had previously found that $\mathbf{AB}_x = \mathbf{AC}$ and $\mathbf{AB}_y = 6000$ lb. Combining these equations with those above, we can solve for $\mathbf{AB}$ and $\mathbf{AC}$.

$$\mathbf{AB}_y = \frac{3\,\mathbf{AB}}{5} = 6000 \text{ lb}$$

$$\mathbf{AB} = \frac{5(6000)}{3} = 10\,000 \text{ lb (compression)}$$

$$\mathbf{AB}_x = \frac{4\,\mathbf{AB}}{5} = \mathbf{AC}$$

$$\mathbf{AC} = \frac{4(10\,000)}{5} = 8000 \text{ lb (tension)}$$

also,      $\mathbf{AB}_x = \mathbf{AC} = 8000$ lb

Since member $AB$ is seen to be pushing on joint $A$, joint $A$ is pushing back on member $AB$, and the member is in compression.

Since member $AC$ is seen to be pulling on joint $A$, joint $A$ is pulling back on member $AC$, and this member is in tension.

The unknown forces at joint $A$ could have been determined by recognizing that the force triangle made by connecting **AB**, **AC**, and $R_a$ (Fig. 4-25c) is similar to triangle $ABG$ of Fig. 4-25a. From the proportionality of these triangles, we find that

$$\frac{\mathbf{AB}}{5} = \frac{\mathbf{AC}}{4} = \frac{R_a}{3}$$

Thus,        $$\mathbf{AB} = \frac{5R_a}{3} = \frac{5(6000)}{3} = 10\ 000\ \text{lb (compression)}$$

and        $$\mathbf{AC} = \frac{4R_a}{3} = \frac{4(6000)}{3} = 8000\ \text{lb (tension)}$$

The use of a force triangle and geometry gives a rapid solution for joints with three forces when one force is known.

A set of free-body diagrams showing joint $A$, member $AB$, and joint $B$ are shown in Fig. 4-26. It is easy to see from these diagrams that if one end of a member acts with a certain force on one joint, it acts with an equal and opposite force on the other joint to which it is connected.

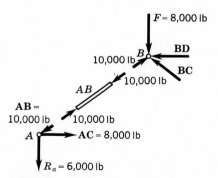

FIGURE 4-26   Free-body diagrams of member *AB* and joints *A* and *B*.

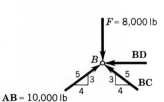

FIGURE 4-27   Free-body diagram of joint *B*.

Knowing the value of force **AB** acting upon joint $B$ is important, since it leaves us only two unknown forces acting on the joint, **BC** and **BD** (Fig. 4-27).

These forces can be determined using $\Sigma F_x = 0$ and $\Sigma F_y = 0$. Note that it is not necessary to calculate the components of **AB** again. As can be seen from Fig. 4-28, since the forces are equal and opposite, the components are also equal and opposite.

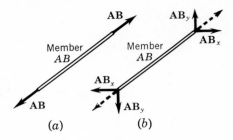

FIGURE 4-28   (a) Free-body diagram of member AB. (b) Components of force acting on member AB.

With respect to joint $B$, $\mathbf{AB}_x = 8000$ lb and $\mathbf{AB}_y = 6000$ lb, as before. Then $\Sigma F_x = 0$ gives

$$\mathbf{AB}_x - \mathbf{BD} - \mathbf{BC}_x = 0$$
$$\mathbf{BC}_x = -\mathbf{BD} + 8000$$

and $\Sigma F_y = 0$ gives

$$\mathbf{AB}_y + \mathbf{BC}_y - 8000 = 0$$
$$\mathbf{BC}_y = 8000 - 6000 = 2000 \text{ lb}$$

From the geometry of the situation, using the similar triangle method as before,

$$\frac{\mathbf{BC}_y}{3} = \frac{\mathbf{BC}_x}{4} = \frac{\mathbf{BC}}{5}$$

from which

$$\mathbf{BC}_x = \frac{4\mathbf{BC}_y}{3} = \frac{4(2000)}{3} = 2670 \text{ lb}$$

Substituting and solving for $\mathbf{BD}$, we obtain

$$\mathbf{BC}_x = 2670 = -\mathbf{BD} + 8000$$
$$\mathbf{BD} = 5330 \text{ lb (compression)}$$

Also,

$$\mathbf{BC} = \frac{5\mathbf{BC}_y}{3} = \frac{5(2000)}{3} = 3330 \text{ lb (compression)}$$

We can now proceed to joint $C$ and $D$, using the principles explained before, and finally to joint $E$ (which can be used as a check). Figure 4-29 shows joints $C$, $D$, and $E$ with the forces acting and the values computed thus far. The remainder of this solution is left to the reader. The results of these calculations are:

$$\mathbf{CD} = 3330 \text{ lb (tension)}$$
$$\mathbf{DE} = 3330 \text{ lb (compression)}$$
$$\mathbf{CE} = 2670 \text{ lb (tension)}$$

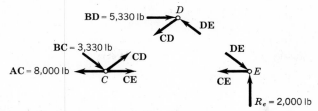

FIGURE 4-29    Free-body diagrams of joints *C, D,* and *E.*

## 4-7  METHOD OF SECTIONS

Suppose we have a truss such as the one loaded as shown in Fig. 4-30. It is desired to find the force in member *FH*.

To do this by the method of joints, it is necessary to solve for the reaction at joint *A* and then solve for the forces involved at joints *A, B, C, D, E,* and *F,* which, of course, is quite involved.

A simpler solution is available through the method of sections. First solve for the reaction at support *A*, using $\Sigma M_n = 0$.

$\Sigma M_n = 0$ gives

$$R_a(48) - 4000(36) - 8000(4) = 0$$
$$48R_a = 144\,000 + 32\,000$$
$$R_a = 3670\,\text{lb}$$

Note that it is not necessary to solve for both reactions in this method, although $R_n$ can be calculated easily using $\Sigma F_y = 0$.

Take a section through the truss, as shown in Fig. 4-31. Note that this section passes through the unknown member *FH* and through two other members *FG* and *EG*. Now draw a free-body diagram of the truss to the left of the section, as in Fig. 4-32. This free-body diagram is not complete because there are forces acting upon the portions of members *FH, FG,* and *EG* which are part of the free body. The existence of these forces can be easily understood from the following reasoning. If the truss

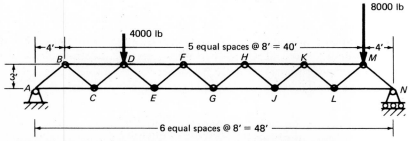

FIGURE 4-30    Truss diagram to illustrate method of sections.

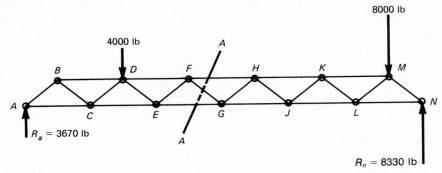

FIGURE 4-31   Free-body diagram of truss showing section plane cutting unknown member *FH*.

existed only as shown in Fig. 4-32 without forces acting at the "cut" members, it would fall down, as can be determined by inspection of the moment situation. Something must be helping to hold the truss up. Obviously, the rest of the truss is doing it and can only do it by means of forces acting at points 1, 2, and 3. The only things which can supply these forces are the remainder of members *FH, FG,* and *EG,* which are part of the rest of the truss. Members *FH, FG,* and *EG* are either in tension or compression. Take, for example, *FH* in compression. Since *FH* is in equilibrium, force *P* must also be acting at point 1 (see Fig. 4-33). A similar argument can be extended to the other members.

The correct free-body diagram then looks like Fig. 4-34. The directions of force **FH, FG,** and **EG** could have been selected arbitrarily. However, examination of the situation shows that we shall need an upward component to help balance the 4000-lb force at *D* (hence, **FG** in the direction as shown), and since **FG** will probably be small (the amount required for $\Sigma F_y = 0$ is only 330 lb), we shall also need a counterclockwise moment to balance the clockwise moment caused by the 3670- and 4000-lb forces. Hence, **FH** and **EG** will be in the directions as shown in Fig. 4-34.

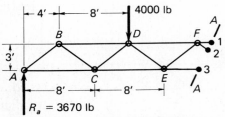

FIGURE 4-32   Incomplete free-body diagram of portion of truss to left of Section A-A.

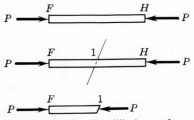

FIGURE 4-33   Equilibrium of a member cut by a section plane.

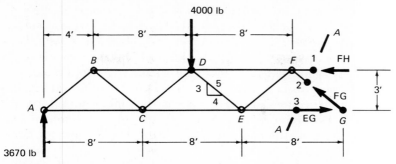

FIGURE 4-34  Complete free-body diagram of portion of truss to left of Section A-A.

If readers desire to circumvent this reasoning exercise, they may select arbitrary directions for the vectors **FH**, **FG**, and **EG**. If an incorrect assumption is made, the answer for that force will be negative and all that need be done is simply to reverse its direction.

The free-body diagram is now complete and we can proceed to solve. Note that we have three unknowns; but we also have three equations available, namely, $\Sigma F_x = 0$, $\Sigma F_y = 0$, and $\Sigma M = 0$.

Since the sum of the moments about any axis is zero, let us try to select an axis which will simplify our calculation to the greatest extent. Note that forces **FG** and **EG** both pass through joint $G$ (located in Fig. 4-34). If we take moments about point $G$, forces **FG** and **EG** have zero moment about $G$, which leaves only the 4000-lb, 3670-lb, and **FH** forces with moments about $G$.

$\Sigma M_g = 0$ gives

$$3670(24) - 4000(12) - 3\mathbf{FH} = 0$$
$$88\,000 - 48\,000 - 3\mathbf{FH} = 0$$
$$3\mathbf{FH} = 40\,000$$
$$\mathbf{FH} = 13\,300 \text{ lb (compression)}$$

Note that it was not necessary to use $\Sigma F_x = 0$ and $\Sigma F_y = 0$ to solve for **FH**.

This problem did not require a calculation for forces **FG** and **EG**, but if they were needed, several approaches are possible. For example, **EG** could be computed directly by taking moments about joint $F$, eliminating the unknown force **FG**. Then **FG** can be obtained by taking moments about joint $E$, or, if preferred, the components of **FG** can be calculated by $\Sigma F_x = 0$ and $\Sigma F_y = 0$. These calculations will give:

$$\mathbf{EG} = 13\,800 \text{ lb (tension)}$$
$$\mathbf{FG} = 550 \text{ lb (compression)}$$

An extremely useful modification of the method of sections allows direct calculation of the forces in slanted internal truss members. The procedure employs $\Sigma F_y = 0$ and is often called the *method of shears*. Suppose it is necessary to calculate **FG** in Fig. 4-34 without finding **FH** or **EG**. It is again necessary first to calculate the support reactions and then to choose a section through the truss which cuts no vertical members and no other slanted member except the one whose force is to be calculated. This requirement is met by the section taken in Fig. 4-34. Now, simply apply the condition that $\Sigma F_y = 0$ and solve for the vertical component of **FG**.

$$3670 - 4000 + \mathbf{FG}_y = 0$$
$$\mathbf{FG}_y = 330 \text{ lb (upward)}$$

From the geometry of the truss,

$$\frac{\mathbf{FG}_y}{\mathbf{FG}} = \frac{3}{5}$$
$$\mathbf{FG} = \frac{330(5)}{3}$$
$$= 550 \text{ lb (compression)}$$

This method can be applied conveniently to members *AB, BC, CD, DE, EF,* and any other slanted member of the truss.

**Sample Problem 5** Find the forces in all members of the truss shown in Fig. 4-35. The method of joints is to be used.

**Solution:** First determine the reactions at the supports. Since the loads are symmetrically placed, the support reactions are each one-half of the total load, or

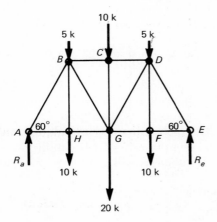

FIGURE 4-35  Diagram for Sample Problem 5.

$\Sigma F_y = 0$

$$R_a + R_e = 5 + 10 + 5 + 10 + 20 + 10 = 60$$
$$R_a = R_e = 30 \text{ kip}$$

At joint $A$ (Fig. 4-36),

$\Sigma F_y = 0$

$$R_a - \textbf{AB} \sin 60° = 0$$

$$\textbf{AB} = \frac{30}{\sin 60°} = \frac{30}{0.866} = 34.6 \text{ kip (compression)}$$

$\Sigma F_x = 0$

$$\textbf{AH} - \textbf{AB} \cos 60° = 0$$
$$\textbf{AH} = 34.6(0.50) = 17.3 \text{ kip (tension)}$$

FIGURE 4-36    Free-body diagram of joint $A$.

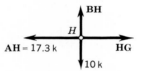

FIGURE 4-37    Free-body diagram of joint $H$.

Now, we cannot go to joint $B$ next since there are three unknown forces, **BH**, **BG**, and **BC**. At joint $H$, however, there are only two unknowns; therefore, we choose joint $H$ next. At joint $H$ (Fig. 4-37),

$\Sigma F_y = 0$

$$\textbf{BH} - 10 = 0$$
$$\textbf{BH} = 10 \text{ kip (tension)}$$

$\Sigma F_x = 0$

$$\textbf{HG} - 17.3 = 0$$
$$\textbf{HG} = 17.3 \text{ kip (tension)}$$

At joint $B$ (Fig. 4-38),

$\Sigma F_y = 0$

$$34.6 \cos 30° - 5 - 10 - \textbf{BG} \cos 30° = 0$$

$$\textbf{BG} = \frac{15}{0.866} = 17.3 \text{ kip (tension)}$$

$\Sigma F_x = 0$

$$34.6 \sin 30° + \textbf{BG} \sin 30° - \textbf{BC} = 0$$
$$\textbf{BC} = 17.3 + 8.7$$
$$\textbf{BC} = 26 \text{ kip (compression)}$$

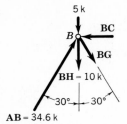

FIGURE 4-38   Free-body diagram of joint *B*.

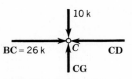

FIGURE 4-39   Free-body diagram of joint *C*.

At joint *C* (Fig. 4-39),

$\Sigma F_y = 0$

$$\mathbf{CG} - 10 = 0$$
$$\mathbf{CG} = 10 \text{ kip (compression)}$$

$\Sigma F_x = 0$

$$26 - \mathbf{CD} = 0$$
$$\mathbf{CD} = 26 \text{ kip (compression)}$$

At joint *G* (Fig. 4-40),

$\Sigma F_y = 0$

$$17.3 \sin 60° + \mathbf{GD} \sin 60° - 10 - 20 = 0$$
$$\mathbf{GD}(0.866) = 30 - 15 = 15$$
$$\mathbf{GD} = \frac{15}{0.866} = 17.3 \text{ kip (tension)}$$

$\Sigma F_x = 0$

$$\mathbf{GF} + 17.3 \cos 60° - 17.3 \cos 60° - 17.3 = 0$$
$$\mathbf{GF} = 17.3 \text{ kip (tension)}$$

The reader will find that the remaining forces can be predicted from those already calculated because the loading is symmetrical, as are the truss members. Thus, **AB** = **DE**, **AH** = **FE**, **BH** = **DF**, **BG** = **GD**, **HG** = **GF**, and **BC** = **CD**. If the loading were not symmetrical, it would be

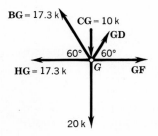

FIGURE 4-40   Free-body diagram of joint *G*.

necessary to proceed from joint to joint until all the forces were evaluated. There will always be one joint that has not been used. Summing forces about this "leftover" joint provides a check on the calculations.

The results for this problem are tabulated below:

| Member | Force, kip | Type |
|--------|-----------|------|
| AB | 34.6 | C |
| BC | 26.0 | C |
| CD | 26.0 | C |
| DE | 34.6 | C |
| EF | 17.3 | T |
| FG | 17.3 | T |
| GH | 17.3 | T |
| HA | 17.3 | T |
| HB | 10.0 | T |
| BG | 17.3 | T |
| CG | 10.0 | C |
| GD | 17.3 | T |
| DF | 10.0 | T |

**\*Sample Problem 6**  Find the forces in members $CD$ and $DL$ in the truss shown in Fig. 4-41.

**Solution:**  The method of sections applied conveniently to this problem. The support reactions will share the total load equally, owing to the symmetrical arrangement.

$$\Sigma F_y = 0$$

$$R_a + R_g - 112.5 - 225 = 0$$

$$R_a = R_g = 168.8 \text{ kN}$$

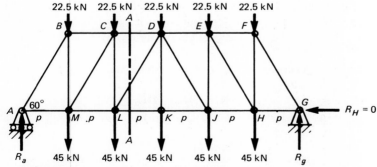

FIGURE 4-41   Diagram for Sample Problem 6.

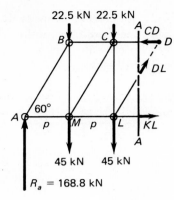

FIGURE 4-42   Free-body diagram of portion of truss to left of Section A-A.

Take a section through the truss which cuts the member to be solved. In this case, both *CD* and *DL* can be cut by the same section. Set out the left portion of the sectioned truss as a free body (Fig. 4-42). Take moments about joint *L* to obtain a direct solution for **CD**. Note that the widths of all panels are equal (but unknown, $p$).

$$\Sigma M_l = 0$$
$$168.8(2p) - 22.5(p) - 45(p) - \mathbf{CD}(\sqrt{3}p) = 0$$
$$\mathbf{CD}(\sqrt{3})p = 337.6p - 67.5p = 270.1p$$
$$\mathbf{CD} = \frac{270.1p}{\sqrt{3}p} = 156 \text{ kN (compression)}$$

Now, sum forces vertically to find $\mathbf{DL}_y$.

$$\Sigma F_y = 0$$
$$168.8 + \mathbf{DL}_y - 22.5 - 22.5 - 45 - 45 = 0$$
$$\mathbf{DL}_y = -168.8 + 135 = -33.8$$

The direction of **DL** was incorrectly chosen in Fig. 4-42. **DL** should act downward, putting the member in compression instead of tension. Thus,

$$\mathbf{DL}_y = 33.8 \text{ kN (downward)}$$

But from the direction of **DL** (60° with the horizontal), we note that

$$\frac{\mathbf{DL}}{2} = \frac{\mathbf{DL}_y}{\sqrt{3}} = \frac{\mathbf{DL}_x}{1}$$

Then        $$\mathbf{DL} = \frac{2\mathbf{DL}_y}{\sqrt{3}} = \frac{2(33.8)}{\sqrt{3}} = 39 \text{ kN (compression)}$$

Incidentally, the direction of **KL** in Fig. 4-42 is correct, which can be verified by inspecting the moment situation about joint $D$.

## PROBLEMS

**4-1.**  Three smooth cylinders, each 18 in in diameter and each weighing 30 lb, are placed in a box as shown in Fig. Prob. 4-1. What are the forces at all points of contact?

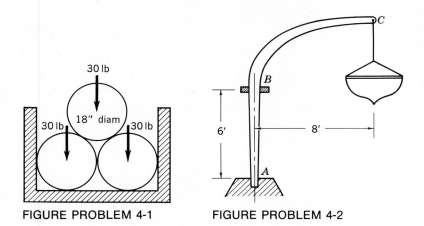

FIGURE PROBLEM 4-1                    FIGURE PROBLEM 4-2

**4-2.**  An ocean liner has an arrangement for supporting lifeboats and for lowering them over the side, as shown in Fig. Prob. 4-2. There is a socket at $A$ and a smooth hole through the deck rail at $B$. If the boat and its load weigh 2000 lb, what are the reactions at $A$ and $B$? Two identical davits support each lifeboat.

**\*4-3.**  A member of uniform cross section having a mass of 68.8 kg is hinged at its upper end to a beam and is held in the position shown in Fig. Prob. 4-3 by means of the horizontal brace $BD$. What will be the reaction at $C$ and the force in $BD$?

**4-4.**  The jointed frame in Fig. Prob. 4-4 carries loads as shown. What are the vertical and horizontal components of the pin reactions?

**\*4-5.**  A brace is hinged at one end to a vertical wall and at the other end to a beam 3.6 m long. The beam has a mass of 183.5 kg and is also hinged to a vertical wall as shown in Fig. Prob. 4-5. The beam carries a load of 2.7 kN at the free end. What will be the compressive force in the brace, and what will be the values of the vertical and horizontal components of the reaction at hinge $A$?

**\*4-6.**  A timber of uniform cross section, having a mass of 90 kg, is hinged at its lower end and held at an angle of 60° with the horizontal by a rod attached as shown in Fig. Prob. 4-6. A cylinder

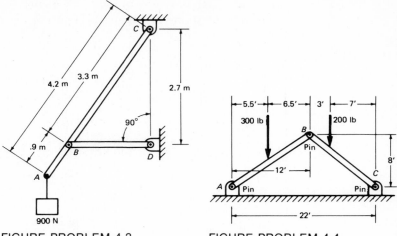

FIGURE PROBLEM 4-3          FIGURE PROBLEM 4-4

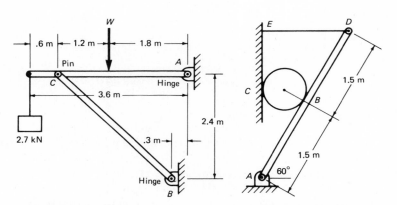

FIGURE PROBLEM 4-5                    FIGURE PROBLEM 4-6

which has a mass of 30 kg is placed between the timber and the wall. What are the horizontal and vertical components of the reaction at $A$? What is the force between the cylinder and the timber?

**\*4-7.** A gate has a mass of 140 kg, which may be considered as uniformly distributed (Fig. Prob. 4-7). A small child, having a mass of 40 kg, climbs up on the gate at the point $B$. What will be the reactions at the hinges?

**4-8.** In an irrigation project, it was found necessary to cross low ground or else swing the canal to the left by cutting into the solid rock. It was decided to run the canal as a flume and support it on a

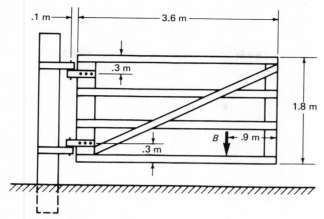

FIGURE PROBLEM 4-7

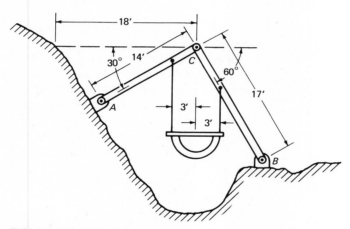

FIGURE PROBLEM 4-8

number of frames, as shown in Fig. Prob. 4-8. The two members rest in sockets in solid rock at points $A$ and $B$. These sockets may be considered as hinges. What will be the vertical and horizontal components of the reactions at $A$ and $B$? The weight of the water in the flume supported by each frame is estimated as 18 200 lb.

**4-9.**   The A-frame in Fig. 4-13 has an additional load of 1200 lb at the center of $BD$. All members weigh 90 lb per foot of length. What are the pin reactions at $B$, $C$, and $D$?

**4-10.**   Figure Problem 4-10 shows a wooden brace hinged to a floor at $A$ and held by means of the rod $BC$. The weight of the brace is 300 lb and may be assumed to be applied as shown. What will be

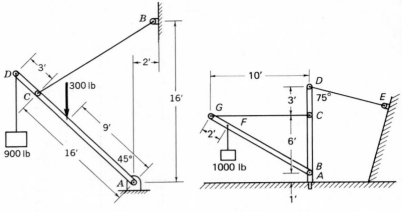

FIGURE PROBLEM 4-10          FIGURE PROBLEM 4-11

the horizontal and vertical components of the force at $A$? What is the force in $BC$?

**4-11.** In the crane shown in Fig. Prob. 4-11, the member $BG$ weighs 600 lb, which is considered to act midway between $B$ and $F$. If other weights of members are neglected, what will be the forces in the two tension rods? What will be the horizontal and vertical components of the reactions at $A$ and $B$?

**4-12.** In an iron foundry, a jib crane, shown in Fig. Prob. 4-12, was used to handle the molten metal. With the moving load in the position shown, what will be the forces in the members $A$, $B$, and $C$? What is the horizontal component of the reaction at the upper end of the vertical member?

**4-13.** For the Pratt truss shown in Fig. Prob. 4-13, find the magnitude and type of forces in members $AB$, $AH$, $BH$, $BC$, $HG$, $BG$, and $CG$.

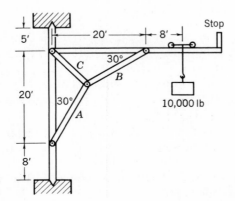

FIGURE PROBLEM 4-12

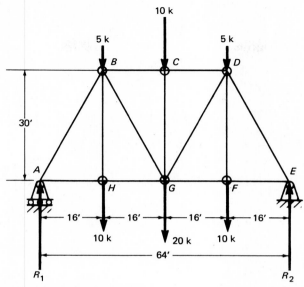

FIGURE PROBLEM 4-13

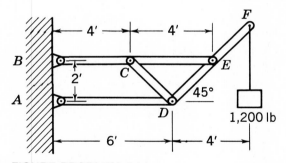

FIGURE PROBLEM 4-14

**4-14.** A cantilever truss, shown in Fig. Prob. 4-14, carries a load of 1200 lb. Find the reactions at $A$, $B$, $C$, and $E$.

**4-15.** Figure Problem 4-15 shows a truss supported at $A$ and $B$ and carrying a load of 4000 lb. Find the external reactions and the forces in all the members, assuming that all joints are pin-connected.

**\*4-16.** Determine the magnitude and direction of the reactions at roller $A$ and pin $B$ in Fig. Prob. 4-16.

**4-17.** Find the pin reaction for the link shown in Fig. Prob. 4-17.

**\*4-18.** Find the forces in the members of the truss shown in Fig. Prob. 4-18.

**4-19.** Find all the forces in the Warren truss shown in Fig. Prob. 4-19, all angles being equal to 60°.

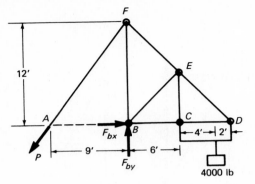

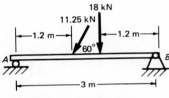

FIGURE PROBLEM 4-15

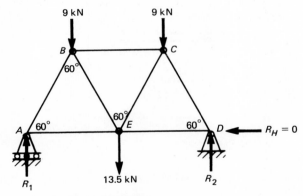

FIGURE PROBLEM 4-16

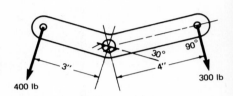

FIGURE PROBLEM 4-17

FIGURE PROBLEM 4-18

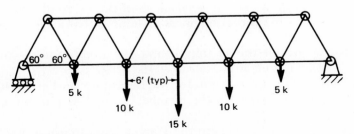

FIGURE PROBLEM 4-19

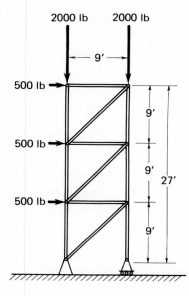

FIGURE PROBLEM 4-20

**4-20.** The members in one plane of a four-cornered tank tower are shown in Fig. Prob. 4-20. The weights given are due to the tank and water above. The horizontal forces are due to the wind. Find the forces in the members.

# CHAPTER
# 5
# Concurrent-Noncoplanar Forces

## 5-1 RESULTANT OF CONCURRENT-NONCOPLANAR FORCES

When concurrent forces are not in the same plane they are called *noncoplanar*. To find the resultant of such forces, it is best to resolve each force into components along three axes making angles of 90° with each other.

Let $x$, $y$, and $z$ be the axes and $O$ the point through which the concurrent forces pass. Take **OE** as one of the forces, making the angles $\alpha$, $\beta$, and $\gamma$ with the axes (Fig. 5-1). Then the components, or projections, of **OE** $= F$ are $F_x = F \cos \alpha$, $F_y = F \cos \beta$, and $F_z = F \cos \gamma$. But the three projections **OG**, **OA**, and **OC** are the three edges of a rectangular parallelopiped of which the force $F$ is the diagonal. From geometry, the square of the diagonal is equal to the sum of the squares of the three edges. Then

$$F^2 = F_x{}^2 + F_y{}^2 + F_z{}^2 = F^2(\cos^2 \alpha + \cos^2 \beta + \cos^2 \gamma)$$

But this cannot be true unless

$$\cos^2 \alpha + \cos^2 \beta + \cos^2 \gamma = 1 \tag{5-1}$$

Angles $\alpha$, $\beta$, and $\gamma$ are called the *direction angles of the force F*, and the foregoing equation gives the condition that their cosines must satisfy. For example, if $\alpha = 60°$ and $\beta = 45°$,

$$\cos^2 60° + \cos^2 45° + \cos^2 \gamma = 1$$
$$0.25 + 0.5 + \cos^2 \gamma = 1$$
$$\cos^2 \gamma = 0.25$$
$$\cos \gamma = 0.5$$
$$\gamma = 60°$$

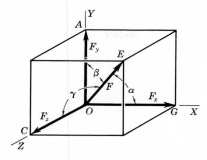

FIGURE 5-1   *x, y,* and *z* components of a force.

If each of the concurrent systems of forces is resolved into its $x$, $y$, and $z$ components, the original system is replaced by three sets of forces acting along the $x$, $y$, and $z$ axes, respectively. Since all the forces along the $x$ axis are in the same straight line, they may be added algebraically and their sum represented by $\Sigma F_x$. Similarly, those along the $y$ axis and the $z$ axis may be added algebraically and their sums represented by $\Sigma F_y$ and $\Sigma F_z$, respectively. Their resultant, the diagonal of the parallelopiped formed by $\Sigma F_x$, $\Sigma F_y$, and $\Sigma F_z$, is equal to

$$R = \sqrt{(\Sigma F_x)^2 + (\Sigma F_y)^2 + (\Sigma F_z)^2} \tag{5-2}$$

**Sample Problem 1**   The force $F$ in Fig. 5-1 is 100 lb. What will be the components along the three axes if $\alpha = 60°$, $\beta = 45°$, and $\gamma = 60°$?

**Solution:**

$$F_x = 100 \cos \alpha = 100 \cos 60° = 100 \times 0.5 = 50 \text{ lb}$$
$$F_y = 100 \cos \beta = 100 \cos 45° = 100 \times 0.707 = 70.7 \text{ lb}$$
$$F_z = 100 \cos \gamma = 100 \cos 60° = 100 \times 0.5 = 50 \text{ lb}$$

**\*Sample Problem 2**   A system of concurrent forces has been reduced to the following components:

$$\Sigma F_x = 45 \text{ N} \qquad \Sigma F_y = 90 \text{ N} \qquad \text{and} \qquad \Sigma F_z = 67.5 \text{ N}$$

Find the magnitude of the resultant and its direction angles.

**Solution:**   The resultant of the three forces may be represented as the diagonal of a parallelopiped whose sides are the three given components. Thus, by Eq. (5-2),

$$R = \sqrt{45^2 + 90^2 + 67.5^2} = 121 \text{ N}$$
$$\cos \alpha = \frac{\Sigma F_x}{R} = \frac{45}{121} = 0.372 \qquad \alpha = 68.2°$$

$$cos\ \beta = \frac{\Sigma F_y}{R} = \frac{90}{121} = 0.744 \qquad \beta = 41.9°$$

$$cos\ \gamma = \frac{\Sigma F_z}{R} = \frac{67.5}{121} = 0.558 \qquad \gamma = 56.1°$$

Of course, once $\alpha$ and $\beta$ are known, $\gamma$ may be found by Eq. (5-1).

## 5-2   CONDITIONS FOR EQUILIBRIUM

When forces are in equilibrium, their resultant is zero; that is,

$$\sqrt{(\Sigma F_x)^2 + (\Sigma F_y)^2 + (\Sigma F_z)^2} = 0$$

But each quantity under the radical sign is positive, being a square. The sum of positive numbers cannot be zero unless each one is equal to zero. Then

$$\Sigma F_x = 0 \qquad \Sigma F_y = 0 \qquad \Sigma F_z = 0 \tag{5-3}$$

This result gives three equations of condition, from which three unknowns may be determined.

   If the force polygon were drawn for a set of concurrent-noncoplanar forces that were in equilibrium, the polygon would close. Then the projection of this polygon on any plane would be a closed-plane polygon. But since it is a closed figure, the forces that form it are in equilibrium. We then have coplanar forces in equilibrium, and the principles of Chap. 2 may be applied.

   $\Sigma F_x = 0$ and $\Sigma F_y = 0$ are the equations of condition that must be satisfied when concurrent forces in equilibrium are projected on the $xy$ plane. Also, $\Sigma F_z = 0$ is the equation that the projection of such forces on the $z$ axis must satisfy.

   The forces might also be projected on the $xz$ plane and the $y$ axis or the $yz$ plane and the $x$ axis. The conclusion may be summarized as follows: When a set of concurrent-noncoplanar forces is in equilibrium, project all the forces on a convenient plane and also on an axis perpendicular to the plane. Then, form $\Sigma F_x = 0$, $\Sigma F_y = 0$, and $\Sigma F_z = 0$, and solve simultaneously.

*Sample Problem 3   A body having a mass of 46 kg is suspended from the ceiling by three cords of equal length. The cords make angles of 30° with the vertical, and they are attached to the ceiling at points equally spaced on the arc of a circle. Find the tensions in the cords (Fig. 5-2).

Solution:   Let the tensions in $OA$, $OB$, and $OC$ be $F_1$, $F_2$, and $F_3$. Set point $O$ out as a free body (Fig. 5-3). Now, project all the forces on a horizontal

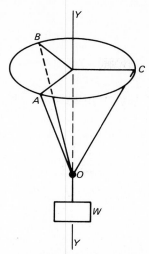

FIGURE 5-2    Diagram for Sample
Problem 3.

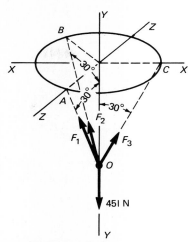

FIGURE 5-3    Free-body diagram of
point O for Sample Problem 3.

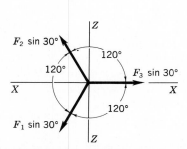

FIGURE 5-4    Forces at point O
projected on horizontal (xz) plane.

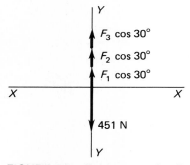

FIGURE 5-5    Forces at point O
projected on vertical (y) axis.

plane $xz$ and a vertical axis $y$ (Figs. 5-4 and 5-5), after calculating the
weight of the body. $W = m \cdot g = (46 \text{ kg})(9.81 \text{ m/s}^2) = 451 \text{ N}$.

$$\Sigma F_x = 0 \qquad -F_1 \sin 30° \cos 60° - F_2 \sin 30° \cos 60° + F_3 \sin 30° = 0$$
$$\Sigma F_y = 0 \qquad (F_1 + F_2 + F_3) \cos 30° - 451 = 0$$
$$\Sigma F_z = 0 \qquad -F_1 \sin 30° \sin 60° + F_2 \sin 30° \sin 60° = 0$$

From the equation for $\Sigma F_z$,

$$F_1 = F_2$$

From the equation for $\Sigma F_x$,

$$F_1 = F_2 = F_3$$

From the equation for $\Sigma F_y$,

$$3F(0.866) = 451$$
$$F = \frac{451}{2.6} = 173 \text{ N (tension)}$$

**Sample Problem 4** A 200-lb weight is suspended from the wall by means of a bracket, as shown in Fig. 5-6. Find the forces in *OA*, *OB*, and *OD*.

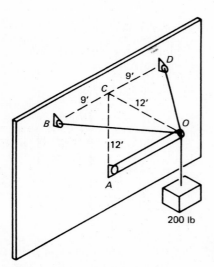

FIGURE 5-6  Diagram for Sample Problem 4.

**Solution:** Let the forces be represented by $F_1$, $F_2$, and $F_3$ (Fig. 5-7a). Take point $O$ as a free body. First project the forces onto a vertical axis through $O$. Since $F_2$ and $F_3$ are perpendicular to this axis, their projections are zero. Then

$$F_1 \sin \measuredangle COA - 200 = 0$$

But
$$\measuredangle COA = 45°$$
$$F_1 = \frac{200}{0.707} = 283 \text{ lb (compression)}$$

Now, project the forces on a horizontal plane through $O$ (Fig. 5-7b). Since $F_2$ and $F_3$ make equal angles with $F_1$, they are equal to each other.

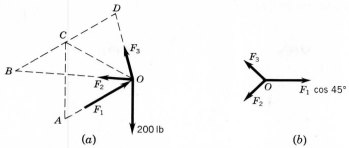

FIGURE 5-7   (a) Free-body diagram of point O for Sample Problem 4. (b) Forces at point O projected on horizontal plane.

$$\Sigma F_x = 0$$
$$-2F_2 \cos \angle COD + 283 \cos 45° = 0$$

But     $DO = \sqrt{9^2 + 12^2} = 15$ ft     and     $\cos \angle COD = \frac{12}{15} = \frac{4}{5}$

Then     $-(2)\frac{4}{5}F_2 + 200 = 0$
$$F_2 = F_3 = 125 \text{ lb (tension)}$$

*Sample Problem 5  An object having a mass of 230 kg is suspended from the wall by means of a bracket, as shown in Fig. 5-8. Find the forces in the members OA, OB, and OD.

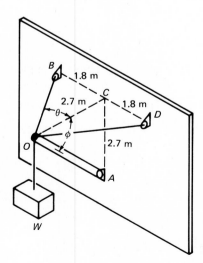

FIGURE 5-8   Diagram for Sample Problem 5.

Solution:  $W = m \cdot g = (230 \text{ kg})(9.81 \text{ m/s}^2) = 2260 \text{ N} = 2.26 \text{ kN}$.  Represent the forces by $P_1, P_2, P_3$. Take BOD as a free body. The forces acting on

*BOD* are $P_1$, 2.26 kN, and the reactions at *B* and *D*. Taking moments about line *BD*, we have

$$P_1 \times 2.7 \sin \phi - 2.26 \times 2.7 = 0$$

But $\phi = 45°$.

$$P_1 = \frac{2.26 \times 2.7}{2.7 \sin 45°} = \frac{2.26}{0.707} = 3.20 \text{ kN (compression)}$$

Now, project the three forces on a horizontal plane, with the point *O* as the free body (Fig. 5-9).

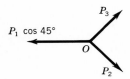

FIGURE 5-9   Forces at point *O* projected on horizontal plane.

$$\angle BOC = \angle COD = \theta$$
$$\tan \theta = \frac{1.8}{2.7} = 0.67$$
$$\theta = 33.7°$$
$$P_3 = P_2$$

$\Sigma F_x = 0$

$$2P_2 \cos \theta = P_1 \cos 45°$$
$$2P_2(0.832) = 3.20(0.707)$$
$$P_2 = P_3 = 1.36 \text{ kN (tension)}$$

**Sample Problem 6**   The boom of the derrick shown in Fig. 5-10 can swing from the horizontal to 25° from the vertical and through the larger angle *CDA*. With the boom horizontal and bisecting the angle *CDA*, find the forces in all members of the derrick.

**Solution:**   Since all the members are two-force members, the lines of action of all forces are known. At point *B*, there are four forces in equilibrium. Since they are noncoplanar, there are but three equations of condition that can be formed, and the solution cannot start at this point.

At point *E*, there are three coplanar forces, one of which is known. Since for this case two equations exist, the solution is possible. Set point *E* out as a free body (Fig. 5-11*a*) and draw the force triangle *LMN* (Fig. 5-11*b*). Since ∢ *BDE* is 90°, side *BE* = 50 ft because *DBE* is a 3-4-5 right triangle. Hence,

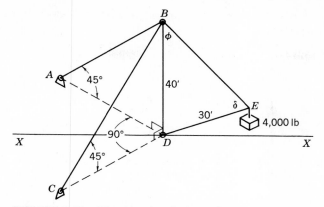

FIGURE 5-10    Diagram for Sample Problem 6.

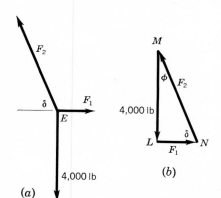

(b)

(a)

FIGURE 5-11    (a) Free-body diagram of
point $E$ for Sample Problem 6.
(b) Force triangle for point $E$.

$$\sin \delta = \tfrac{40}{50} = 0.80$$
$$\delta = 53.13° \text{ or } 53°8'$$

and

$$\phi = 90° - \delta = 36.87° \text{ or } 36°52'$$

From Fig. 5-11b,

$$\sin \delta = \frac{4000}{F_2}$$

$$F_2 = \frac{4000}{0.80} = 5000 \text{ lb (tension)}$$

$$\sin \phi = \frac{F_1}{F_2}$$

$$F_1 = 5000(0.60) = 3000 \text{ lb (compression)}$$

The results for $F_1$ and $F_2$ could have been obtained by use of similar triangles $DBE$ and $LMN$.

Point $B$ may now be set out as a free body, since the force in $BE$ is known and there are but three unknowns left (Fig. 5-12).

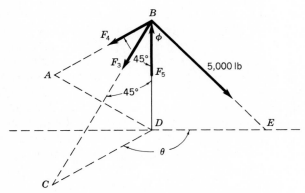

FIGURE 5-12    Free-body diagram of point $B$.

First, project the forces on a horizontal plane through $B$, as shown in Fig. 5-13$a$. The projection of $F_5$ is zero since it is perpendicular to the plane

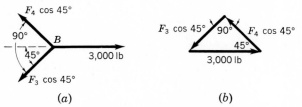

FIGURE 5-13    ($a$) Forces at point $B$ projected on horizontal plane with boom horizontal and central. ($b$) Force triangle for point $B$.

of projection. Now, draw the force triangle (Fig. 5-13$b$). Since it is an isosceles right triangle, $F_3$ equals $F_4$.

$$\cos 45° = \frac{F_4 \cos 45°}{3000}$$
$$F_4 = 3000 \text{ lb} = F_3 \text{ (tension)}$$

Now, project all forces at point $B$ onto a vertical line through $B$.

$\Sigma F_z = 0$

$$F_3 \cos 45° + F_4 \cos 45° + 5000 \cos \phi - F_5 = 0$$

But      $F_3 = F_4$     and     $\cos \phi = 0.80$

Then      $2[3000(0.707)] + 4000 = F_5$

$$F_5 = 8240 \text{ lb (compression)}$$

When a derrick is to be designed, the maximum forces that exist in every member are to be found for a given load while the boom moves through its entire range. Since the force triangle $LMN$ is similar to $DBE$, two of the sides of which are of fixed lengths, the force in $BE$ will be a maximum when $BDE$ is a right angle; that is, when the boom is horizontal (assuming $BDE$ may not exceed 90°).

Now the maximum forces will occur in $AB$, $CB$, and $BD$ when the force in $BE$ is a maximum. It now remains to find what angle boom $DE$, when it is horizontal, must make with $CD$ to give maximum forces in the mast and legs. Let $\angle CDE = \theta$. Set point $B$ out as a free body. Now, project all the forces onto a horizontal plane through point $B$ (Fig. 5-14$a$). Next, draw the force triangle (Fig. 5-14$b$), from which

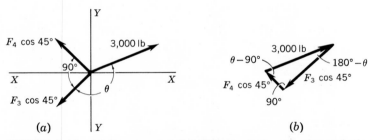

FIGURE 5-14   (a) Forces at point $B$ projected on horizontal plane with boom horizontal and positioned for maximum tension in leg $CB$. (b) Force triangle for point $B$.

$$\sin (\theta - 90°) = \frac{F_3 \cos 45°}{3000}$$

$$F_3 = \frac{3000 \sin (\theta - 90°)}{\cos 45°}$$

Now, $F_3$ is greatest when $\sin (\theta - 90°)$ is greatest.

$$\sin (\theta - 90°) = 1$$
$$\theta - 90° = 90°$$
$$\theta = 180°$$

This makes the plane of the boom perpendicular to the vertical plane through the leg $AB$. Although this case is special, in that the planes of the legs are perpendicular to each other, the same method of procedure for any angle $CDA$ will lead to the same conclusion.

The maximum tension is found in one leg of a derrick when the plane of the boom is perpendicular to the plane of the other leg. Substituting $\theta = 180°$ in the equation for $F_3$, we have

$$F_3 = \frac{3000 \sin 90°}{0.707} = 4240 \text{ lb (tension)}$$

$$F_4 = \frac{3000 \sin 0°}{0.707} = 0$$

The analysis for maximum tension could be applied as well to maximum compression, if it were possible to swing the plane of the boom into the smaller angle $CDA$. However, the maximum compression will be developed in either leg when the plane of the boom coincides with the plane of that leg and is on the same side of the mast. Again, set point $B$ out as a free body, with the boom horizontal and in the position for maximum compression in leg $CB$, and project the forces onto a horizontal plane (Fig. 5-15). By inspection,

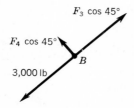

$F_3 \cos 45°$

$F_4 \cos 45°$

$B$

3,000 lb

FIGURE 5-15   Forces at point $B$ projected on horizontal plane with boom horizontal and positioned for maximum compression in leg $CB$.

and

$$F_4 = 0$$
$$F_3 \cos 45° = 3000$$
$$F_3 = 4240 \text{ lb (compression)}$$

To obtain the force in $DB$ for each of the foregoing cases, project onto the vertical axis $DB$. Then

$$5000 \cos \phi + 4240 \cos 45° = F_5$$
$$4000 + 3000 = F_5$$
$$F_5 = 7000 \text{ lb (compression)}$$

This result is less than the value obtained when the plane of the boom bisected angle $CDA$. To prove this, let $\angle CDE = \theta$ (Fig. 5-14a). Then, by Fig. 5-14b,

$$F_3 = \frac{3000 \sin (\theta - 90°)}{\cos 45°} = -\frac{3000 \cos \theta}{0.707}$$

$$F_4 = \frac{3000 \sin (180° - \theta)}{\cos 45°} = \frac{3000 \sin \theta}{0.707}$$

From the vertical projection on $DB$,

$$F_5 = 5000 \cos \phi + F_3 \cos 45° + F_4 \cos 45°$$
$$= 4000 + 3000(-\cos \theta + \sin \theta)$$

Now, $F_5$ is a maximum when $\sin \theta - \cos \theta$ is greatest. This condition occurs when both functions are numerically equal and $\cos \theta$ is negative; i.e., when $\theta = 135°$. When angle $CDA$ is not 90°, the same line of reasoning will show maximum force in the mast when the plane of the boom bisects the angle between the legs.

**Sample Problem 7** In Sample Problem 6, in order to lift the weight vertically with the boom stationary and horizontal, it is connected to $E$ by a block and tackle. Four lengths of cable run from $E$ to $W$. One end is fastened at $E$, and the other runs along $DE$ over a pulley at $D$ and then to a winch. Between $B$ and $E$, there are five strands of another cable, the free end of which runs down $BD$ to a pulley at $D$ and then to a second winch (see Fig. 5-16). Find the maximum forces in all members and the reaction at $D$.

**Solution:** Set $W$ out as a free body (Fig. 5-17). If we neglect friction on the pulleys, the tension in the cable is the same at all points, and each of the upward forces is $T_1$. *For vertical equilibrium,*

$$4T_1 = 4000$$
$$T_1 = 1000 \text{ lb (tension)}$$

Now, set point E out as a free body (Fig. 5-18). By taking $\Sigma F_y = 0$, $F_2 = 5000$ lb, as in Sample Problem 6; but this is carried by five cables, and

$$5T_2 = 5000$$
$$T_2 = 1000 \text{ lb (tension)}$$

From $\Sigma F_x = 0$,

$$F_1 = 1000 + 3000 = 4000 \text{ lb (compression)}$$

where 3000 lb is the horizontal component of $F_2$.

Now, set point $B$ out as a free body (Fig. 5-19). Projecting the forces on a horizontal plane, we see that $F_3$ and $F_4$ have the same values as in

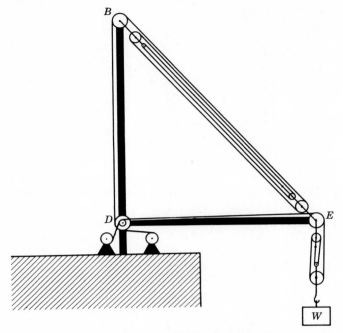

FIGURE 5-16   Diagram for Sample Problem 7.

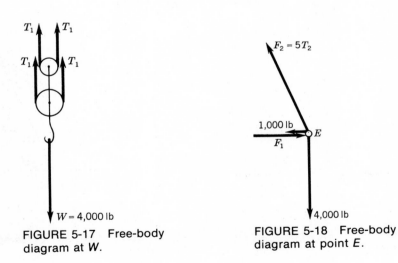

FIGURE 5-17   Free-body diagram at W.

FIGURE 5-18   Free-body diagram at point E.

Sample Problem 6. By summing all vertical components along $BD$ and setting them equal to zero, we obtain $F_5 = 9300$ lb (compression), or 1060 lb greater than in the preceding example.

For determining the reaction at $D$, set out $D$ as a free body (Fig.

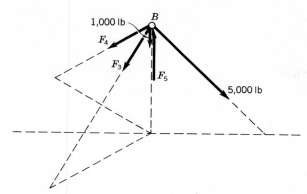

FIGURE 5-19  Free-body diagram at point B.

5-20). The reaction not being known, the components are used. By
$\Sigma F_y = 0$ and $\Sigma F_x = 0$,

$$F_{dy} = 9300 + 1000 - 1000 = 9300 \text{ lb}$$
$$F_{dx} = 4000 - 1000 - 1000 = 2000 \text{ lb}$$
$$F_d = \sqrt{9300^2 + 2000^2} = 9510 \text{ lb}$$
$$\tan \theta = \frac{9300}{2000} = 4.65$$
$$\theta = 77.85° \text{ or } 77°50'$$

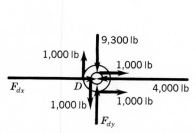

FIGURE 5-20  Free-body diagram
at point D.

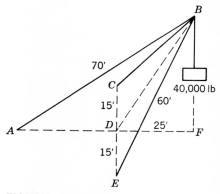

FIGURE 5-21  Diagram for Sample
Problem 8.

**Sample Problem 8** Figure 5-21 shows a shear-leg crane lifting a
40 000-lb load. The legs are 60 ft long and 30 ft apart at their lower ends.
The backstay is 70 ft long All members are pin-connected, and A, E, and
C are in the same horizontal plane. Find the forces in the members.

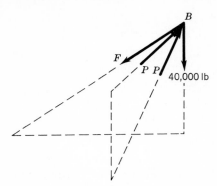

FIGURE 5-22   Free-body diagram at point *B*.

**Solution:**   All members are two-force members and concurrent at *B*. Set *B* out as a free body (Fig. 5-22). Project all forces onto a vertical plane *ABF*. Since the two forces *P* are in a plane *BCE*, perpendicular to the plane of projection, they project into *P'*, their resultant (Fig. 5-23*a*).

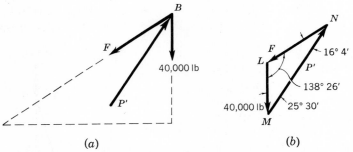

(*a*)                                    (*b*)

FIGURE 5-23   (*a*) Forces at point *B* projected on vertical plane *ABF*. (*b*) Force triangle at point *B*.

Next, form the force triangle *LMN* (Fig. 5-23*b*). Angles *M*, *L*, and *N* must now be found.

$$BD^2 = 60^2 - 15^2 = 3375$$
$$BD = 58.1 \text{ ft}$$
$$BF^2 = BD^2 - 25^2 = 2750$$
$$BF = 52.4 \text{ ft}$$
$$\cos \sphericalangle ABF = \frac{52.4}{70} = 0.7490$$
$$\sphericalangle ABF = 41.57° \text{ or } 41°34'$$
$$\sphericalangle L = 180° - 41°34' = 138°26'$$
$$\tan \sphericalangle DBF = \tan \sphericalangle M = \frac{25}{52.4} = 0.477$$

$$\measuredangle M = 25.5° \text{ or } 25°30'$$
$$\measuredangle N = 180° - (138°26' + 25°30') = 16°4'$$

From the triangle $LMN$,

$$\frac{F}{\sin 25°30'} = \frac{40\ 000}{\sin 16°4'}$$
$$F = 62\ 200 \text{ lb (tension)}$$
$$\frac{P'}{\sin 138°26'} = \frac{40\ 000}{\sin 16°4'}$$

Since $P'$ is the resultant of the thrusts $P$, it will equal the sum of the components along $BD$, or

$$P' = 95\ 900 \text{ lb}$$
$$2P \cos \measuredangle DBC = 95\ 900$$
$$P = \frac{95\ 900}{2(58.1/60)} = 49\ 500 \text{ lb (compression)}$$

*Sample Problem 9  A tripod has legs 3.6 m long. The lower ends of the legs are placed at the corners of an equilateral triangle 1.8 m on a side. What are the forces in the legs when a load of 4.5 kN is placed on top?

Solution:  Figure 5-24 shows an elevation of the tripod and a bottom view of the base. The point $D'$ is the projection of the vertex of the tripod on the plane of the base. Since the sides of the base are equal in length, this projection falls at the geometric center of the triangle. The angle that each leg makes with the vertical can be determined from the triangle $ADD'$. From the geometry of the triangles,

$$AD' = \tfrac{2}{3}AE = \tfrac{2}{3}\sqrt{1.8^2 - 0.9^2}$$
$$= \tfrac{2}{3}(1.56) = 1.04 \text{ m}$$
$$DD' = \sqrt{(AD)^2 - (AD')^2} = 3.45 \text{ m}$$

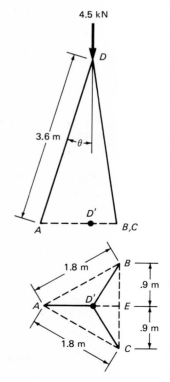

FIGURE 5-24  Diagram for Sample Problem 9.

Since the legs are symmetrically arranged and make equal angles with the vertical, then from $\Sigma F_y = 0$,

$$3F \cos \theta = 4.5$$

$$3F\left(\frac{3.45}{3.6}\right) = 4.5$$

$$F = \frac{3.6(4.5)}{3(3.45)} = 1.57 \text{ kN say, } F = 1.6 \text{ kN (compression)}$$

## PROBLEMS

**5-1.** Figure Problem 5-1 shows a tripod used in unloading heavy machinery. What is the force in each of the legs if the load is 4 tons?

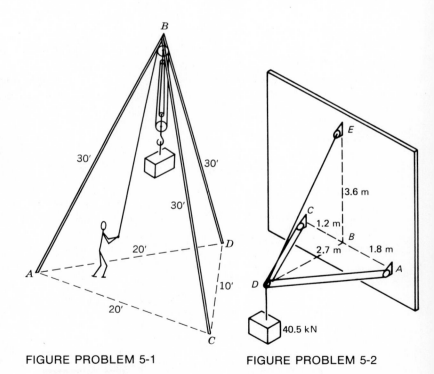

FIGURE PROBLEM 5-1         FIGURE PROBLEM 5-2

**\*5-2.** Find the forces in all members in the bracket shown in Fig. Prob. 5-2. Members $AD$ and $CD$ are horizontal.

**5-3.** What is the force in member *BD* in the derrick in Fig. Prob. 5-3, with boom *AC* in the position shown?

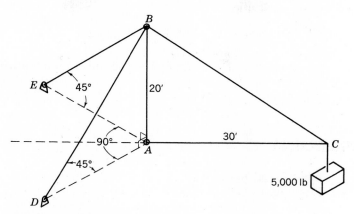

FIGURE PROBLEM 5-3 and 5-4

**5-4.** For the derrick in Fig. Prob. 5-4, what would be the force in *BD* if boom *AC* were rotated into plane *ABD*?

**5-5.** What are the forces in all parts of the shear-leg crane, shown in Fig. Prob. 5-5, if a naval gun weighing 50 tons is lifted?

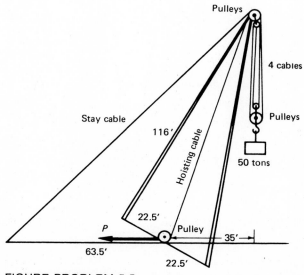

FIGURE PROBLEM 5-5

**\*5-6.**   Figure Problem 5-6 shows a derrick used in a quarry. The mast and boom are pivoted at the top and the bottom so that they can be rotated through an angle $\alpha$ of 360°. The angle $\theta$ that the boom makes with the vertical is also adjustable. If a load having a mass of 9.20 metric tons is raised, what is the force in the boom when $\theta = 45$°? The length of the mast $DF$ is 7.2 m and the boom $EF$ is 5.4 m (1 metric ton = 1000 kg).

**5-7.**   The derrick in Fig. Prob. 5-7 is stabilized by three guy wires $AD$, $BD$, and $CD$. In erecting the derrick, wires $AD$ and $BD$ are each pulled taut to an initial tensile force of 2500 lb by means of turnbuckles. What should the initial tensile force in wire $CD$ be, so that forces in a horizontal plane through point $D$ are fully balanced, if $\alpha = 0$°, $\beta = 60$°, and $\gamma = 75$°? (The derrick is not lifting a load.)

**5-8.**   What force do the guy wires in Prob. 5-7 impart to member $DF$?

**5-9.**   The derrick in Fig. Prob. 5-9 lifts a load of 18 tons. Members $DF$ and $EF$ are 30 and 20 ft long, respectively. The various angles are $\alpha = 15$°, $\beta = 60$°, $\gamma = 70$°, and $\theta = 60$°. Wires $AD$ and $BD$ were initially tightened to balance the horizontal initial pull of $CD$. The tension in guy wire $CD$ is 5000 lb after the derrick is loaded. Find (a) force in $DE$, (b) resultant force in $DF$, (c) resultant forces in the guy wires.

**\*5-10.**   A load having a mass of 9.20 metric tons is to be lifted by means of a contractor's derrick, as shown in Fig. Prob. 5-10. If the vertical plane containing the boom bisects the angle $ADB$, what reactions must be provided at points $A$ and $B$ to keep the derrick

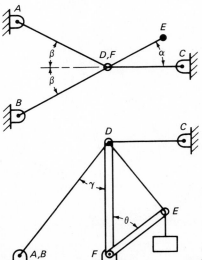

FIGURE PROBLEM 5-6 through 5-9

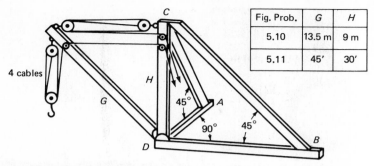

| Fig. Prob. | G | H |
|---|---|---|
| 5.10 | 13.5 m | 9 m |
| 5.11 | 45' | 30' |

FIGURE PROBLEM 5-10 and 5-11

from tipping? The boom has a mass 1.84 metric tons and makes an angle of 45° with the horizontal. (1 metric ton = 1000 kg.)

**5-11.** When the boom of the derrick in Fig. Prob. 5-11 swings into the plane *BCD*, what will be the force in leg *BC*? The load is 7 tons and the boom weighs 3000 lb and makes an angle of 60° with the horizontal.

**\*5-12.** Both of the 900 N forces in Fig. Prob. 5-12 are acting in a horizontal plane and are applied to the top of pole *AC* at point *A*. To eliminate bending of the pole, a guy wire *AB* is used. Determine the force acting in the pole and the guy wire.

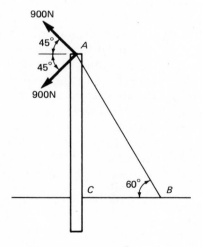

FIGURE PROBLEM 5-12

# CHAPTER
# 6
# Static and Kinetic Friction

## 6-1   MAXIMUM STATIC FRICTION

Friction was defined in Sec. 2-17 as a force between two surfaces in contact acting to resist the motion of one surface over the other. Friction is a passive force. It acts only when there is an effort made to move one body over the other.

In Fig. 6-1, let a 100-lb body rest on a rough plane and be acted on by a horizontal force $P$. If the body does not move, the forces are in equilibrium. Then $P = F$, and $N = 100$ lb; i.e., when $P = 10$ lb, $F = 10$ lb; and when $P = 20$ lb, $F = 20$ lb, etc.

But if $P = 30$ lb and the body is just on the point of moving, $F = 30$ lb is its greatest, or maximum, static friction. Any value of $P$, say 31 lb or more, will cause motion.

Experiments have shown that within certain limits the maximum value of friction depends on the normal force on the plane and the materials in the two rubbing surfaces. So long as the normal force per square inch does not exceed the elastic limit of the bearing surface, the friction is independent of the area of contact between the two surfaces, but does depend on the normal force between the surfaces.

Experiments have also shown that, for practical purposes, the ratio of the maximum friction to the normal force on the plane is a constant for any given material. This ratio is defined as the *coefficient of static friction*. Let $f$ represent the coefficient of friction and $F_m$ the maximum friction force. Then

$$f = \frac{F_m}{N} \tag{6-1}$$

Thus, if $F_m = 30$ lb, for Fig. 6-1,

$$f = \tfrac{30}{100} = 0.3$$

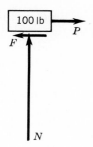

FIGURE 6-1    Forces acting on body on rough plane.

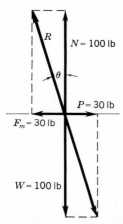

FIGURE 6-2    Resultant of normal force $N$ and maximum friction force $F_m$.

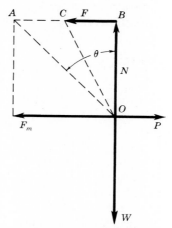

FIGURE 6-3    Resultant of normal force $N$ and any static friction force $F$, including impending motion when $F = F_m$.

The resultant of $F_m$ and $N$ when motion is impending is a single force $R$, as shown in Fig. 6-2. Since $P$, $W$, and $R$ are in equilibrium, they form a triangle such as $OAB$ (Fig. 6-3).

Now, let $\theta$ be the angle that $R$ makes with the normal to the plane. Then $\tan \theta = \frac{30}{100} = 0.3$; that is, when motion is impending, the resultant $R$ makes an angle whose tangent is the coefficient of friction $f$. $R$ is always drawn on the opposite side of the normal from the direction of motion. That is, if motion would take place to the right, $R$ bears to the left. This principle is evident from the fact that $F$ is always resisting motion.

In Fig. 6-3, $\angle AOB = \theta$, the maximum angle of friction. Suppose $P$ is less than $F_m$; then the friction $F$ is less than $F_m$, say **BC**. Then **OC** is the resultant making an angle $COB$ with the normal. Motion cannot take place until $P$ is greater than **BA**.

From Eq. (6-1), $F_m = fN$. If $f = 0.25$ and $N = 50$ lb, then $F_m = 12.5$ lb. If $f = 0.2$ and $N = 75$ lb, then $F_m = 15$ lb.

Every surface has depressions on its face. These depressions vary with the material and are more evident in a rough surface than in one with a smooth surface. When one body rests on another, the elevations on one tend to fill the depressions on the other. Before there can be motion, one body must be lifted out of the depressions, or the surface leveled or smoothed. Although this illustration does not explain all the action of friction forces, it at least furnishes an explanation of why a body moves more easily over ice than over a brick sidewalk and also why it is easier to move a weight by pulling at an angle to give a small vertical or lifting component.

The force necessary to continue uniform motion is usually less than the force necessary to start it. The friction of motion in rotating shafts, drums on brake shoes, etc., is called *kinetic friction*.

**Sample Problem 1** A 75-lb body rests on a horizontal plane for which $f = 0.4$. The body is acted on by a force of 20 lb at 30° with the horizontal, as shown in Fig. 6-4. Will the body move?

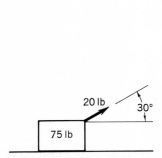

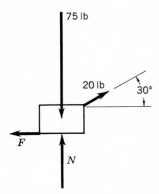

FIGURE 6-4   Diagram for Sample Problem 1.

FIGURE 6-5   Free-body diagram for Sample Problem 1.

**Solution:**   Set the object out as a free body (Fig. 6-5). The forces are in equilibrium vertically.

$$\Sigma F_y = 0$$
$$N - 75 + 20 \sin 30° = 0$$
$$N = 65 \text{ lb}$$

By Eq. (6-1), maximum friction

$$F_m = 0.4N = 0.4(65) = 26 \text{ lb}$$

For equilibrium horizontally,

$$\Sigma F_x = 0$$

$$-F + 20 \cos 30° = 0$$
$$F = 17.32 \text{ lb}$$

But the body will not move until the maximum friction of 26 lb is overcome. Since only a force of 17.32 lb is necessary to prevent motion, the body remains at rest. Friction equals 17.32 lb.

*Sample Problem 2   A body having a mass of 46 kg rests on a horizontal plane for which $f = 0.4$. A force $P$ acts on the body at an angle of 20° with the horizontal. Find its magnitude for impending motion (Fig. 6-6).

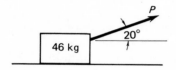

FIGURE 6-6   Diagram for Sample Problem 2.

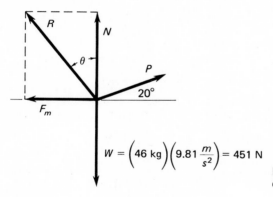

$$W = \left(46 \text{ kg}\right)\left(9.81 \frac{m}{s^2}\right) = 451 \text{ N}$$

FIGURE 6-7   Free-body diagram for Sample Problem 2.

Solution:   When motion is impending, it will often be found convenient to use the resultant $R$ of $N$ and $F_m$, which must make the ∢ $\theta$ with the normal (tan $\theta = f$). Setting out the free body (Fig. 6-7), we find there are three forces in equilibrium, $R$, $P$, and $W$.

$$\tan \theta = f = 0.4$$
$$\theta = 21.8° \text{ or } 21°48'$$

This problem may be solved by using the summation method. Since the components of the forces in the horizontal direction are in equilibrium at the instant of impending motion,

$$\Sigma F_x = 0$$
$$P \cos 20° - R \sin \theta = 0$$
$$P \cos 20° = R \sin \theta$$

$$\Sigma F_y = 0$$
$$R \cos \theta + P \sin 20° - 451 = 0$$
$$R = \frac{451 - P \sin 20°}{\cos \theta}$$
$$P \cos 20° = \frac{\sin \theta}{\cos \theta}(451 - P \sin 20°)$$
$$= \tan \theta (451 - P \sin 20°)$$
$$P(\cos 20° + 0.4 \sin 20°) = 0.4(451) = 180$$
$$P = \frac{180}{0.9397 + 0.1368} = 167 \text{ N}$$

It can be seen that the force $P$ exerts a lifting component and reduces the amount of the normal pressure; thus, $N = 394$ N.

Table 6-1 gives several values of coefficient of friction. The degree of smoothness of the surfaces affects the value of $f$ very materially, and this fact accounts for the range of values. The kinetic coefficients are usually from 20 to 40 percent lower than the values of static $f$.

**TABLE 6-1**  COEFFICIENTS OF FRACTION ($f$)*

| Material | Static | | Sliding | |
|---|---|---|---|---|
| | Dry | Lubricated | Dry | Lubricated |
| Steel on steel | 0.78 | 0.23 | 0.42 | 0.08 |
| On babbitt | 0.42 | 0.17 | 0.35 | 0.14 |
| On Teflon | 0.04 | | | 0.04 |
| On lead | 0.95 | 0.50 | 0.95 | 0.30 |
| On aluminum | 0.61 | | 0.47 | |
| On copper | 0.53 | | 0.36 | 0.18 |
| On brass | 0.51 | | 0.44 | |
| Cast iron on cast iron | 1.10 | | 0.15 | 0.07 |
| On copper | 1.05 | | 0.29 | |
| On zinc | 0.85 | | 0.21 | |
| Aluminum on aluminum | 1.05 | | 1.40 | |
| Nickel on nickel | 1.10 | | 0.53 | 0.12 |
| Wood on wood | 0.62 | | 0.48 | 0.16 |
| Leather on wood | 0.61 | | 0.52 | |
| On cast iron | | | 0.56 | 0.13 |
| Teflon on Teflon | 0.04 | | | 0.04 |
| Glass on glass | 0.94 | 0.01 | 0.40 | 0.09 |
| Laminated plastic on steel | | | 0.35 | 0.05 |

* Baumeister, Theodore, Eugene A. Avalone, and Theodore Baumeister III: *Marks' Standard Handbook for Mechanical Engineers*, 8th ed., p. 3–26, McGraw-Hill Book Company, Inc., New York, 1978.

## 6-2   FRICTION ON AN INCLINED PLANE

Let a body weighing $W$ lb rest on a rough inclined plane, Fig. 6-8. Suppose the weight is just on the point of slipping down the plane. Then friction is a maximum $F_m$. From Sec. 2-17, the component of $W$ down the plane is $W \sin \alpha$, and that normal to the plane is $W \cos \alpha$.

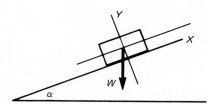

FIGURE 6-8   Body on an inclined plane.

$$F_m = W \sin \alpha$$
$$N = W \cos \alpha$$

Dividing,
$$\frac{F_m}{N} = \frac{W \sin \alpha}{W \cos \alpha} = \tan \alpha$$

But $F_m/N = f = \tan \theta$, where $\theta$ is the angle of friction, or the angle that the resultant makes with the normal. Then $\tan \theta = \tan \alpha$ and $\theta = \alpha$.

When a body rests on an inclined plane, if the angle of friction $\theta$ is less than $\alpha$, the angle of the plane, the body will slip. If $\theta > \alpha$, the body will not slide down the plane.

**Sample Problem 3** A body weighing 50 lb rests on a 20° plane. Let $f = 0.2$. Will the body slide down the plane?

**Solution:**
$$\tan \theta = 0.2$$
$$\theta = 11°20'$$

Since $\theta < \alpha$, the body will slide. Alternatively,

$$N = W \cos \alpha = 50 \cos 20° = 47 \text{ lb}$$
$$F_m = 0.2N = 0.2(47) = 9.4 \text{ lb}$$

The component of $W$ down the plane is

$$W \sin \alpha = 50 \sin 20° = 17.1 \text{ lb}$$

Since 17.1 lb > 9.4 lb, the body must move.

**Sample Problem 4** In the preceding example let $f = 0.3$ and $\alpha = 15°$. Will the body move? How large is the frictional force?

**Solution:**

$$\tan \theta = 0.3$$
$$\theta = 16°42'$$

Since $\theta > \alpha$, the body will not move. Now,

$$N = W \cos \alpha = 50 \cos 15° = 48.3 \text{ lb}$$
$$F_m = 0.3(48.3) = 14.49 \text{ lb}$$

But the force trying to move the body down the plane is

$$W \sin \alpha = 50 \sin 15° = 12.9 \text{ lb}$$

Since 12.9 lb < 14.49 lb, it is evident at the body will not move. But friction is a passive force. Only enough resistance is offered to prevent motion. Therefore, the frictional force actually developed is 12.9 lb.

## 6-3 FORCE NECESSARY TO MOVE A BODY UP A ROUGH PLANE

Figure 6-9 shows a body on a plane with a force $P$ acting on the body parallel to the plane. If motion is impending up the plane, the resultant $R$ of $F_m$ and $N$ makes an angle $\theta$ ($\tan \theta = f$) with the normal to the plane and is drawn on the left-hand side. There are then three concurrent forces in equilibrium. Construct the triangle of forces by drawing **AB** equal to $W$, the only known force. From $B$, draw a line parallel to $P$. Since the figure closes, $R$ must end at $A$. Then, from $A$ draw a line parallel to $R$, forming the triangle $ABC$ (Fig. 6-10). Since the normal makes an angle $\alpha$ with the vertical, angle $A$ of the triangle is $\theta + \alpha$.

$$\angle B = 90° - \alpha$$
$$\angle C = 180° - (90° - \alpha + \theta + \alpha) = 90° - \theta$$

This is easily seen from Figs. 6-9 and 6-10.

By the law of sines, the force $P$ necessary to just start the body up the plane is

$$\frac{P}{\sin (\theta + \alpha)} = \frac{W}{\sin (90° - \theta)} = \frac{W}{\cos \theta}$$
$$P = \frac{W \sin (\theta + \alpha)}{\cos \theta}$$

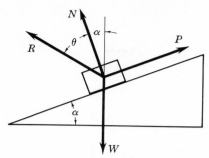

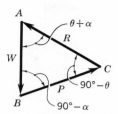

FIGURE 6-9   Body with motion impending (but not yet moving) up a rough plane.

FIGURE 6-10   Force triangle for impending motion up a rough plane.

*Sample Problem 5** Find the force required to move a body having a mass of 46 kg up a 20° inclined plane if $f = 0.3$.

$$W = (46 \text{ kg})(9.81 \text{ m/s}^2) = 451 \text{ N}$$

**Solution a:**   From Sample Problem 4, $\theta = 16°42' = 16.7°$.

$$P = \frac{451 \, \sin\,(20° + 16.7°)}{\cos 16.7°} = \frac{451(0.5976)}{0.9580} = 281 \text{ N}$$

**Solution b:**   Choose axes as in Fig. 6-11.

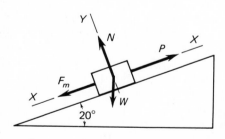

FIGURE 6-11   Diagram for Sample Problem 5.

$\Sigma F_x = 0$

$$P - F_m - W \sin 20° = 0$$

$\Sigma F_y = 0$

$$N - W \cos 20° = 0$$
$$N = 451(0.94) = 424 \text{ N}$$
$$F_m = 0.3(424) = 127 \text{ N}$$
$$P = 127 + 451 \sin 20°$$
$$= 127 + 154 = 281 \text{ N}$$

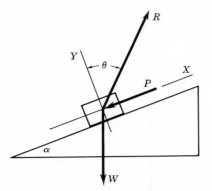

FIGURE 6-12   Diagram for Sample Problem 6.

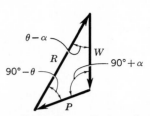

FIGURE 6-13   Force triangle for Sample Problem 6.

**Sample Problem 6** Find the force $P$ parallel to the plane (Fig. 6-12) needed to just start the body down the plane.

**Solution:**   From Fig. 6-13 and the law of sines,

$$\frac{P}{\sin(\theta - \alpha)} = \frac{W}{\sin(90° - \theta)} = \frac{W}{\cos\theta}$$
$$P = \frac{W\sin(\theta - \alpha)}{\cos\theta}$$

## 6-4  LEAST FORCE

Let a body $W$ rest on a horizontal plane (Fig. 6-14). A force $P$ acts at an angle $\gamma$. Find the least force to just move the body. $R$, the resultant reaction, makes an angle $\theta$ with the normal. There are three concurrent forces in equlibrium. Draw **AB** equal to $W$ (Fig. 6-15). Since the direction of $P$ is unknown, we draw a line parallel to $R$ from $A$. The diagram must close, the $P$ must be drawn from $B$ to some point on **AD** and is to be the least force, represented by the shortest line. But the shortest line from $B$ is a perpendicular to **AD**, which in this case is line **BC**.

Since $P$ is perpendicular to $R$, $P$ makes an angle $\theta$ with the horizontal; $\gamma = \theta$. But $ABC$ of Fig. 6-15 is a right triangle.

$$P = W \sin\theta$$

**Sample Problem 7** Let $W = 50$ lb and $f = 0.3$. Find the least horizontal force needed to move the body. Find the least force for any direction (Fig. 6-14).

**Solution:**   If $P$ is horizontal, $P = F$.

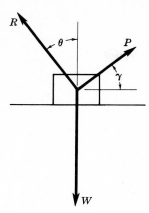

FIGURE 6-14  Free-body diagram for finding least force to move a body.

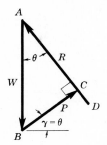

FIGURE 6-15  Force triangle for finding least force.

$$N = W \qquad f = 0.3$$
$$F_m = Nf = 50(0.3) = 15 \text{ lb}$$
$$\therefore P = 15 \text{ lb}$$

But, for least force,

$$P = W \sin \theta = 50 \sin 16°42' = 14.37 \text{ lb (at } 16°42')$$

## 6-5  CONE OF FRICTION

If the line **OA** of Fig. 6-3 making the maximum angle of friction $\theta$ with the normal is revolved about **OB** as an axis, the cone generated is called the *cone of friction*.

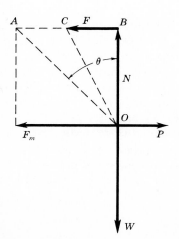

FIGURE 6-3  Repeated.

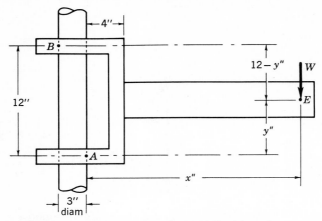

FIGURE 6-16   Diagram for Sample Problem 8.

If the resultant $R$ of the normal reaction and the friction falls within the cone of friction, the forces acting on the body are not great enough to cause motion.

This principle is used in self-locking mechanisms and also in taper pins. For example, if the angle of the taper pin is less than the angle of friction, no force at a right angle to the axis of the pin could cause it to move in the direction of its axis. This principle will be illustrated by applying it in the solution of a problem.

**Sample Problem 8**   A lift slides on a vertical shaft 3 in in diameter. Find the greatest distance from the shaft at which any load $W$ can be placed and still permit the lift to slide on the shaft. Let $f = 0.3$. Disregard the weight of the lift (Fig. 6-16).

**Solution a:**   When the load is placed on the lift, force is exerted perpendicular to the shaft, on the right side at the bottom and on the left side at the top. Let $A$ and $B$ be the points of application of these resultant horizontal forces $F_{ax}$ and $F_{bx}$. Friction at $A$ and $B$ is called $F_m$ in the direction indicated by vectors. Since motion is taking place, or impending, maximum friction is developed and the resultant reaction at $A$ is along $AE$ at the angle $\theta = \tan^{-1}(0.3)$ with the normal $AH$. At $B$, it is along $BE$ at the same angle with the horizontal (Fig. 6-17).

For impending motion, the lift is in equilibrium under the action of three forces. By the principle of concurrence they must meet at the point (point $E$) at which $AE$ and $BE$ meet. From triangles $KBE$ and $HAE$,

$$\tan \theta = \frac{y}{x} = 0.3$$
$$y = 0.3x$$

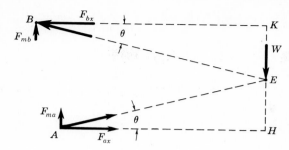

FIGURE 6-17   Forces acting on lift in Sample Problem 8.

From the dimensions in Fig. 6-16,

$$\frac{12 - y}{x + 3} = 0.3$$
$$12 - y = 0.3x + 0.9$$

Substituting for $y$,

$$12 - 0.3x = 0.3x + 0.9$$
$$0.6x = 11.1$$
$$x = 18.5 \text{ in}$$

That is, with $W$ less than 18.5 in from the edge of the shaft, the lift will slide.

**Solution b (Summation Method):**   The conditions for equilibrium give

$\Sigma F_y = 0$
$$F_{ma} + F_{mb} - W = 0$$

$\Sigma F_x = 0$
$$F_{ax} - F_{bx} = 0$$

$\Sigma M_a = 0$
$$F_{mb}(3) + Wx - F_{bx}(12) = 0$$

Since $F_{ax} = F_{bx}$,

$$F_{ma} = F_{mb}$$

From $\Sigma F_y$ above,

$$2F_m = W \qquad \text{or} \qquad F_m = \frac{W}{2}$$

Since $F_m = fN$ and $N = F_{ax} = F_{bx}$,

$$N = \frac{F_m}{f} = \frac{W}{2f}$$

Substituting in the expression for $\Sigma M_a = 0$, we have

$$\frac{W}{2}(3) + Wx - \frac{W}{2f}(12) = 0$$

Dividing by $W$ produces

$$\frac{3}{2} + x - \frac{12}{2(0.3)} = 0$$

$$x = \frac{6}{0.3} - 1.5 = 20 - 1.5 = 18.5 \text{ in}$$

## 6-6  WEDGE ACTION

A wedge is a piece of wood or metal used in lifting heavy loads, transmitting power, etc. It may be made with one or two inclined faces. Figure 6-18 shows a load of 3000 lb to be lifted by means of two wedges with faces inclined at an angle of 15° with the horizontal. Coefficient of friction for all rubbing surfaces is 0.3. Find the forces $P$ needed to just lift the load.

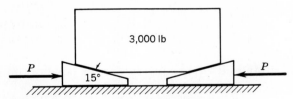

FIGURE 6-18   Lifting a load by wedges.

First, set the load out as a free body (Fig. 6-19$a$). It is in equilibrium under the action of the 3000-lb weight and the two resultant reactions $R_1$ of the two rubbing surfaces. The direction of $R_1$ is determined by first drawing a line normal to each of the inclined surfaces at their centers. The motion of the load relative to the wedge is upward. Therefore, the force $R_1$ must act inclined to the normal so that it will resist this motion, as shown in the figure. The angle that $R_1$ makes with the normal is $\theta = \tan^{-1}(0.3) = 16°42'$. Now, draw the force triangle, Fig. 6-19$b$. By the law of sines,

$$\frac{R_1}{3000} = \frac{\sin 31°42'}{\sin 116°36'}$$
$$R_1 = 1760 \text{ lb}$$

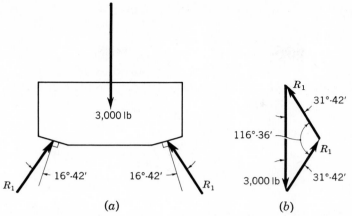

FIGURE 6-19   (a) Free-body diagram of load for impending
upward motion. (b) Force triangle for load.

Now, set the wedge out as a free body (Fig. 6-20a). It is in equilibrium
under the action of force $P$ and the resultant reactions of the two rubbing
surfaces. Since motion is impending, $R_1$ and $R_2$ must make the angle of
friction with the normals at the center of the rubbing surfaces. The
motion of the wedge is to the right. This motion is resisted by the two
surfaces in contact with the wedge. Therefore, $R_1$ and $R_2$ are drawn to
the right of and at an angle of 16°42' with their respective normals. Now,
draw the force triangle (Fig. 6-20b). By the law of sines,

$$\frac{P}{R_1} = \frac{\sin 48°24'}{\sin 73°18'}$$
$$P = 1370 \text{ lb}$$

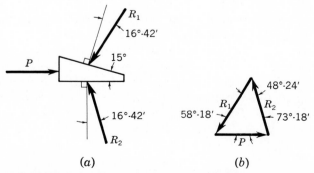

FIGURE 6-20   (a) Free-body diagram of wedge for
impending upward motion. (b) Force triangle for wedge.

The two free-body diagrams show that $R_1$ in one figure is equal, parallel, and opposite to $R_1$ in the other. This fact is easily understood because action is equal and opposite to reaction.

For this type of problem, the solution by means of the force triangle is probably the simplest. However, for purposes of comparison, the summation method will now be applied. For the load (Fig. 6-19a),

$$\Sigma F_y = 0$$
$$2R_1 \cos 31°42' - 3000 = 0$$
$$R_1 = \frac{1500}{\cos 31°42'} = 1760 \text{ lb}$$

For the wedge (Fig. 6-20a),

$$\Sigma F_y = 0$$
$$R_2 \cos 16°42' - R_1 \cos 31°42' = 0$$
$$R_2 = \frac{1760(0.8508)}{0.9578} = 1560 \text{ lb}$$
$$\Sigma F_x = 0$$
$$P - 1760 \sin 31°42' - 1560 \sin 16°42' = 0$$
$$P = 1760(0.5255) + 1560(0.2873) = 1370 \text{ lb}$$

With the same conditions in Fig. 6-21a, suppose the problem is to find what force applied to the wedge will cause impending downward motion of the load. First, set the wedge (Fig. 6-22a) and load (Fig. 6-21a) out as free bodies. Since the motion of the load is downward, the resultant reaction of the surfaces of the wedges must act on the side of the normal

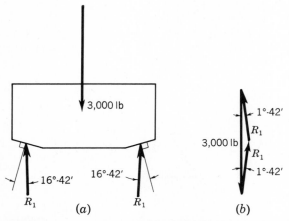

FIGURE 6-21   (a) Free-body diagram of load for impending downward motion. (b) Force triangle for load.

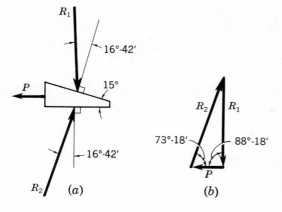

FIGURE 6-22   (a) Free-body diagram of wedge for impending downward motion. (b) Force triangle for wedge.

opposite to that acted on in the preceding problem. Also, since the wedge tends to move to the left, the reactions must act on the left side of the normals. The force triangles are now drawn. The one for the wedge shows $P$ acting to the left. The mechanism is self-locking, and it requires a force to pull the wedge from under the load.

## 6-7   JOURNAL FRICTION

When a shaft or axle rotates in its bearings, a suitable amount of lubrication is usually applied in order to reduce the effect of friction. However, since there is sliding between surfaces in contact, a force of friction is present and must be recognized. Owing to the lubrication, the value of $f$ is substantially reduced compared to the value of $f$ for a similar unlubricated surface. The effect of the friction is to cause wear on the bearings as well as to increase the turning moment needed to continue rotation. Figure 6-23 shows a shaft and bearing. The resultant $R$ of the friction $F$ and the normal reaction $N$ will be equal and opposite to the resultant of the weight $W$ and the force $P$. The friction will therefore be

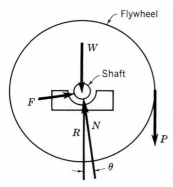

FIGURE 6-23   Force acting in a journal bearing.

at a slight angle with the horizontal in order to satisfy the requirements for equilibrium. In other words, the shaft will at first climb up on the bearing until a point is reached where the forces are in equilibrium.

*Sample Problem 9** A flywheel having a mass of 920 kg is 1.8 m in diameter. The coefficient of friction is 0.01. What tangential force applied to the rim will cause rotation? The diameter of the shaft on which the flywheel is mounted is 150 mm = 0.15 m.

**Solution:**

$$W = m \cdot g = (920 \text{ kg})(9.81 \text{ m/s}^2) = 9020 \text{ N}$$
$$F = 0.01(9020) = 90.2 \text{ N}$$

Taking moments about the axis of the shaft,

$$P(0.9) = 90.2(0.075)$$
$$P = \frac{90.2(0.075)}{0.9} = 7.5 \text{ N}$$

This is the force that may be assumed to be sufficient to turn the wheel. There is a slight omission in that the actual normal force is equal to $9020 + P$. But, in this case, it can be seen that there is no appreciable error.

**Sample Problem 10** A pulley 27 in in diameter weighs 80 lb. It is mounted on a 2-in axle. A weight of 1000 lb is to be lifted. If $f = 0.1$, what vertical pull will be needed? The value of $f$ implies poor lubrication.

**Solution:**

$$N = 80 + 1000 + P$$
$$F = 0.1(80 + 1000 + P)$$

Taking moments about the center of the axle,

$$0.1(80 + 1000 + P)1 = P(13.5) - 1000(13.5)$$
$$80 + 1000 + P = (P - 1000)(135) = 135P - 135\,000$$
$$135P - P = 135\,000 + 1080 = 136\,080$$
$$134P = 136\,080$$
$$P = \frac{136\,080}{134} = 1015 \text{ lb}$$

## 6-8   JACKSCREW

A jackscrew has a threaded screw which turns in a stationary frame or base (Fig. 6-24). As the force is applied to the lever, the screw turns and

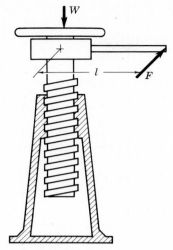

FIGURE 6-24   Jackscrew.

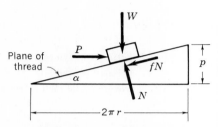

FIGURE 6-25   Free-body diagram of a jackscrew thread.

moves upward, lifting the load. The thread in the frame is an inclined plane. The load supported by the screw is transmitted to the inclined plane. As shown in Fig. 6-25, a single thread is set out as a free body.

From $\Sigma F_x = 0$

$$P = N \sin \alpha + fN \cos \alpha$$

From $\Sigma F_y = 0$

$$N \cos \alpha - fN \sin \alpha = W$$
$$\frac{P}{W} = \frac{\sin \alpha + f \cos \alpha}{\cos \alpha - f \sin \alpha} = \tan (\alpha + \theta)$$

To produce motion, the screw must be rotated by the force $F$. Then

$$Fl = Pr = Wr \tan (\alpha + \theta)$$

**\*Sample Problem 11**   A jackscrew with a mean radius of 18 mm has a pitch of 6 mm. With $f = 0.1$, what force applied to the end of a 500 mm lever will lift 1.84 metric tons (*Note:* 1 metric ton = 1000 kg)?

**Solution:**

$$W = m \cdot g = (1840 \text{ kg})(9.81 \text{ m/s}^2) = 18.05 \text{ kN}$$
$$\tan \alpha = \frac{6}{2\pi(18)} = 0.053$$
$$\alpha = 3°2' = 3.03°$$

$$\tan \theta = 0.1$$
$$\theta = 5°42' = 5.7°$$
$$F(500) = 18(18.05) \tan 8.73°$$
$$F = \frac{18(18.05)(0.1536)}{500} = 0.1 \text{ kN} = 100 \text{ N}$$

## PROBLEMS

**6-1.** Find the force required, parallel to the plane, to just start a body weighing '100 lb down an inclined plane. The slope of the plane is 12° with the horizontal, and $f = 0.3$.

**\*6-2.** Find the force required, parallel to the plane, to start a body having a mass of 46 kg down an inclined plane. The slope of the plane is 20° with the horizontal, and $f = 0.3$.

**6-3.** If, in Fig. 6-9, $P$ is a horizontal force, prove that for motion up the plane, $P = W \tan (\alpha + \theta)$. See Sec. 6-3.

**6-4.** If, in Fig. 6-9, $P$ is a horizontal force, prove that for motion down the plane, $P = W \tan (\alpha - \theta)$. See Sec. 6-3.

**6-5.** Show that the least force $P$ necessary to start a body up a plane whose slope is $\alpha$ is given by the equation $P = W \sin (\theta + \alpha)$. This result shows that the least force is required when it acts at an angle $\theta$ with the direction of impending motion.

**6-6.** Parcels are to slide down a sheet-metal chute. If the coefficient of friction is 0.15, find the minimum angle of inclination of the chute.

**6-7.** A body weighing 100 lb rests on a plane making an angle of 10° with the horizontal. A horizontal force of 30 lb is just sufficient to cause motion up the plane. What is the coefficient of friction?

**\*6-8.** A body having a mass of 46 kg rests on an inclined plane. A horizontal force of 135 N is just sufficient to cause motion up the plane. If $f = 0.1$, what angle does the plane make with the horizontal?

**6-9.** Will the bodies in Fig. Prob. 6-9 slide? What is the tension in the cord connecting the two bodies? What force is needed to start the 200-lb body to the right? The bodies are brass and the planes are steel.

**\*6-10.** In the friction drive shown in Fig. Prob. 6-10, the wheel $A$ transmits

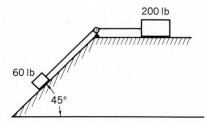

200 lb

60 lb

45°

FIGURE PROBLEM 6-9

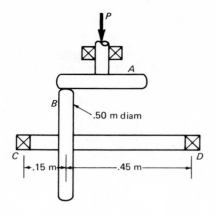

FIGURE PROBLEM 6-10

a turning moment of 22 N·m to the wheel $B$. If the coefficient between the surfaces of the wheels in contact is 0.35, what must be the force $P$ between the wheels, and what are the vertical reactions at the bearings $C$ and $D$?

**6-11.** A load of 1000 lb is to be raised by means of a wedge which is moved by a horizontal force $P$ (Fig. Prob. 6-11). The load is compelled to move up the slope, because of the horizontal force $F$. What are the amounts of the horizontal forces $P$ and $F$ if all surfaces are wood?

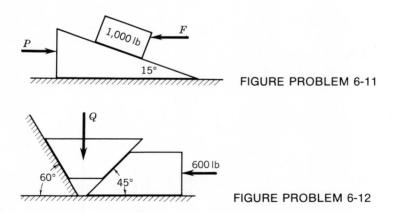

FIGURE PROBLEM 6-11

FIGURE PROBLEM 6-12

**6-12.** What load $Q$ can be supported by the force of 600 lb if all surfaces are dry steel (Fig. Prob. 6-12)?

**\*6-13.** A ladder 3 m long and having a mass of 18 kg is placed with the lower end resting on a rough floor and the upper end against a smoooth wall at a point 2.4 m above the floor. If the value of $f =$ 0.2, will the ladder remain in position? The weight of the ladder is assumed to act at the midpoint of its length.

**6-14.** A turbine-generator rotor weighing 5 tons on a 14-in diameter shaft can be turned with a force of 80 lb applied at a radius of 4 ft. Find the coefficient of friction. There are two bearings.

**6-15.** Block $A$, weighing 250 lb, rests on a surface whose coefficient of friction is 0.25. A force $F = 68$ lb is applied to a cable which passes over a 4-in diameter sheave (with no slip), as in Fig. Prob. 6-15. The sheave weighs 2 lb and is mounted on a 1-in diameter shaft with a friction coefficient of 0.03. Will block $A$ be moved?

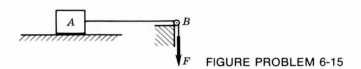

FIGURE PROBLEM 6-15

**\*6-16.** The mean radius of the screw of a square-threaded jackscrew is 25 mm. The pitch of the threads is 7.5 mm. If the coefficient of friction is 0.12, what force applied to the end of a lever 600 mm long is needed to raise a load having a mass of 1.84 metric tons?

# CHAPTER
# 7
# Simple Stresses

## 7-1   STRESS

When a restrained body is subjected to external forces there is a tendency for the shape of the body to be deformed or changed. Since materials are not perfectly rigid, the applied forces will cause the body to deform. The *internal resistance* to deformation of the fibers of a body is called *stress*.

Stress can be considered either as total stress or unit stress. *Total stress* represents the total resistance to an external effect, and is expressed in pounds, kips, newtons, or other force dimension. *Unit stress* represents the resistance developed by a unit area of cross section, and is expressed in pounds per square inch (psi), kips per square inch (ksi), newtons per square meter, megapascals (1 MPa = $10^6$ Pa = $10^3$ kN/m²), or other force per unit area dimension. For the remainder of this text, the word *stress* will be used to signify *unit stress*.
The various types of stress may be classified as:

1. Simple or direct stress
   (*a*)   Tension*
   (*b*)   Compression*
   (*c*)   Shear
2. Indirect stress
   (*a*)   Bending (see Chap. 12)
   (*b*)   Torsion (see Chap. 13)
3. Combined stress (see Chap. 14)
   (*a*)   Any possible combination of types 1 and 2

This chapter deals with simple stresses only.

* See Sec. 2-8.

## 7-2  SIMPLE STRESS

Simple sress is often called *direct stress* because it develops under direct loading conditions. That is, simple tension and simple compression occur when the applied force (called *load*) is in line with the axis of the member (*axial* loading) (see Figs. 7-1 and 7-3); and simple shear occurs when equal, parallel, and opposite forces tend to cause a surface to slide relative to the adjacent surface (see Fig. 7-5).

There are many loading situations in which the stresses that develop are not simple stresses. For example, referring to Fig. 11-7, the member is subjected to a load which is perpendicular to the axis of the member (*transverse* loading). This will cause the member to bend, resulting in deformation of the material and stresses being developed internally to resist the deformation. All three types of stresses—tension, compression, *and* shear—will develop, but they will not be simple stresses since they were not caused by direct loading.

When any type of *simple* stress develops, we can calculate the magnitude of that stress by

$$s = \frac{F}{A} \tag{7-1}$$

where

$s$ = average unit stress, psi; Pa (usually expressed as $10^6$ Pa or MPa)
$F$ = external force causing stress to develop, lb; N
$A$ = area over which stress develops, in²; m²

If a member is in tension due to the action of force $F$, as in Fig. 7-1, the stress on each cross section is the same since the cross-sectional areas $A_1$, $A_2$, $A_3$, etc., are all equal. However, the stresses on the various cross sections in Fig. 7-2 are different since the areas are different. The largest stress occurs where the force must be resisted by the smallest area. In this case, section 3 carries the largest stress. That is, $s_1 = F/A_1 < s_2 = F/A_2 < s_3 = F/A_3$.

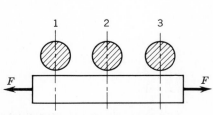

FIGURE 7-1   Tension member— uniform cross section.

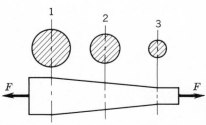

FIGURE 7-2   Tension member— varying cross section.

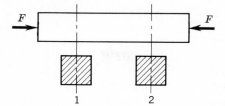

FIGURE 7-3  Compression member—uniform cross section.

A similar analysis applies to members in compression. Figure 7-3 shows a compression member in which the same stress exists on all cross sections.

The compression member shown in Fig. 7-4 carries different stresses at different levels because of the changing areas. Of course, the stress is maximum where the area is minimum, since the same load must be transmitted by all sections.

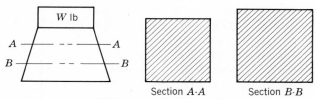

Section *A-A*    Section *B-B*

FIGURE 7-4  Compression member—varying cross section.

It should be noted that in cases of either simple tension or simple compression, the areas which resist the load are *perpendicular* to the direction of the forces.

When a member is subjected to simple shear, the resisting area is parallel to the direction of the force. Common situations causing shear stresses are shown in Figs. 7-5 and 7-6.

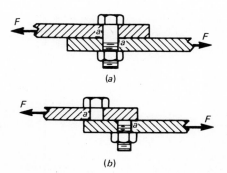

(a)

(b)

FIGURE 7-5  (*a*) Bolt resisting shear.
(*b*) Bolt failure due to shear.

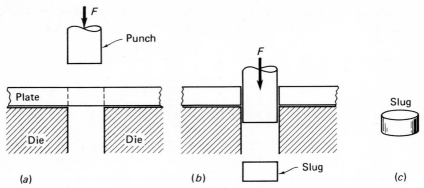

FIGURE 7-6 (a) Punch approaching plate. (b) Punch shearing plate. (c) Slug showing sheared area.

A bolt connecting two flat plates, as in Fig. 7-5a, is subjected to shear stress on the circular area $aa$. Thus, the magnitude of this stress is $s_s = F/A$. If the bolt material is not strong enough to provide the necessary shear stress, the failure might occur as shown in Fig. 7-5b. A similar situation is encountered in riveted joints.

Stamping and blanking operations depend upon the material to fail cleanly in shear. Figure 7-6a shows a circular blanking punch approaching the work. The extent of the slug to be sheared out is indicated by broken lines. Figure 7-6b shows the punch penetrating the plate with the slug falling out below. Figure 7-6c indicates the shear area as the cylindrical surface of the slug.

**Sample Problem 1** A steel rod 1 in in diameter is subjected to a pull of 6000 lb. What is the average stress?

**Solution:**

$$F = 6000 \text{ lb}$$

$$A = \frac{\pi(1^2)}{4} = 0.785 \text{ in}^2$$

$$s = \frac{F}{A} = \frac{6000}{0.785} = 7640 \text{ psi (tension)}$$

**\*Sample Problem 2** A 100- by 100-mm bar, 150 mm long, is under a compression force of 45 kN. Find the average stress.

**Solution:**

$$F = 45 \text{ kN} = 45(10^3) \text{ N}$$
$$A = (0.1 \text{ m})(0.1 \text{ m}) = 0.01 \text{ m}^2$$
$$s = \frac{F}{A} = \frac{45(10^3)}{0.01} = 4\,500\,000 \text{ Pa} = 4.5(10^6) \text{ Pa} = 45 \text{ MPa (compression)}$$

**Sample Problem 3**  In Fig. 7-7, what will be the average shearing stress on the area $ABCD$? $F = 2000$ lb, $AB = 8$ in, and $BC = 4$ in.

**Solution:**

$$P = 2000 \text{ lb}$$
$$A = 32 \text{ in}^2$$
$$s = \frac{F}{A} = \frac{2000}{32} = 62.5 \text{ psi (shear)}$$

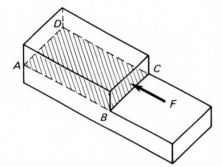

FIGURE 7-7   Diagram for Sample Problem 3.

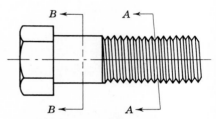

FIGURE 7-8   Diagram for Sample Problem 4.

**Sample Problem 4**  Figure 7-8 shows a common type of bolt. If a tensile load is applied to the bolt, the greatest tensile stress will be developed across the root area (section $AA$) rather than the shank area (section $BB$) since the root area is smaller.

A 1″-8 UNC hex-head bolt carries a tensile load of 6000 lb. Find the stresses on the root and shank areas.

**Solution:**   From App. B, Table 3, the root diameter is found to be 0.8466 in.

$$F = 6000 \text{ lb}$$
$$\text{Root area} = \frac{\pi}{4}(0.8466)^2 = 0.563 \text{ in}^2$$
$$s = \frac{6000}{0.563} = 10\ 660 \text{ psi (tension at root)}$$
$$\text{Shank area} = \frac{\pi(1)^2}{4} = 0.785 \text{ in}^2$$
$$s = \frac{6000}{0.785} = 7640 \text{ psi (tension on shank)}$$

\***Sample Problem 5**  The rod and yoke shown in Fig. 7-9 are subjected to a tensile load of 29 kN. The pin diameter is 12 mm. The rod diameter

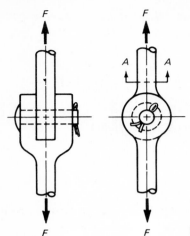

FIGURE 7-9   Diagram for Sample Problem 5.

at section $AA$ is 22 mm. Find the average shearing stress in the pin and the average tensile stress in the rod at section $AA$.

**Solution:**   The pin is in double shear; that is, it tends to fail simultaneously on two separate areas. The total area resisting shear is

$$A_s = 2\left(\frac{\pi d^2}{4}\right) = 2\left[\frac{\pi (12)^2}{4}\right] = 2(113) = 226 \text{ mm}^2 = 226(10^{-6}) \text{ m}^2$$

The average shearing stress in the pin is

$$s_s = \frac{F}{A_s} = \frac{29(10^3)}{226(10^{-6})} = 128(10^6) \text{ Pa} = 128 \text{ MPa}$$

At section $AA$, the area resisting tension is

$$A_t = \frac{\pi d^2}{4} = \frac{\pi (22)^2}{4} = 380 \text{ mm}^2 = 380(10^{-6}) \text{ m}^2$$

The average tensile stress in the rod is

$$s_t = \frac{F}{A_t} = \frac{29(10^3)}{380(10^{-6})} = 76(10^6) \text{ Pa} = 76 \text{ MPa}$$

**\*Sample Problem 6**  A circular brass rod is to carry a tensile load of 45 kN, and its average stress is not to exceed 52 MPa. What diameter rod is required?

**Solution:**

$$s = \frac{F}{A} \quad \text{or} \quad A = \frac{F}{s}$$

$$A = \frac{45(10^3)}{52(10^6)} = 0.86(10^{-3}) \text{ m}^2 = 0.86(10^3) \text{ mm}^2$$

$$= 860 \text{ mm}^2 \text{ (required tensile area)}$$

$$A = \frac{\pi}{4}d^2 \quad \text{or} \quad d = \sqrt{\frac{4A}{\pi}}$$

$$d = \sqrt{\frac{4(860)}{\pi}} = \sqrt{1090} = 33 \text{ mm}$$

*__Sample Problem 7__ If the diameter found in the previous problem is doubled, what tensile load can be carried?

**Solution:**

$$s = \frac{F}{A} \quad \text{or} \quad F = As$$

$$A = \frac{\pi d^2}{4} = \frac{\pi(66)^2}{4} = 3420 \text{ mm}^2 = 3420(10^{-6}) \text{ m}^2$$

$$F = (3420 \cdot 10^{-6})(52 \cdot 10^6) = 178 \text{ kN}$$

(*Note:* If the calculations had not been rounded off, $F$ would equal 180 kN, because a doubling of the diameter results in an increase of 4 times the load.)

## PROBLEMS

**7-1.** A bar of steel 1 by $1\frac{1}{2}$ in in cross section and 6 in long is subjected to a pull of 6000 lb. What is the average tensile stress in the bar?

**\*7-2.** A short rod of steel 50 mm in diameter supports a load of 9 kN. What is the average compressive stress?

**7-3.** What load $F$ will develop an average shearing stress of 8800 psi in the shank of the $\frac{3}{4}''$-10 UNC bolt shown in Fig. 7-5a?

**\*7-4.** A structure having a mass of 180 metric tons is supported equally by 24 short posts, 150 mm in diameter. What average compressive stress will be developed?

**\*7-5.** In Prob. 7-4, what diameter posts may be used if an average compressive stress of 7.5 MPa is permitted?

**7-6.** A 1''-8 UNC hex-head bolt is subjected to a tensile load of 10 000 lb. The height of the head is $\frac{3}{4}$ in. Find the average tensile stress in the bolt and the average shearing stress in the bolt head (Fig. Prob. 7-6).

**7-7.** In Prob. 7-6, what height of head is necessary so that the bolt will

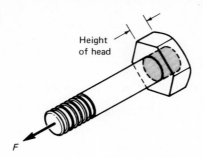

Height
of head

F

FIGURE PROBLEM 7-6

be stressed twice as much in tension as the head is stressed in shear?

**7-8.** Select an American Standard fine-thread bolt to carry a tensile load of 1000 lb. An average tensile stress of 10 000 psi must not be exceeded.

*7-9. What force is required to punch a hole 50 mm in diameter through a 16-mm-thick aluminum-alloy plate? An average stress of 265 MPa will cause this material to shear.

*7-10. A steel rod 30 mm in diameter and 18 m long is suspended vertically. Calculate the average stress in the rod due to its own weight at a section 9 m from the top. What is the maximum average stress in the rod due to its weight, and on what section does it act? ($\rho = 7830$ kg/m³ for steel.)

**7-11.** A platform 16 by 24 ft is to support a uniformly distributed load of 480 psf of surface. How many square timber posts whose nominal size is 4 by 4 in (see App. B, Table 11, for actual sizes) must be used to support the platform if an average stress of 1200 psi in compression is permitted? Show how you would arrange these posts.

**7-12.** The lower one-third of a wall consists of concrete, and the upper two-thirds consists of brick. What total height of wall would cause an average stress of 100 psi at the base? ($\rho = 150$ lb/ft³ for concrete; $\rho = 120$ lb/ft³ for brick.)

**7-13.** The maximum gas pressure in an 8-in diameter diesel-engine cylinder is 750 psi. There are six $\frac{7}{8}$"-9 UNC cap screws fastening the cylinder head to the engine block. In order to provide a gas-tight joint, the cap screws have been tightened so that each screw has an initial tensile load of 14 000 lb. Find the maximum average tensile stress in the cap screws when the engine is operating. Ignore the effect of the gasket.

*7-14. What maximum load $F$ can the link shown in Fig. Prob. 7-14 withstand if the average tensile stress must not exceed 75 MPa? All dimensions are in millimeters.

*7-15. A hollow, cylindrical, short steel compression member is required to support an axial load of 180 kN. Specifications limit the average compressive stress to 95 MPa. The cylinder is 200 mm long.

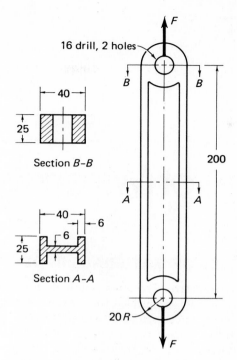

Section B-B

Section A-A

FIGURE PROBLEM 7-14

a. Find the inside and outside diameters if the outside diameter is $1\frac{1}{2}$ times the inside diameter.

b. Find the inside and outside diameter if the wall thickness is 6 mm.

**7-16.** For the machine-part connection shown in Fig. Prob. 7-16, describe (in words) the possible ways in which failure might occur; indicate what kind of simple stress causes the failures described; and calculate (using the symbols in the figure) the area that resists each failure.

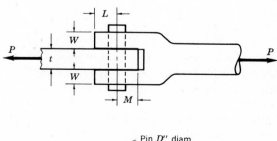

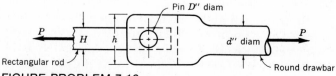

FIGURE PROBLEM 7-16

# CHAPTER
# 8
# Properties of Materials

## 8-1 USE OF PROPERTIES OF MATERIALS IN DESIGN

Designers of machinery and structures must know both the action of the forces that are applied and the strength of the various members working together to resist the forces successfully. It is essential for the members to be strong enough; but any part that is unnecessarily strong means a waste of material. It is therefore highly important for the designer to know the mechanical and physical properties of the ferrous (e.g., steel) and nonferrous (e.g., aluminum) metals, timber, concrete, plastics, brick, stone, etc., with which he or she works. In the solution of problems a knowledge of the laws of mechanics alone is insufficient and must be accompanied by a substantial knowledge of the various properties of the materials.

Many experiments are conducted in the laboratory to determine required properties. One of the most important tests which gives the designer many properties of a material is the tension test. A typical tension test is described below.

## 8-2 TENSION TEST

In order to determine the tensile properties of metals, standard tests have been devised by the American Society for Testing and Materials (ASTM) and other groups. Figure 8-1 shows the dimensions of a steel test specimen.

The specimen is mounted in a testing machine and an axial tensile load is applied. The length increases as the load is increased. This change in length is measured by an extensometer which is attached to the specimen. The data in Fig. 8-2a were obtained in a typical tension test.

With reference to the data, some explanation is in order. Under Initial data, the material is specified as AISI 1020 cold-rolled steel. The

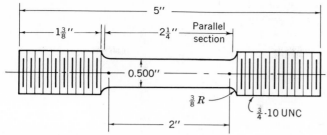

FIGURE 8-1    Tension test specimen.

American Iron and Steel Institute (AISI) uses a code, similar to the Society of Automotive Engineers (SAE) code, to identify various steels. A steel designated as AISI 1020 is a low-carbon steel containing approximately 0.20 percent carbon. If a steel is designated as AISI 1045, it is a medium-carbon steel containing approximately 0.45 percent carbon. Also under Initial data, the hardness of the steel specimen is given as Rockwell B75 (commonly written as RB75). The Rockwell test establishes the hardness of a material by forcing a specific-shaped indentor below the surface of the specimen. The depth of penetration is then measured and specified. The hardness test can be conducted with a combination of different magnitude of forces and indentor shapes, resulting in different Rockwell scales, such as A, B, C, etc. Rockwell B75 indicates that the B scale was used. Note that the hardness under Final data is Rockwell B76, which indicates that the specimen became harder as the test progressed.*

## 8-3    STRESS AND STRAIN

In order to permit comparison with standard values, the data must be converted to a unit basis. The load is expressed in terms of load per unit area, which is unit stress (henceforth called *stress*). The deformation (in this case elongation) is expressed in terms of deformation per unit length, which is unit strain (henceforth called *strain*).

Strain can be calculated by

$$\epsilon = \frac{\delta}{l} \tag{8-1}$$

where $\quad\quad\quad\quad\quad \epsilon$ = unit strain, in/in; m/m
$\delta$ = deformation, in; m
$l$ = original length, in; m

* For a complete discussion of hardness testing and the classification systems used for steels and other metals, refer to Sidney H. Avner, *Introduction to Physical Metallurgy*, 2d ed., McGraw-Hill Book Company, New York, 1974.

**TENSION TEST**

| Laboratory data (a) | | | Calculated data (b) | |
|---|---|---|---|---|
| Initial data:<br>  Material AISI 1020<br>  cold-rolled steel | Load<br>kip | Elongation<br>in 2 in,<br>in × 10⁻⁴ | Stress,<br>ksi | Strain,<br>in/in<br>× 10⁻⁴ |
| Diameter: 0.503 in<br>Gage length: 2.000 in | $F$ | $\delta$ | $s$ | $\epsilon$ |
| Hardness: Rockwell B75 | 0.5 | 0 | 2.5 | 0 |
| Speed of testing: | 1.5 | 3 | 7.5 | 1.5 |
|   0.025 in/min | 2.5 | 6 | 12.5 | 3.0 |
|  | 3.5 | 10 | 17.5 | 5.0 |
|  | 4.5 | 13 | 22.5 | 6.5 |
|  | 5.5 | 16 | 27.5 | 8.0 |
|  | 6.5 | 20 | 32.5 | 10.0 |
|  | 7.5 | 23 | 37.5 | 11.5 |
|  | 8.5 | 26 | 42.5 | 13.0 |
|  | 9.5 | 30 | 47.5 | 15.0 |
|  | *10.0 | 32 | 50.0 | 16.0 |
|  | 10.5 | 34 | 52.5 | 17.0 |
|  | 11.0 | 36 | 55.0 | 18.0 |
|  | 11.5 | 39 | 57.5 | 19.5 |
|  | 12.0 | 43 | 60.0 | 21.5 |
|  | 12.5 | 46 | 62.5 | 23.0 |
|  | 13.0 | 51 | 65.0 | 25.5 |
|  | 13.5 | 60 | 67.5 | 30.0 |
|  | 14.0 | 73 | 70.0 | 36.5 |
|  | 14.5 | 88 | 72.5 | 44.0 |
|  | 15.0 | 118 | 75.0 | 59.0 |
| Ultimate | 15.83 |  | 79.15 |  |
| Breaking | 12.50 |  | 62.50 |  |
| Final data:<br>  Diameter: 0.312 in<br>  Gage length: 2.328 in<br>  Hardness: Rockwell B76<br>  Character of fracture:<br>    ¾ cup-cone | | | Elastic-limit stress:<br>  52 500 psi<br>Yield stress: 72 500 psi<br>Ultimate stress:<br>  79 150 psi<br>Percent elongation in<br>  2 in: 16.4%<br>Percent reduction in<br>  area: 62.0%<br>Modulus of elasticity<br>  $E = 29.7 \times 10^6$ psi | |

FIGURE 8-2  Data for typical tension test.

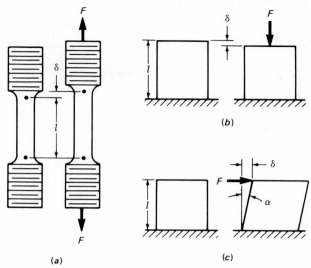

FIGURE 8-3    (a) Tensile specimen before loading and after a load F is applied. (b) Short compression member before loading and after a load F is applied. (c) Member before loading and after a shearing load F is applied.

(Note: $\epsilon = \dfrac{\delta}{l} = \tan \alpha$)

In Fig. 8-3, three applications of the use of Eq. (8-1) for calculating unit strain, $\epsilon$, are shown. Strains related to the three simple stresses—tension, compression, and shear—are shown. Note that in all cases, the deformation, $\delta$, is measured in the same direction as the applied force.

The following sample calculations will demonstrate conversions from load and deformation to stress and strain. The values used correspond to the starred data in Fig. 8-2.

$$\text{Stress} = \frac{\text{load}}{\text{area}} \qquad s = \frac{F}{A} \qquad\qquad \text{Strain} = \frac{\text{elongation}}{\text{original length}} \qquad \epsilon = \frac{\delta}{l}$$

$$A = \frac{\pi d^2}{4} = \frac{\pi (0.503)^2}{4} = 0.20 \text{ in}^2 \qquad\qquad \epsilon = \frac{0.0032}{2.000} = 0.0016 \text{ in/in}$$

$$s = \frac{10\,000}{0.20} = 50\,000 \text{ psi} = 50.0 \text{ ksi}$$

By following this procedure, the test data are converted to stress and strain, as tabulated in Fig. 8-2b.

These results are represented graphically by plotting stress vs. strain (Figs. 8-4 and 8-5). The American Society for Testing and Materials (ASTM) test specifications require that an initial load be applied to the specimen after which the extensometer dial is set to zero. This ensures that the specimen is securely gripped by the machine before data are

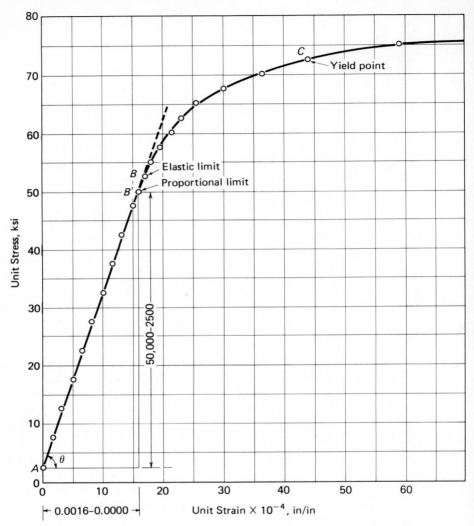

FIGURE 8-4   Partial stress-strain graph for test data of Fig. 8-2.

recorded. For this reason, point $A$ is plotted at the initial stress (corresponding to the initial load) and zero strain.

The straight-line portion of the curve $AB'$ indicates a direct proportion between stress and strain in this range. Point $B'$ represents the limit of proportionality, and thus is called the *proportional limit*.

At any stress up to point $B$, the specimen will return to its original dimensions upon removal of the load. At stresses higher than $B$, the specimen will not return to its original dimensions. Point $B$ represents the limit of elasticity since a permanent deformation, or *set*, occurs when the

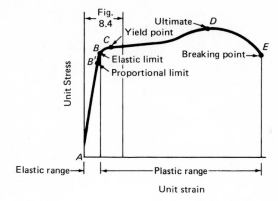

FIGURE 8-5 Complete stress-strain diagram for data of Fig. 8-2.

material is stressed beyond this point. Thus, point $B$ is called the *elastic limit*.

For ductile materials, the elastic limit $B$ and the proportional limit $B'$ are sufficiently close so that they may be assumed to be located at the same point without serious error. Henceforth, this point will be referred to as the elastic limit. However, if it is necessary to establish the accurate location of the elastic limit, the specimen will have to be loaded and unloaded until the condition described in the previous paragraph is reached.

In the course of testing the material beyond the elastic limit, a point is reached where, without a significant increase in the load, the specimen continues to elongate. This is point $C$ on the stress-strain curve (Figs. 8-4 and 8-5), and is called the *yield point*. For materials which do not show an apparent yielding, this point may also be found by the 0.2 percent offset method which is discussed in Sec. 8-7.

The stress at point $D$ (Fig. 8-5), is the *ultimate strength* of the material and is the maximum stress which the specimen can withstand. In the neighborhood of this point the specimen begins to *neck-down*, that is, its cross-sectional area decreases rapidly (Fig. 8-6), and failure occurs at point $E$.

# 8-4 DEFINITIONS

The following terms are among those which are associated with properties of materials.

*Hardness:* The ability of a material to resist wear or penetration; the strength of a metal varies directly as its hardness.

*Malleability:* The ability of a material to deform appreciably in a hammering or rolling operation before rupture.

*Toughness:* The ability of a material to withstand high-impact (shock) loads. The area under the stress-strain curve is a measure of toughness.

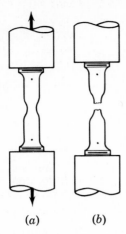

FIGURE 8-6   (a) Tension specimen necking down before failure. (b) Tension specimen after failure.

(a)          (b)

*Ductility:*   The ability of a material to deform appreciably under a tensile load before rupture; high percent elongation and percent reduction of area indicates ductility.

*Brittleness:*   The tendency of a material to fail with very little deformation; low percent elongation and percent reduction of area indicates brittleness.

*Elasticity:*   The ability of a material to return to its original size and shape upon removal of the load; high elastic limits indicate materials with good elasticity.

*Plasticity:*   The ability of a material to deform nonelastically without rupturing; large plastic ranges indicate materials with good plasticity.

*Stiffness:*   The abilty of a material to resist deformation; stiffness of a material varies directly as its modulus of elasticity.

*Alloy:*   A mixture of two or more elements, at least one being a metal, exhibiting metallic properties.

## 8-5   MODULUS OF ELASTICITY

The slope of the *straight-line* portion of the stress-strain curve is the ratio of stress to strain.

$$\text{Slope of } AB = \frac{s}{\epsilon} = \tan \theta$$

This ratio is often considered a constant for a material and is called the *modulus of elasticity*, represented by $E$. Actually, $E$ varies with temperature, but such variation will not be considered in this book.

$$\tan \theta = E = \frac{s}{\epsilon} \qquad \text{or} \qquad s = E\epsilon \qquad (8\text{-}2)$$

The units of modulus of elasticity are pounds per square inch (psi) and Pa. In U.S. customary units, $E$ is usually expressed in terms of $10^6$ psi. In SI metric units, $E$ is usually expressed in terms of $10^9$ Pa or GPa.

Another form of the equation can be obtained by substituting $F/A$ for $s$ and $\delta/l$ for $\epsilon$.

$$E = \frac{s}{\epsilon} = \frac{F/A}{\delta/l} = \frac{Fl}{A\delta} \tag{8-3}$$

The reader should recall that the ASTM specifications require an initial load on the specimen before elongation is mesured. Thus, the initial stress is $s_a$ at zero strain. Therefore, to determine the modulus of elasticity from the stress-strain curve (Fig. 8-4),

$$\tan \theta = E = \frac{s_b{}' - s_a}{\epsilon_b{}' - \epsilon_a} \tag{8-4}$$

In the application of Eq. (8-4), points $A$ and $B'$ may be any two widely separated points on the *straight-line* portion of the stress-strain curve.

From the data in Fig. 8-2b, $\epsilon_a$ is zero, and

$$E = \frac{50\,000 - 2500}{0.0016 - 0} = \frac{47\,500}{0.0016} = 29\,700\,000 \text{ psi}$$

This value of $E$ is within the expected range for steel, as indicated in App. B., Table 1.

The modulus of elasticity ($E$) for a material is the same in tension and compression. However, the modulus of elasticity in shear, called the *modulus of rigidity* and designated as $G$ (not to be confused with G, which is used in the SI metric system to represent the prefix giga), is smaller in value. The values of $G$ are found experimentally and are indicated in App. B, Table 1.

## 8-6   DUCTILITY

In manufacturing processes, such as rolling, drawing, extrusion, etc., it is important to know the relative ductility of the material. Percent reduction in area and percent elongation are measures of ductility. These terms are defined by the following equations.

$$\text{Percent reduction in area*} = \frac{\text{original area} - \text{final area}}{\text{original area}}(100) \tag{8-5}$$

* An alternate form of this equation may be written:

$$\text{Percent reduction in area} = \frac{(\text{original diameter})^2 - (\text{final diameter})^2}{(\text{original diameter})^2}(100)$$

$$\text{Percent elongation} = \frac{\text{change in gage length}}{\text{original gage length}} (100) \qquad (8\text{-}6)$$

Using Eqs. (8-5) and (8-6) with previous test data, we obtain

$$\text{Original area} = 0.20 \text{ in}^2$$

$$\text{Final area} = \frac{\pi}{4}(0.312)^2 = 0.076 \text{ in}^2$$

$$\text{Percent reduction in area} = \frac{0.20 - 0.076}{0.20}(100) = 62.0 \text{ percent}$$

$$\text{Percent elongation} = \frac{2.328 - 2.000}{2.000}(100) = 16.4 \text{ percent}$$

This elongation is an average value for the 2-in gage length. Actually, the elongation is more severe near the point of failure.

## 8-7  BRITTLE METALS

Members made of brittle materials, such as cast iron, do not readily deform before breaking. Therefore, they do not show an apparent yield point in physical tests. Figure 8-7 is a stress-strain diagram for a brittle specimen in tension. Since the yield strength is often necessary in design, it can be arbitrarily determined by either the 0.20 percent offset method*

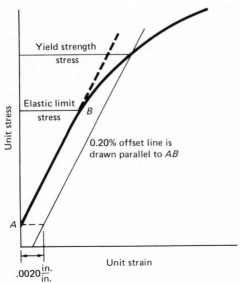

FIGURE 8-7  Stress-strain diagram for tension test of brittle material.

---

* Theodore Baumeister, Eugene A. Avalone, and Theodore Baumeister III: *Marks' Standard Handbook for Mechanical Engineers*, 8th ed., p. 5-2, McGraw-Hill Book Company, Inc., New York, 1978.

or the apparent elastic limit method.* The offset method establishes the yield strength by drawing a line parallel to the straight-line portion of the stress-strain curve as indicated in Fig. 8-7. Note that 0.20 percent of 1 in of length results in a strain of 0.0020 in/in (if metric units are used, the offset is 0.0020 m/m).

## 8-8  ALLOWABLE STRESSES

The strength properties of a material can be determined from physical tests, such as the tension test described above. In most design situations it is not only important to prevent failure of the material but also to avoid permanent deformation. This means that the maximum stress that is considered safe for a material to carry must fall within the elastic range. The values used depend upon the applicaton to which the building materials or machine parts are to be put. These stresses, called *allowable, safe working,* or *design stresses*, are decided upon by designers and engineers through years of experience and experiment.

## 8-9  FACTOR OF SAFETY

Allowable stresses depend primarily upon the type of loading and the material used. Values of the physical properties of materials given in various tables are, at best, average values. In addition, the actual loading may vary somewhat from predicted conditions. Therefore, to ensure a safe design the allowable stress must be made low enough to cover these uncertainties. Allowable stresses are based either on the ultimate strength or the yield stress. The *factor of safety* ($N_u$, when based on the ultimate) is defined as the ratio of the ultimate strength to the allowable stress.

$$N_u = \frac{\text{ultimate stress}}{\text{allowable stress}}$$

or $\qquad$ Allowable stress $= \dfrac{\text{ultimate stress}}{N_u}$ $\qquad$ (8-7)

Similarly, the factor of safety ($N_y$) based on the yield strength is given by

$$N_y = \frac{\text{yield stress}}{\text{allowable stress}}$$

or $\qquad$ Allowable stress $= \dfrac{\text{yield stress}}{N_y}$ $\qquad$ (8-8)

* O. W. Eshbach, *Handbook of Engineering Fundamentals*, 3d ed., p. 409, John Wiley & Sons, Inc., New York, 1975.

Values of $N_u$ and $N_y$ for various conditions are given in App. B, Table 2. In certain fields, allowable stresses are directly specified by codes which have been established by supervising organizations or by legislation.

## 8-10 POISSON'S RATIO

One of the interesting properties of materials is the deformation that takes place in a transverse direction when the piece is subjected to a force in a longitudinal direction. If a tensile force is applied, the piece will elongate and the transverse dimensions will decrease (Fig. 8-8).

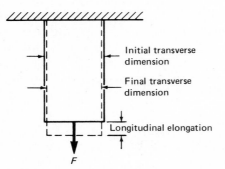

FIGURE 8-8   Dimensional changes due to tensile load.

Conversely, a comprehensive force will shorten the piece and increase the transverse dimensions. The ratio of the transverse strain to the longitudinal strain is known as *Poisson's ratio* ($\mu$). Table 8-1 gives average values of Poisson's ratio for some materials.

**TABLE 8-1    POISSON'S RATIO ($\mu$)**

| Material | $\mu$ |
|---|---|
| Aluminum alloy | 0.330 |
| Brass | 0.340 |
| Bronze | 0.350 |
| Cast iron | 0.270 |
| Concrete | 0.200 |
| Copper | 0.355 |
| Glass | 0.244 |
| Monel | 0.315 |
| Steel | 0.288 |
| Stainless steel | 0.305 |
| Wrought iron | 0.278 |

The modulus of rigidity $G$ (see Sec. 8-5) is related to the modulus of elasticity $E$ through Poisson's ratio, as given by Eq. (8-9).

$$G = \frac{E}{2(1 + \mu)} \qquad (8\text{-}9)$$

**Sample Problem 1** A steel rod, 2 in in diameter, is subjected to a pull of 60 000 lb. What are the longitudinal and transverse strains?

**Solution:**

$$F = 60\ 000 \text{ lb}$$
$$A = \frac{\pi(2)^2}{4} = 3.14 \text{ in}^2$$
$$E = 30 \times 10^6 \text{ psi}$$

*Longitudinal:*

$$s = \frac{F}{A} = \frac{60\ 000}{3.14} = 19\ 100 \text{ psi (tension)}$$

$$E = \frac{s}{\epsilon}$$

$$\epsilon = \frac{s}{E} = \frac{19\ 100}{30\ 000\ 000} = 0.000637 \text{ in/in}$$

*Transverse:*

$$\mu = \frac{\epsilon_{\text{transverse}}}{\epsilon_{\text{longitudinal}}}$$

$$\epsilon_{\text{transverse}} = \mu \epsilon_{\text{longitudinal}} = 0.288(0.000637) = 0.000183 \text{ in/in}$$

## 8-11 THERMAL EXPANSION

All materials tend to change their dimensions when subjected to changes in temperature. Table 8-2 lists values of *coefficients of thermal linear expansion* ($\alpha$) for several common materials in units of inches per inch per degree Fahrenheit and meters per meter per degree Celsius. That is, $\alpha$ is the change in length for each unit of original length for each degree of temperature change. For example, a structural steel bar 8 ft 4 in long (100 in) subjected to a *change in temperature* of 200°F will change its length by $6.5 \times 10^{-6}$ in/in/°F, or the total change in length for the bar will be $6.5 \times 10^{-6} \times 100 \times 200 = 0.13$ in. If the change in temperature was an *increase* in temperature, the length of the bar would *increase* by 0.13 in. Similarly, if the temperature change was a *decrease* of 200°F, the length of the bar would have been *decreased* (shortened) by 0.13 in. Equation (8-10) gives the expression for the total deformation or change in length due to a temperature change.

$$\delta = \alpha l\, \Delta t \qquad (8\text{-}10)$$

**TABLE 8-2**  COEFFICIENTS OF LINEAR EXPANSION ($\alpha$)

| Material | $10^{-6}$ in/in/°F | $10^{-6}$ m/m/°C |
|---|---|---|
| Aluminum alloy | 12.8 | 23.0 |
| Brass | 10.4 | 18.7 |
| Bronze | 10.0 | 18.0 |
| Cast iron | 6.3 | 11.3 |
| Concrete | 5.5 | 9.9 |
| Copper | 9.2 | 16.6 |
| Glass | 5.0 | 9.0 |
| Monel | 7.8 | 14.0 |
| Steel | 6.5 | 11.7 |
| Wrought iron | 6.4 | 11.5 |

where
$\delta$ = total deformation, in; m
$\alpha$ = coefficient of thermal linear expansion,
    in/in/°F; m/m/°C (Table 8-2)
$l$ = original length, in; m
$\Delta t$ = change in temperature, °F; °C ($\Delta t = t_2 - t_1$)

Equation (8-10) gives the actual total deformation for members that are *free* to change their dimensions. When a member deforms freely due to a temperature change, no stresses occur in the material due to $\Delta t$. If, however, a member is *prevented* from deforming freely when a temperature change occurs, the restraining forces which prevent the deformation will produce stresses in the material. Stresses due to restrained deformation will be discussed below, after the following applications of *free* thermal expansion.

**Sample Problem 2**  Steel rails, 60 ft long, are to be laid with a small gap between the end of one section and the beginning of the next. How large should the gap be at 20°F so that the rail ends will just touch when the temperature is 110°F?

**Solution:**  Assume that the rails are free to expand longitudinally. Since each rail length will undergo equal expansion, and will expand at both ends, one rail may be assumed to close half the gap at each end. However, an adjacent rail will close the other half of the gap so that we may calculate the elongation of one rail as if it closed one full gap rather than two halves of a gap. From Eq. (8-10),

$\delta = \alpha l \, \Delta t$
$\alpha = 6.5 \times 10^{-6}$ in/in/°F (Table 8-2)
$l = 60 \, \cancel{ft} \times 12\dfrac{\text{in}}{\cancel{ft}} = 720$ in

$$\Delta t = t_2 - t_1 = 110 - 20 = 90°F$$
$$\delta = (6.5 \times 10^{-6})(720)(90) = 0.42 \text{ in}$$
$$\text{(total elongation of one 60-ft rail section)}$$
$$\therefore \text{ Gap should be 0.42 in or about 27/64 in}$$

*Sample Problem 3 A continuous Monel shaft is to be designed for connecting two power units in a ship. The shaft will be supported by suitable bearings and will be 25 m long. The system is to be designed for a minimum temperature of $-32°C$ and a maximum temperature of $77°C$. Determine the amount of contraction and the amount of expansion that must be provided for if the system is to be installed at $27°C$.

**Solution:**

*Contraction:*

$$\Delta t = t_2 - t_1 = (-32) - 27 = -59°C \text{ (temperature drop)}$$
$$\alpha = 14.0 \times 10^{-6} \text{ m/m/°C (Table 8-2)}$$
$$l = 25 \text{ m}$$
$$\delta = \alpha l \, \Delta t = (14.0 \times 10^{-6})(25)(-59) = ^-0.021 \text{ m} = -21 \text{ mm}$$
$$\text{(total contraction from installed position)}$$

The minus $(-)$ sign for $\delta$ indicates negative expansion (i.e., contraction).

*Expansion:*

$$\Delta t = 77 - 27 = 50°C \text{ (temperature rise)}$$
$$\delta = (14.0 \times 10^{-6})(25)(50) = 0.018 \text{ m} = 18 \text{ mm}$$
$$\text{(total expansion from installed position)}$$

Note that the total change in shaft length that must be designed for is 21 + 18 = 39 mm.

## 8-12 THERMAL STRESSES

In Sec. 8-11, the total deformation of a member due to a temperature change was defined by Eq. (8-10). It was further stated that the value of $\delta$ given by Eq. (8-10) would represent the actual change in length of a member that is *free* to change its dimensions. There are many situations, however, when a member undergoing a temperature change is *prevented* from expanding or contracting. When thermal expansion or contraction is restrained, the restraining forces cause new stresses in the member. These stresses are just as real as if equivalent externally applied loads were acting instead of the restraining forces. Let us develop an expression for these new stresses (or changes in stresses) due to a temperature change of a restrained member.

From Eq. (8-2),

$$s = E\epsilon, \quad \text{where} \quad \epsilon = \frac{\delta}{l}$$

From Eq. (8-10),

$$\frac{\delta}{l} = \alpha \, \Delta t$$

Therefore,

$$\epsilon = \alpha \, \Delta t$$

Substituting this expression into Eq. (8-2), we obtain

$$s = E\alpha \, \Delta t \tag{8-11}$$

where $s$ = thermal stress due to prevented deformation, psi; Pa
$E$ = modulus of elasticity, psi; Pa
$\alpha$ = coefficient of thermal linear expansion, in/in/°F; m/m/°C
$\Delta t$ = temperature change °F; °C

In the cases considered here, the thermal stress will be assumed to be axial in either tension or compression; *tension* occurring when the member should contract but is prevented from doing so; *compression* occurring when the member should expand but is prevented from doing so.

**Sample Problem 4**   A Class 20 cast-iron pipe 12 ft long is rigidly fastened between two walls. Find the stress induced in the pipe due to a temperature rise of 25°F.

**Solution:**
$$E = 11 \times 10^6 \text{ psi (App. B, Table 1)}$$
$$\alpha = 6.3 \times 10^{-6} \text{ in/in/°F (Table 8-2)}$$
$$\Delta t = 25°F$$

By Eq. (8-11),

$$s = E\alpha \, \Delta t = 11 \times 10^6 (6.3 \times 10^{-6})(25) = 1730 \text{ psi (compression)}$$

**\*Sample Problem 5**   A steel tie rod is subjected to an axial pull of 18 kN. The allowable tensile stress is specified as 160 MPa.

(a) Find the minimum diameter required.

(b) A temperature drop of 38°C is anticipated. If the ends of the tie rod are fixed, what is the minimum required diameter?

**Solution a:**

$$A = \frac{F}{s} = \frac{18(10^3)}{160(10^6)} = 0.112 \times 10^{-3}\,\text{m}^2 = 112 \times 10^{-6}\,\text{m}^2 = 112\,\text{mm}^2$$

$$A = \frac{\pi d^2}{4}$$

$$d = \sqrt{\frac{4A}{\pi}} = \sqrt{\frac{4(112)}{\pi}} = \sqrt{143}$$

$$= 11.96\,\text{mm} \qquad \text{use 12-mm rod}$$

**Solution b:**

$$E = 200 \times 10^9\,\text{Pa (App. B, Table 1)}$$
$$\alpha = 11.7 \times 10^{-6}\,\text{m/m/°C (Table 8-2)}$$
$$s = E\alpha\,\Delta t = (200 \times 10^9)(11.7 \times 10^{-6})(38)$$
$$= 89\,\text{MPa (tension)}$$

Since the allowable stress = 160 MPa, then 160 − 89 = 71 MPa is the maximum stress which may be developed by the 18-kN load.

$$A = \frac{F}{s} = \frac{18(10^3)}{71(10^6)} = 0.254 \times 10^{-3}\,\text{m}^2 = 254 \times 10^{-6}\,\text{m}^2 = 254\,\text{mm}^2$$

$$d = \sqrt{\frac{4A}{\pi}} = 18\,\text{mm} \qquad \text{use 18-mm rod}$$

# 8-13 MEMBERS COMPOSED OF TWO MATERIALS IN PARALLEL

The term *parallel* as used in this section and the term *series* as used in Sec. 8-14 refer to the similarity between the paths of force transmission and the corresponding current flow in simple electrical circuits.

Consider a tension member composed of two different materials (Fig. 8-9). Let $F$ be the tensile load carried by this member. Since materials $A$ and $B$ act as a single member, their elongation must be the same. Therefore, $\delta_a = \delta_b$, and since the length is the same for both materials, then $\epsilon_a = \epsilon_b$.

But
$$\epsilon_a = \frac{s_a}{E_a} \qquad \epsilon_b = \frac{s_b}{E_b}$$

Therefore,
$$\frac{s_a}{E_a} = \frac{s_b}{E_b} \tag{8-12}$$

FIGURE 8-9   Member composed of two different materials in parallel.

Also the total load $F = F_a + F_b$, the sum of the force carried by $A$ and the force carried by $B$. Therefore,

$$F = A_a s_a + A_b s_b. \tag{8-13}$$

The use of Eqs. (8-12) and (8-13) is demonstrated in the example below.

**Sample Problem 6**   A short, 8- by 8-in timber post is reinforced with two 7- by $\frac{1}{2}$-in steel plates bolted to the post so that the parts act as a single member (Fig. 8-10). Note that the dressed size of an 8 by 8 in is actually $7\frac{1}{2}$ by $7\frac{1}{2}$ in (see App. B, Table 11).

Find the compressive load that this post can carry if the stress in the timber is not to exceed 1000 psi. Assume that $E = 1\,600\,000$ psi for timber.

**Solution:**   Use $E = 30 \times 10^6$ psi for steel. By Eq. (8-12),

$$s_{\text{steel}} = s_{\text{timber}} \frac{E_{\text{steel}}}{E_{\text{timber}}} = 1000 \left( \frac{30 \times 10^6}{1.6 \times 10^6} \right)$$
$$= 18\,750 \text{ psi} \quad \text{(stress in steel when timber is stressed to 1000 psi)}$$

Applying Eq. (8-13),

$$F = (As)_{\text{timber}} + (As)_{\text{steel}}$$
$$= (7.5)(7.5)(1000) + (2)(7)(0.5)(18\,750)$$
$$= 56\,200 + 131\,000 = 187\,200 \text{ lb} \quad \text{(safe compressive load)}$$

Say 187 000 lb (safe compressive load)

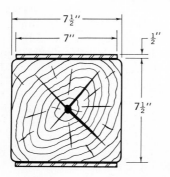

FIGURE 8-10   Diagram for Sample Problem 6.

## 8-14  MEMBERS COMPOSED OF TWO MATERIALS IN SERIES

Consider a member in compression composed of two different materials, as in Fig. 8-11. Since the load $F$ must be transmitted from one end of the member to the other, $F = F_a = F_b$. When the cross-sectional areas of the two materials are equal, then $s_a = s_b$.

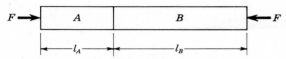

FIGURE 8-11   Member composed of two different materials in series.

The total strain (shortening) is shared by the materials, so that $\delta = \delta_a + \delta_b$. But

$$\delta_a = \frac{Fl_a}{A_aE_a} \quad \text{and} \quad \delta_b = \frac{Fl_b}{A_bE_b}$$

Therefore,

$$\delta = F\left(\frac{l_a}{A_aE_a} + \frac{l_b}{A_bE_b}\right) \tag{8-14}$$

Also, the total load

$$F = A_as_a = A_bs_b. \tag{8-15}$$

The use of Eqs. (8-14) and (8-15) is demonstrated in the following example.

**Sample Problem 7**  A steel bar, 1 in square and 8 in long, is set end to end with a cast-iron (CI) Class 40 bar, 2 in square and 4 in long, between two immovable supports. What stress will develop in each material due to a temperature rise of 50°F?

**Solution:**  From App. B, Table 1,

$$E_s = 30 \times 10^6 \text{ psi (for steel)}$$
$$E_c = 16 \times 10^6 \text{ psi (for CI Class 40)}$$

From Table 8-2,

$$\alpha_s = 6.5 \times 10^{-6} \text{ in/in/°F (for steel)}$$
$$\alpha_c = 6.3 \times 10^{-6} \text{ in/in/°F (for CI)}$$

*If the member were not restrained*, each material would elongate by amounts of

$$\delta_s = \alpha_s l_s \, \Delta t = 6.5 \times 10^{-6} \, (8)(50) = 0.00260 \text{ in}$$
$$\delta_c = \alpha_c l_c \, \Delta t = 6.3 \times 10^{-6} \, (4)(50) = 0.00126 \text{ in}$$

The total elongation, $\delta$, that would occur would equal $\delta = \delta_s + \delta_c = 0.00260 + 0.00126 = 0.00386$ in.

*Since the member is not free to elongate*, the materials are placed in compression by a force $F$, which can be found from Eq. (8-14).

$$F = \frac{\delta}{\dfrac{l_s}{A_s E_s} + \dfrac{l_c}{A_c E_c}} = \frac{0.00386}{\dfrac{8}{(1)(30 \times 10^6)} + \dfrac{4}{(4)(16 \times 10^6)}}$$

$$F = \frac{0.00386}{0.27 \times 10^{-6} + 0.06 \times 10^{-6}} = \frac{0.00386}{0.33 \times 10^{-6}} = 11\ 700 \text{ lb}$$

Applying Eq. (8-15),

$$s_s = \frac{F}{A_s} = \frac{11\ 700}{1} = 11\ 700 \text{ psi} \qquad \text{(compression in steel)}$$

$$s_c = \frac{F}{A_c} = \frac{11\ 700}{4} = 2\ 900 \text{ psi} \qquad \text{(compression in cast iron)}$$

## PROBLEMS

**8-1.** Find the total elongation and the stress in a 1-in-diameter AISI 1045 steel rod 12 ft long under a pull of 12 000 lb.

**\*8-2.** The total elongation of a square wrought-iron bar 50 mm on a side and 3 m long is 1.5 mm. What is the load?

**\*8-3.** What is the strain of Monel when stressed in tension to 196 MPa? If the original length was 992 mm, what is the final length?

**8-4.** A wrought-iron rod 0.80 in² in cross section is elongated 0.06 in by a load of 16 000 lb. What was the original length?

**\*8-5.** A bar of AISI 1020 structural steel 30 mm in diameter is under successive tensile loads of 198, 210, and 270 kN. Calculate the stress that develops at each load. What are the strains for each load? Are they all true strains? If not, why not?

**8-6.** The following data were obtained from a tension test of 6061-T6 aluminum alloy.

   *a.* Plot a stress-vs.-strain diagram for these data.

   *b.* Estimate the elastic-limit stress from the stress-vs.-strain diagram.

   *c.* From the stress-vs.-strain diagram, determine the modulus of elasticity.

| Load, lb | Elongation, in | Data | |
|---|---|---|---|
| 500 | 0 | Original gage length | 2.000 in |
| 1 000 | 0.0004 | Final gage length | 2.400 in |
| 1 500 | 0.0009 | Original diam | 0.5038 in |
| 2 000 | 0.0013 | Final diam | 0.3850 in |
| 2 500 | 0.0018 | Yield point | 9140 lb |
| 3 000 | 0.0022 | Ultimate load | 11 940 lb |
| 3 500 | 0.0027 | Rockwell | B53.8 before |
| 4 000 | 0.0033 | Rockwell | B57.2 after |
| 4 500 | 0.0038 | | |
| 5 000 | 0.0043 | | |
| 5 500 | 0.0048 | | |
| 6 000 | 0.0053 | | |
| 6 500 | 0.0058 | | |
| 7 000 | 0.0065 | | |
| 7 500 | 0.0073 | | |
| 8 000 | 0.0084 | | |
| 8 500 | 0.0103 | | |

d. Calculate the yield stress and the ultimate stress.
e. Calculate the percent elongation and the percent reduction in area.

**8-7.** The following data were obtained from a tension test of yellow brass.

| Load, lb | Elongation, in | Data | |
|---|---|---|---|
| 500 | 0 | Original gage length | 2.000 in |
| 1 000 | 0.0004 | Final gage length | 2.438 in |
| 1 500 | 0.0007 | Original diam | 0.5048 in |
| 2 000 | 0.0011 | Final diam | 0.3660 in |
| 2 500 | 0.0015 | Yield point | 11 600 lb |
| 3 000 | 0.0018 | Ultimate load | 12 300 lb |
| 3 500 | 0.0022 | Rockwell | B59.2 before |
| 4 000 | 0.0026 | Rockwell | B59.6 after |
| 4 500 | 0.0030 | | |
| 5 000 | 0.0034 | | |
| 5 500 | 0.0039 | | |
| 6 000 | 0.0044 | | |
| 6 500 | 0.0049 | | |
| 7 000 | 0.0055 | | |
| 7 500 | 0.0063 | | |
| 8 000 | 0.0072 | | |
| 8 500 | 0.0081 | | |
| 9 000 | 0.0092 | | |
| 9 500 | 0.0140 | | |

a. Plot a stress-vs.-strain diagram for these data.

b. Estimate the elastic-limit stress from the stress-vs.-strain diagram.

c. From the stress-vs.-strain diagram, determine the modulus of elasticity.

d. Calculate the yield stress and the ultimate stress.

e. Calculate the percent elongation and the percent reduction in area.

*8-8. In an experiment in the testing laboratory, a specimen 20 mm in diameter and 250 mm long reached the elastic limit under a tensile load of 120 kN and failed under a load of 175 kN. The length of the bar at its elastic limit was 251.2 mm, and at failure was 333.8 mm. Compute the stress at the elastic limit, the ultimate tensile stress, and the strain for the elastic limit.

8-9. A round, Class 20 cast-iron post 2 ft long, 8 in OD, and 6 in ID supports a load of 25 tons. What is the total decrease in length?

*8-10. How much will an alloy steel punch 25 mm in diameter and 75 mm long be compressed while punching a hole through a 12-mm AISI 1045 steel plate?

*8-11. A wire made of copper alloy is 4.5 m long and 1 mm in diameter. It is stretched 14.5 mm by a load of 340 N. What is the modulus of elasticity of this alloy?

8-12. An AISI 1020 steel rod 32 ft long is used as a hanger for a theater balcony. This rod has to support 52 000 lb. What should be the diameter of the rod, if its total elongation does not exceed 0.15 in and the stress does not exceed 12 000 psi?

8-13. A short, round AISI 1020 bar has a cross-sectional area of 4.20 $in^2$ and carries a compressive load of 20 tons. What is the factor of safety $N_u$?

*8-14. The shearing force on a hard-steel (AISI 1095) pin 30 mm in diameter is 25 kN. What is the factor for safety $N_u$? The pin is in single shear.

8-15. Find the diameter of a Monel rod in tension required to carry a shock load of 7500 lb:

a. Based on ultimate strength

b. Based on yield strength

*8-16. A Monel bar is 60 by 35 mm in cross section. Using a safety factor $N_u$ of 5, what tensile load will the bar support?

8-17. A square timber member carries a compressive load of 73 000 lb. What should be the nominal cross-sectional dimensions using a safety factor of 2.5 based on the elastic limit of 2500 psi? (Use App. B, Table 11, to select a proper member.)

8-18. How many 2- by 3-in wrought-iron bars are required to carry a steady tensile load of 255 000 lb? Determine the allowable stress based on $N_y$.

*8-19. What space must be left between ends of AISI 1095 steel rails, each 9.9 m long and at a temperature of 10°C, so that the ends may be in contact at a temperature of 38°C? If these rails had been in contact at 10°C, what stress would result from the rise in temperature?

**8-20.** An AISI 1045 steel tie rod 24 ft long and 1 in in diameter is used to tie together two walls of a building. The rod is screwed up to a tension of 9000 lb. What will be the stress in the rod if the temperature rises 30°F? If it falls 30°F?

**8-21.** How wide a gap should be allowed between the 50-ft-long paving slabs of concrete in a street so that the slabs will touch each other at a temperature of 90°F? The pavement is laid at a temperature of 65°F. What would be the compressive stress in the slabs if the temperature rose to 110°F? ($E$ for concrete is 3 000 000 psi.)

**8-22.** A surveyor's steel measuring tape is exactly 100 ft long at 60°F when subjected to a pull of 10 lb. Determine the correction to be made in reading the tape if it is used at 100°F under a pull of 20 lb. The tape is $\frac{1}{32}$ in thick and $\frac{3}{8}$ in wide. Express the correction in inches per foot of length and indicate whether it is added to or subtracted from the reading.

**\*8-23.** A 40-mm-diameter bronze rod is 1.8 m long and is subjected to a temperature rise of 60°C. $E = 100$ GPa.
  *a.* Find the total elongation and the stress if the ends are not fixed.
  *b.* Find the total elongation and the stress if the ends are rigidly fixed.
  *c.* For the conditions in part *b*, determine the force in the rod.

**\*8-24.** A wrought-iron bar 100 mm wide and 25 mm thick is placed between two steel bars, each 100 mm wide and 12 mm thick. If the bars act as a unit:
  *a.* What stresses are produced by a pull of 450 kN applied to the combination?
  *b.* What is the elongation due to the load if the member is 1.5 m long?

**8-25.** A short, CI Class 40 square column (Fig. Prob. 8-25) is filled with plain concrete. A load $F$ is applied so as to cause the parts to act as a single member. If the modulus of elasticity of concrete is 2 000 000 psi, what is the compressive load that this short column can carry if the stress in the concrete is not to exceed 900 psi?

**8-26.** A concrete cylinder 4 in in diameter and 2 ft long supports a structural steel rod 1 in in diameter and 6 in long. What is the maximum compressive load that can be applied to the top of the steel rod such that the total shortening of the two cylinders does not exceed 0.005 in? Use $E = 3.0 \times 10^6$ psi for concrete.

**8-27.** A 6061-T6 aluminum-alloy rod 4 in long is supported at one end so that its other end provides a gap of 0.002 in with the end of a yellow brass rod 6 in long supported in a like manner and in line

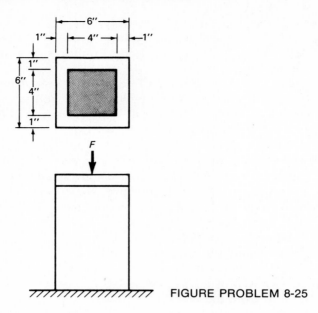

FIGURE PROBLEM 8-25

with it. The cross-sectional area of each rod is 2 in². It is anticipated that the rods will be subjected to a maximum temperature increase of 100°F.

a. What maximum stress will develop in each of the rods if the supports are immovable?

b. If both rods are made of aluminum alloy, what stress will develop in each of the rods?

c. If both rods are made of yellow brass, what stress will develop in each of the rods?

*8-28. A surveyor's steel measuring tape is exactly 30 m long at 15°C when subjected to a pull of 45 N. Determine the correction to be made in reading the tape if it is used at 35°C under a pull of 70 N. The tape is 0.8 mm thick and 10 mm wide. Express the correction in millimeters per meter of length and indicate whether it is added to or subtracted from the reading.

# CHAPTER
# 9

## Bolted, Riveted, and Welded Joints, and Thin-Walled Pressure Vessels

### 9-1 INTRODUCTION

In the construction of a steam generator or in the erection of a building or bridge, the individual members or parts must be securely connected in order to safely carry and transmit forces and moments due to the loadings. Such connections are achieved by means of bolts, rivets, pins, and welds.

This chapter will be concerned with the analysis and design of several types of bolted, riveted, and welded joints which occur in the construction and mechanical fields. However, the emphasis will be on bolted and welded joints, since rivets are seldom used in these fields today.

### 9-2 BOLTED JOINTS

Bolted joints are designated either lap or butt joints. When two plates are placed with the ends overlapping, they form a lap joint. Figure 9-1 represents the top and sectional views of a lap joint with a single row of bolts, and Fig. 9-2 represents one with a double row of bolts. Rows of bolts are perpendicular to the applied load.

When two plates are placed edge to edge and are connected by bolts, as in Fig. 9-3, they form a butt joint. Since there is one row of bolts on either side of the centerline of the joint, Fig. 9-3 is called a single-bolted butt joint. Similarly, Fig. 9-4 shows a double-bolted butt joint (two rows

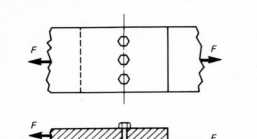

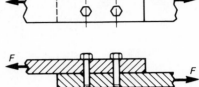

FIGURE 9-1  Single bolted lap joint.     FIGURE 9-2  Double bolted lap joint.

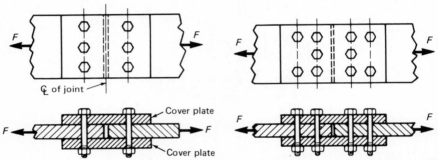

FIGURE 9-3  Single bolted butt joint.     FIGURE 9-4  Double bolted butt joint.

of bolts on each side of the joint). It should be noted that bolted butt joints require the use of cover plates.

## 9-3  TYPES OF FAILURE IN BOLTED JOINTS

The actual stresses developed in bolted joints are quite complex, but in practice they are treated as *simple* tension, compression, and shearing stresses.

The analysis of bolted joints involves the continued use of

$$s = \frac{F}{A} \tag{9-1}$$

where $s$ represents the particular stress to be determined.

It is usually assumed that each bolt will carry its proportional share of the total load on the joint. In the lap joint shown in Fig. 9-2, each bolt transmits one-fourth of the load from one plate to the other. The transmission of load by the bolts may be represented as in Fig. 9-5.

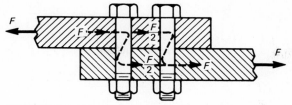

FIGURE 9-5   Force transmission in lap joint—
Fig. 9-2.

A butt joint is composed of two identical halves. Each half is designed to carry the full load. Therefore, in analyzing butt joints, only one-half of the joint is considered. The joint shown in Fig. 9-4 is analyzed as in Fig. 9-6. In this butt joint each bolt carries one-fifth of the load and transmits one-half of it to each cover plate (Fig. 9-6).

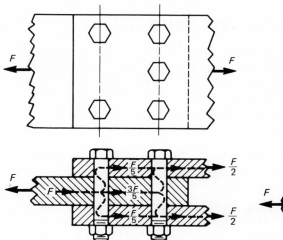

FIGURE 9-6   Force transmission in
butt joint—Fig. 9.4.

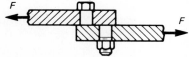

FIGURE 9-7   Bolt shear failure
in lap joint—single shear.

To demonstrate the types of failure, consider the lap joint shown in Fig. 9-1. The joint may fail by shearing the bolts, as shown in Fig. 9-7. Another possibility is that the plate may pull apart at its weakest section. This section occurs on the centerline of the row of bolt holes in Fig. 9-1, and the failure would appear as in Fig. 9-8.

A third mode of failure may occur by crushing of the bolts or the plate material in contact with the bolts. This type of failure is called a *bearing*, or compression, failure. Such a failure is shown in Fig. 9-9. (*Note:* A bolted joint may be designed so that the possibility of failure in bearing

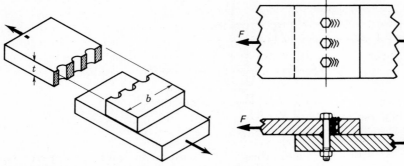

FIGURE 9-8    Plate tension failure in lap joint.

FIGURE 9-9    Bearing failure in lap joint.

need not be considered. This type of joint is referred to as a *friction-type connection* and will be discussed later in the chapter.)

## 9-4  STRESSES IN A BOLTED JOINT

The joints in this chapter involve only direct (axial) loading. Eccentric loading, which tends to cause rotation of the joint, will be dealt with in Chap. 14. Shear, tension, and bearing stresses in axially loaded bolted joints are each represented by

$$s = \frac{F}{A} \tag{9-1}$$

or $$F = As \tag{9-1a}$$

*Shear:*    When computing the shear stress in a bolted joint, Eq. (9-1) gives the relation between the external force $F$, which produces a shearing stress $s_s$, and the total area resisting shear $A_s$. Since the total shear area $A_s$ is composed of a number of circular bolt cross sections, each having a diameter $d$, the special forms of Eq. (9-1) and (9-1a) for shear are

$$s_s = \frac{F}{A_s} = \frac{F}{n\left(\dfrac{\pi d^2}{4}\right)} \tag{9-2}$$

or $$F = \left[ n\left(\frac{\pi d^2}{4}\right) \right] s_s \tag{9-2a}$$

where
$F$ = external force on joint, lb
$s_s$ = shear stress in bolt material, psi
$d$ = bolt diameter at shear plane, in
$n$ = number of shear areas resisting force $F$

In a lap joint, such as Fig. 9-2, each bolt is in *single shear*. Thus, there are four shear areas in this joint, and in Eqs. (9-2) and (9-2a), $n = 4$. Each bolt in one-half of a butt joint is in *double shear*. That is, each bolt tends to fail on two cross sections simultaneously, as shown in Fig. 9-10. Therefore, for the joint in Fig. 9-4, the value of $n$ is 10 in Eqs. (9-2) and (9-2a).

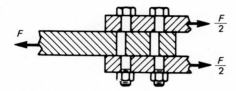

FIGURE 9-10   Bolt shear failure in butt joint—double shear.

***Tension:***   This analysis deals with the possibility of the main plate failing in tension, as in Fig. 9-8. Equation (9-1) applies, where $s_t$ is the tensile stress in the plate on cross-sectional area $A_t$. This area resists a total force $F$. For the lap joint in Fig. 9-8, the effective area of the plate cross section through the center of the bolts is reduced because of the bolt holes.

$$s_t = \frac{F}{A_t} \qquad F = A_t s_t$$

$$A_t = bt - nDt = (b - nD)t$$

$$F = [(b - nD)t]s_t \qquad (9\text{-}3)$$

where    $F$ = external force on joint, lb
   $s_t$ = tensile stress in plate material at cross section in question, psi
   $b$ = width of plate, in
   $t$ = thickness of plate, in
   $D$ = diameter of bolt hole, in
   $n$ = number of holes in cross section in question

Note that $bt$ is the cross-sectional area of the plate without holes, and $Dt$ is the area of material removed from the plate cross section by each hole. Also, $(b - nD)$ represents the effective width of plate remaining after $n$ holes are punched or drilled.

***Bearing:***   With reference to Fig. 9-11, it is evident that when the force $F$ is applied to the plates, each plate is forced against the curved surface of the bolt and tends to crush it and the plate. Thus, a compressive stress is produced which varies in intensity for various points on the curved surface. However, it is customary in figuring this bearing stress to take a section through the axis of the bolt perpendicular to the direction of the applied forces. This section is a rectangle of height $t$, the thickness of the plate, and width $d$, the diameter of the bolt. Then the bearing area for one bolt is $td$. If $s_c$ is the compressive, or bearing, stress,

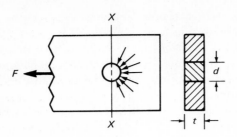

FIGURE 9-11   Bearing forces and bearing area.

$$s_c = \frac{F}{A_c} \qquad F = A_c s_c$$
$$A_c = ntd$$
$$F = [ntd]s_c \qquad\qquad (9\text{-}4)$$

where     $F$ = external force on joint, lb
$s_c$ = bearing (compressive) stress in plate and bolt material, psi
$t$ = thickness of plate, in
$d$ = bolt diameter, in
$n$ = number of bearing areas (usually one per bolt in a given plate)

## 9-5   TERMINOLOGY AND CODES FOR BOLTED JOINTS

The distance between centers of adjacent bolts is called *pitch*.

In the case of a long or continuous bolted joint, subjected to uniform tension for its entire length, the strength of the joint may be computed from a repeating section. A repeating section is the shortest length of joint which contains a complete bolt pattern. In Fig. 9-12 the distance $l$ is the length of the repeating section. Usually, the length of the repeating section is equal to the pitch of the bolts in row 1 (outer row).

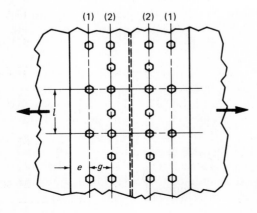

FIGURE 9-12   Continuous bolted joint.

The distance $g$ between rows of bolts is called the gage.

The distance from the center of a bolt to any plate edge is called the *edge distance*. Such a distance is indicated in Fig. 9-12 by $e$.

The design of bolted joints for specific applications is usually governed by codes which have been established by local legislation or technical groups. Two of the most commonly used codes are the American Institute of Steel Construction (AISC) code for connections, and the American Society of Mechanical Engineers (ASME) code for boilers and pressure vessels. Some of the provisions of these codes are given in Tables 9-1 and 9-2. For other codes or more complete information on the above codes, the student is referred to the technical agency concerned.

**TABLE 9-1**  AISC CODE FOR CONNECTIONS (RIVETS, BOLTS, AND THREADED PARTS)

| | Allowable Stress, psi | |
|---|---|---|
| Type of Stress and Material Specification | Friction-Type Connections | Bearing-Type Connections |
| *Shear on fasteners:* | | |
| A 502, Grade 1, Hot-driven rivets | | 15 000 |
| A 502, Grade 2, Hot-driven rivets | | 20 000 |
| A 307, Low-carbon steel bolts | | 10 000 |
| A 325, High-strength bolts; threads occur at the shear plane | 17 500 | 21 000 |
| A 325, High-strength bolts; threads do not occur at the shear plane | 17 500 | 30 000 |
| A 490, High-strength bolts; threads occur at the shear plane | 22 000 | 28 000 |
| A 490, High-strength bolts; threads do not occur at the shear plane | 22 000 | 40 000 |
| Threaded parts of other steels | | $0.30 s_y$ |
| *Bearing on projected area (bearing-type connections only)†:* | | $1.35 s_y$ |
| A 36, Carbon structural steel; thickness 8 in and less ($s_y = 36\ 000$ psi) | | 48 500 |
| A 440, High-strength structural steel; thickness $\frac{3}{4}$ in and less ($s_y = 50\ 000$ psi) | | 67 500 |
| A 441, High-strength low-alloy structural steel; thickness $\frac{3}{4}$ in and less ($s_y = 50\ 000$ psi) | | 67 500 |
| A 572, Grade 50, High-strength low-alloy steel; thickness $1\frac{1}{2}$ in and less ($s_y = 50\ 000$ psi) | | 67 500 |
| A 242, Corrosion-resistant high-strength low-alloy structural steel; thickness $\frac{3}{4}$ in and less ($s_y = 50\ 000$ psi) | | 67 500 |
| A 588, Corrosion-resistant high-strength low-alloy structural steel; thickness 4 in and less ($s_y = 50\ 000$ psi) | | 67 500 |

(continued)

**TABLE 9-1**   AISC CODE FOR CONNECTIONS
(RIVETS, BOLTS, AND THREADED PARTS)—(Continued)

| | Allowable Stress, psi | |
|---|---|---|
| Type of Stress and Material Specification | Friction-Type Connections | Bearing-Type Connections |
| *Tension on net section (bearing- and friction-type connections)†:* | $0.60\,s_y$ | $0.60\,s_y$ |
| A 36, Carbon structural steel; thickness 8 in and less ($s_y = 36\,000$ psi) | 22 000 | 22 000 |
| A 440, High-strength structural steel; thickness $\frac{3}{4}$ in and less ($s_y = 50\,000$ psi) | 30 000 | 30 000 |
| A 441, High-strength low-alloy structural steel; thickness $\frac{3}{4}$ in and less ($s_y = 50\,000$ psi) | 30 000 | 30 000 |
| A 572, Grade 50, High-strength low-alloy steel; thickness $1\frac{1}{2}$ in and less ($s_y = 50\,000$ psi) | 30 000 | 30 000 |
| A 242, Corrosion-resistant high-strength low-alloy structural steel; thickness $\frac{3}{4}$ in and less ($s_y = 50\,000$ psi) | 30 000 | 30 000 |
| A 588, Corrosion-resistant high-strength low-alloy structural steel; thickness 4 in and less ($s_y = 50\,000$ psi) | 30 000 | 30 000 |

† For steels having an $s_y$ other than 36 000 or 50 000 psi, refer to the *Manual of Steel Construction of AISC.*
Additional provisions:
  a.  Minimum preferable pitch (center to center between adjacent fasteners) = 3d to provide clearance for tools in assembling connections.
  b.  Minimum edge distance varies approximately from $1\frac{1}{4}\,d$ to $1\frac{3}{4}\,d$.
  c.  For purpose of design, hole diameters are taken as $\frac{1}{8}$ in larger than fastener diameters; $D = d + \frac{1}{8}''$.
*Note:* The AISC code (*Manual of Steel Construction*) also establishes allowable stresses for rivets, bolts, and threaded parts subjected to direct tension.

Referring to Table 9-1, the AISC code for connections, the allowable shear stress given for most of the listed fasteners distinguishes between *friction-type* and *bearing-type* connections.

The designer of a connection (joint) may use either of the two types. The *friction type* assumes no movement between adjacent surfaces of the plate material and, as a result, bearing stress is not considered. However, *shear* on the fastener and *tension* on the plates are taken into account. The *bearing type* does assume the possibility of movement between adjacent surfaces. Thus, the possibility of failure in shear, bearing, and tension are all considered.

Referring to Table 9-2, the ASME code for bolted connections on boilers and pressure vessels, the use of this code is based on the possibilty of failure in shear, tension, and bearing.

**TABLE 9-2**  ASME BOILER AND PRESSURE VESSEL CODE
FOR BOLTED CONNECTIONS

| Type of Stress and Material Specification | Allowable Stress, psi |
|---|---|
| *Shear on fastener:* | |
| SA 325, Medium-carbon-steel high-strength bolts | 15 400 |
| SA 499, Medium-carbon-steel high-strength bolts | |
| Diameter 1 in and less | 18 400 |
| Diameter from $1\frac{1}{8}$ in to $1\frac{1}{2}$ in | 16 200 |
| Diameter from $1\frac{5}{8}$ in to 3 in | 11 600 |
| *Bearing on fastener:* | |
| SA 325, Medium-carbon-steel high-strength bolts | 30 700 |
| SA 499, Medium-carbon-steel high-strength bolts | |
| Diameter 1 in and less | 36 800 |
| Diameter from $1\frac{1}{4}$ in to $1\frac{1}{2}$ in | 32 300 |
| Diameter from $1\frac{5}{8}$ in to 3 in | 23 200 |
| *Bearing on plate:* | |
| SA 285, Carbon-steel plates | |
| Grade A | 17 900 |
| Grade B | 20 000 |
| Grade C | 21 900 |
| SA 36,  Structural-steel plates | 20 200 |
| SA 612, High-strength steel plates | |
| Grade A | 33 100 |
| Grade B | 32 300 |
| *Tension on net section:* | |
| SA 285, Carbon-steel plates | |
| Grade A | 11 200 |
| Grade B | 12 500 |
| Grade C | 13 700 |
| SA 36,  Structural-steel plates | 12 600 |
| SA 612, High-strength steel plates | |
| Grade A | 20 700 |
| Grade B | 20 200 |

Additional provisions:
  a.  Minimum thickness of main plate for pressure vessel is $\frac{3}{16}$ in.
  b.  Minimum thickess of butt straps (cover plates) for a double-strap connection shall
      be not less than $\frac{2}{3}$ of the main plate. For specific recommended thicknesses, refer
      to the code.
  c.  Hole diameter = bolt diameter + 0.02 in maximum.
  d.  Use bolt diameter for all calculations; $D = d$.
  e.  Minimum gage distance (distance between rows) varies from $1\frac{3}{4}$ to $2\frac{1}{2}$ times the
      bolt diameter $(d)$.
  f.  Minimum edge distance varies from $1\frac{1}{2}$ to $1\frac{3}{4}$ times the bolt diameter $(d)$.

**Sample Problem 1**  A single-bolted lap joint with 12- by $\frac{1}{2}$-in A36 plates
contains four $\frac{7}{8}$-in-diameter A325 bolts (Fig. 9-13). Based on the AISC
code, friction-type connection, determine the largest load the joint can
safely carry.

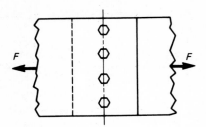

FIGURE 9-13 Diagram for Sample Problem 1.

**Solution:** To find the safe load on the joint, calculate the safe loads in shear and tension. Bearing is not considered because the joint is a friction-type connection. The largest safe load on the joint will be the *least* of the two calculated loads since this load must satisfy both conditions.

***Shear:*** *(Eq. 9-2a):* Note that each bolt is in single shear; $n = 4$.

$$F_s = \left[ n\left(\frac{\pi d^2}{4}\right) \right] s_s = 4\left[ \frac{\pi(0.875)^2}{4} \right] (17\ 500)$$
$$= 4(0.601)(17\ 500) = 42\ 100\ \text{lb}$$

***Tension:*** *(Eq. 9-3):* There are four holes in the plate along the weak section; $n = 4$, $D = d + \frac{1}{8}$ (AISC code).

$$F_t = [(b - nD)t]s_t = [12 - 4(1)]0.5(22\ 000)$$
$$= 8(0.5)(22\ 000) = 88\ 000\ \text{lb}$$

The largest safe load on the joint is 42 100 lb. This is called the *strength of the joint.*

**Sample Problem 2** A double-bolted butt joint, with four 1-in-diameter A490 bolts on each side and two bolts per row, joins 10- by ¾-in A441 plates (Fig. 9-14). The cover plates are ½ in thick. Based on the AISC code, bearing-type connection, find the strength of the joint (maximum safe load). Assume that the bolt threads do not occur at the shear plane.

**Solution:**

***Shear:*** Note that each bolt is in double shear and that only one-half of the joint is considered; $n = 8$.

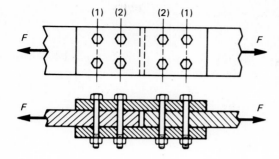

FIGURE 9-14  Diagram for Sample Problem 2.

$$F_s = \left[ n\left( \frac{\pi d^2}{4} \right) \right] s_s = 8 \left[ \frac{\pi (1)^2}{4} \right] (40\ 000)$$
$$= 8(0.785)(40\ 000) = 251\ 000 \text{ lb}$$

***Tension:*** The weak section in the main plate occurs at row 1 (Fig. 9-14). This section must transmit the full load and has two bolt holes. The section at row 2 also has two holes but is required to transmit only half of the external force (see Sec. 9-3). For the cover plates in tension the critical section is at row 2. It is usually assumed that each cover plate transmits one-half of the external force to the other side of the joint. In this problem, it is not necessary to check the safe load on the cover plates since the total cover-plate thickness exceeds that of the main plate and the number of bolt holes in row 2 is the same as for row 1. For the main plate at row 1, $n = 2$, $D = d + \frac{1}{8}$ (AISC code)

$$F_t = [(b - nD)t]s_t = [10 - 2(1\tfrac{1}{8})](0.75)(30\ 000)$$
$$= (7.75)(0.75)(30\ 000) = 174\ 000 \text{ lb}$$

***Bearing:*** Each rivet contributes one bearing area in the main plate; $n = 4$.

$$F_c = [ntd]s_c = 4(0.75)(1)(67\ 500)$$
$$= 3(67\ 500) = 203\ 000 \text{ lb}$$

The strength of the joint is 174 000 lb.

## 9-6  EFFICIENCY OF A BOLTED JOINT

The area of the cross section of a plate is called the *gross area, bt*.

The quotient obtained by dividing the tensile strength of the gross area into the strength of the bolted joint is called the *efficiency* of the joint and is represented by $\eta$ (the Greek letter *eta*). That is, the efficiency is

$$\eta = \frac{\text{strength of the joint}}{\text{tensile strength of the gross area}} \times 100 \qquad (9\text{-}5)$$

In determining the efficiency, the repeating section width of plate (*l*) may be used instead of the entire width.

**Sample Problem 3**   Find the efficiency of a single-bolted lap joint with $\frac{1}{2}$-in-thick plates (A36), $\frac{7}{8}$-in-diameter bolts (A325), and a pitch width (repeating section) of 3 in. Use the AISC code, bearing-type connection; bolt threads do not occur at the shear plate.

**Solution:**

*Shear:*

$$n = 1$$
$$F_s = 1\left[\frac{\pi(0.875)^2}{4}\right](30\,000) = 18\,000\text{ lb}$$

*Tension:*

$$n = 1$$
$$F_t = [3 - 1(1)](\tfrac{1}{2})(22\,000) = 22\,000\text{ lb}$$

*Bearing:*

$$n = 1$$
$$F_c = 1(0.875)(\tfrac{1}{2})(48\,500) = 21\,200\text{ lb}$$

From the above calculations, the strength of the joint is 18 000 lb. Tensile strength of gross area is

$$F_g = (lt)s_t = 3(\tfrac{1}{2})(22\,000) = 33\,000\text{ lb}$$

Therefore,   $$\eta = \frac{18\,000}{33\,000} = 0.545 = 54.5\%$$

**Sample Problem 4**   Find the efficiency of a triple-bolted butt joint with $\frac{3}{4}$-in main plates, $\frac{1}{2}$-in cover plates, and 1-in-diameter SA499 steel bolts as arranged in Fig. 9-15. Use the ASME code. The plates are SA612, Grade A.

**Solution:**   The bolt-hole size of 1 in is used for all calculations (see Table 9-2).

All calculations refer to a 10-in repeating section, as indicated in Fig. 9-15.

*Shear:*

$$n = 12\text{ (6 bolts in double shear)}$$
$$F_s = \left[n\left(\frac{\pi d^2}{4}\right)\right]s_s = (12)(0.785)(1^2)(18\,400) = 173\,000\text{ lb}$$

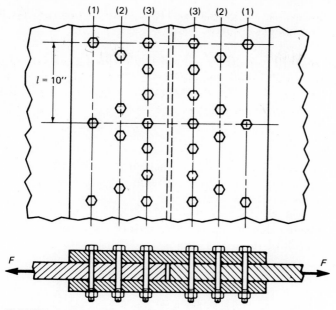

FIGURE 9-15   Diagram for Sample Problem 4.

***Bearing:***

$$n = 6$$
$$F_c = [ntd]s_c = 6(0.75)(1)(33\ 100) = 149\ 000\ \text{lb}$$

(*Note:* The allowable bearing stress is based on the plate allowable because it is less than the fastener allowable of 36 800 psi.)

***Tension*** (*Row* 1):

$$n = 1$$
$$F_t = [(\ell - nD)t]s_t = [10 - 1(1)](0.75)(20\ 700)$$
$$= (9)(0.75)(20\ 700) = 140\ 000\ \text{lb}$$

***Tension*** (*Row* 2):

$$n = 2$$

The purpose of these calculations is to evaluate the external force $F$ which a 10-in repeating section can carry. This force is transmitted by the bolts to the cover plates, assuming that each bolt transmits its share of the force (see Sec. 9-3 and Fig. 9-6). The tensile force on the plate at row 2 is smaller than the external force $F$, by the amount which the bolt in row

1 has transmitted to the cover plates. Thus, the force on row 2 in this problem is $\frac{5}{6}$ of the external force.

$$\tfrac{5}{6}F_t = [(l - nD)t]s_t = [10 - 2(1)](0.75)(20\ 700)$$
$$= 8(0.75)(20\ 700) = 124\ 000$$
$$F_t = \tfrac{6}{5}(124\ 000) = 149\ 000\ \text{lb}$$

**Tension** (*Row* 3):

$$n = 3$$
$$\tfrac{3}{6}F_t = [(l - nD)t]s_t = [10 - 3(1)](0.75)(20\ 700)$$
$$= 7(0.75)(20\ 700) = 109\ 000$$
$$F_t = \tfrac{6}{3}(109\ 000) = 218\ 000\ \text{lb}$$

From the above calculations, the strength of the joint is 140 000 lb. Tensile strength of the gross area is

$$F_g = (lt)s_t = 10(0.75)(20\ 700) = 155\ 000\ \text{lb}$$

and

$$\eta = \frac{140\ 000}{155\ 000} = 0.903 = 90.3\%$$

Check cover-plate size. Assume one-half the external force is transmitted to each cover plate. This load must be transmitted by the cover plate from the inner-row centerline (row 3) to the inner row in the other half of the joint. Since the safe load is 140 000 lb, each cover plate must transmit 70 000 lb across its weakest section (row 3).

$$\text{Allowable } F_{cp} = [(l - nD)t]s_t = [10 - 3(1)](0.5)(20\ 700)$$
$$= 7(0.5)(20\ 700) = 72\ 500\ \text{lb}$$
$$\text{Actual } F_{cp} = 70\ 000\ \text{lb} \qquad \text{therefore OK}$$

## 9-7 BOLTED JOINTS OF MAXIMUM EFFICIENCY

The procedure for design of bolted joints of maximum efficiency is similar for either lap or butt joints. A practical maximum-efficiency joint is designed so that plate tension in the outer row is the limiting factor. The procedure may be outlined as follows.

1. Determine the allowable load on the main plate at row 1 by subtracting one bolt hole from the gross plate section.
2. Determine the allowable shear force on one bolt. Use one area if single shear or two areas if double shear.
3. Determine the allowable bearing force on one bolt.
4. The smaller numerical result from step 2 or 3 is the limiting force on one bolt. Determine the minimum number of bolts required

by dividing the allowable plate load (from step 1) by the limiting bolt force.

$$\text{Minimum number of bolts} = \frac{\text{allowable plate load}}{\text{limiting bolt force}}$$

5.  Select an appropriate bolt pattern.
6.  Check the plate stress at each row in the pattern.
7.  Revise the pattern if the stress on any row exceeds that of row 1.
8.  Calculate the joint efficiency.

Suggested trial patterns for lap and butt joints of maximum efficiency are given in Figs. 9-16 and 9-17, respectively.* The number shown below

* David Singer, *Basic Structural Design*, Pelex Publishers Inc., New York, 1952.

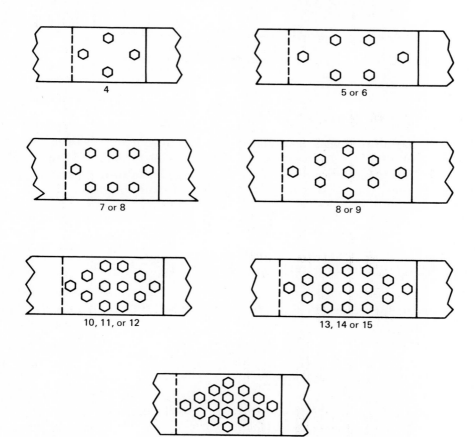

FIGURE 9-16   Suggested trial patterns—lap joints.

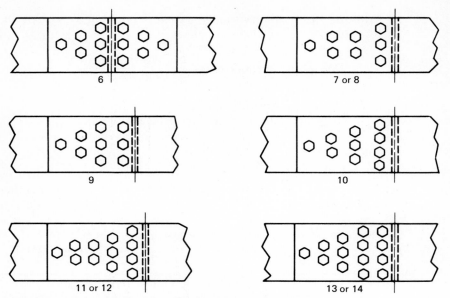

FIGURE 9-17   Suggested trial patterns—butt joints.

each pattern is the number of bolts required to develop the allowable plate stress in row 1.

**Sample Problem 5**   Two plates (A36), 10 in wide by $\frac{1}{4}$ in thick are to be connected by a lap joint with $\frac{5}{8}$-in-diameter bolts (A325). Design a joint of maximum efficiency using the AISC code, bearing-type connection. The bolt threads do not occur at the shear plate.

**Solution:**   Allowable load on main plate at row 1, using $\frac{3}{4}$-in-diameter hole:

$$F_t = [(b - nD)t]s_t = [10 - 1(0.75)](0.25)(22\,000)$$
$$= 9.25(0.25)(22\,000) = 50\,900 \text{ lb}$$

Allowable shear force on one bolt (single shear):

$$F_s = \left[ n\!\left( \frac{\pi d^2}{4} \right) \right] s_s = 1(0.785)(0.625^2)(30\,000)$$
$$= 0.307(30\,000) = 9200 \text{ lb}$$

Allowable bearing force on one bolt:

$$F_c = [ntd]s_c = 1(0.25)(0.625)(48\,500)$$
$$= 7580 \text{ lb}$$

Limiting force on one bolt = 7580 lb.
Minimum number of bolts required:

$$\frac{50\ 900}{7580} = 6.7 \text{ bolts} = 7 \text{ bolts}$$

Referring to Fig. 9-16, select the trial pattern for seven or eight bolts, as in Fig. 9-18. Check the trial pattern.

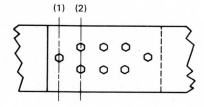

FIGURE 9-18   Diagram for Sample Problem 5.

Force transmitted by each bolt:

$$\frac{\text{Allowable plate load}}{\text{Number of bolts}} = \frac{50\ 900}{8} = 6360 \text{ lb}$$

Tension at row 2:

$$\tfrac{7}{8}F_t = [(b - nD)t]s_t = [10 - 2(0.75)](0.25)(22\ 000)$$
$$= 8.5(0.25)(22\ 000) = 46\ 800$$
$$F_t = \tfrac{8}{7}(46\ 800) = 53\ 500 \text{ lb}$$

The pattern is satisfactory, since the allowable load on the main plate at row 1 is less than on row 2.
The joint efficiency is

$$\eta = \frac{50\ 900}{10(\tfrac{1}{4})(22\ 000)} = \frac{50\ 900}{55\ 000} = 0.925 = 92.5\%$$

## 9-8   RIVETED JOINTS

The procedures discussed earlier in this chapter for investigating and designing bolted joints may be used for riveted joints as well. Check the appropriate code specifications for design procedures and allowable stresses.

With improvements in welding technology and the availability of economical high-strength bolts, riveting for structures and pressure vessels has practically disappeared. While rivets are still used in lightweight and light load applications, adhesives and other fastening materials are successfully competing with rivets.

## 9-9 WELDED JOINTS

For many purposes, welded connections are found to be more suitable than bolted or riveted connections. Welding is a process of joining two pieces of metal by fusion. In the types of welded joints discussed here, the pieces to be joined are heated and molten material is deposited between them. While there are a great variety of welding processes in use today, the heating is most frequently accomplished by means of an electric arc. This section will deal with lap and butt joints (called groove joints), although there are several other types of welded joints of some importance. Figure 9-19 shows what is known as a *side-fillet* lap weld with a corner

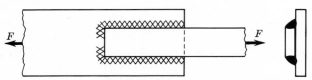

FIGURE 9-19   Side fillet welds (with corner returns).

return. The fused metal is deposited in the angle between the two pieces. Shearing stresses are developed in this weld when either tensile or compressive forces are applied to the long axis of the plates. Figure 9-20 shows a joint using *end* lap welds with corner returns. This type of weld is usually used in conjunction with side welds.

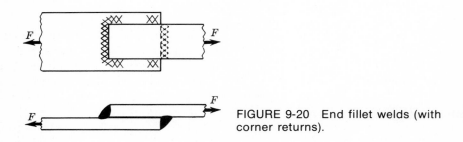

FIGURE 9-20   End fillet welds (with corner returns).

Practically all metals can be welded successfully, although for some, special techniques are required. Commonly encountered materials, such as cast iron, cast steel, structural steel, tool steel, bronze, aluminum, and copper, are all capable of being welded. Figure 9-21 shows various types of butt (groove) welds, most of which require advanced preparation of the plate edges at the joint.

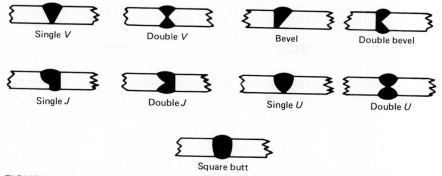

FIGURE 9-21   Types of butt (groove) welds.

The size of fillet welds is designated by the length of the leg, as shown in Fig. 9-22a. Fillet welds resist load on the throat area, which is the minimum cross section of the weld. The throat area is the product of the throat thickness and the effective length of the weld. The throat thickness is determined from Fig. 9-22b as Leg·sin 45° or 0.707 Leg. For

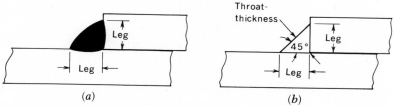

FIGURE 9-22   (a) Cross section of a fillet weld. (b) Simplified cross section of a fillet weld.

1 in of effective weld length, we can determine the allowable load for welds of given sizes at certain allowable shearing stresses. Thus, for $s_s = 21\ 000$ psi, a $\frac{3}{16}$-in weld with an effective length of 1 in can carry an allowable load of $[0.707(\frac{3}{16})(1)](21\ 000) = 2800$ lb. In the same way, allowable loads per inch of effective length may be calculated for various weld sizes.

The AISC code specifies allowable stresses for welds on several types of structural steel by different methods. These specifications conform to American Welding Society (AWS) standards. Both AISC and AWS specifications should be consulted for complete information. Table 9-3 gives appropriate parts of the AISC specifications for the purposes of this book.

Table 9-4 is a listing of allowable loads on welds of various sizes for two different levels of allowable shearing stress.

The total allowable force which a given fillet weld can withstand may be expressed by

$$F = F'(L) \qquad\qquad (9\text{-}6)$$

where   $F$ = total allowable force on entire weld, lb
        $F'$ = allowable force per inch of effective weld length, lb,
from Table 9-4
        $L$ = total effective length of fillet welding, in

**TABLE 9-3**   AISC CODE FOR WELDS (MANUAL ARC METHOD)

| Fillet Welds | |
| --- | --- |
| Type of Steel and Electrode | Allowable Shearing Stress on Throat Area of Weld, psi |
| A36 steel with Class E 60 series electrodes | 18 000 |
| A36, A242, A441, A572, and A588 steels with Class E 70 series electrodes | 21 000 |

Additional provisions:*
    A440 steel is not recommended for welding.
    The effective area of a fillet weld shall be considered as the effective length of the weld times the effective throat thickness.
    Maximum size of a fillet weld applied to a square edge of plate shall be $\frac{1}{16}$ in less than the plate thickness, for plate thickness $\frac{1}{4}$ in or more, and equal to the plate thickness for plate thickness less than $\frac{1}{4}$ in.

Butt (Groove) Welds—Complete Penetration

The allowable stresses permitted for the connected plate material shall apply to complete-penetration butt-weld stresses in tension, compression, bending, and bearing. For shear, use the same allowable shear stress as is used for fillet welds.
    The effective area of a butt weld shall be considered as the effective length of the weld times the effective throat thickness.
    The effective length of a butt weld shall be the width of the part joined.
    The effective throat thickness of a complete-penetration butt weld shall be the thickness of the thinner part joined.

    * The following provisions may be of interest, but have not been uniformly applied to the problems in this book. The reader is referred to the *Steel Construction Manual* of the AISC for full specifications.
    Minimum effective length of a fillet weld shall not be less than 4 times the nominal size of the weld.
    Minimum width of laps on lap joints shall be 5 times the thickness of the thinner part joined and not less than 1 in.
    Side or end fillet welds should be returned continuously around the corners for a distance not less than twice the nominal size of the weld.
    Lap joints joining plates or bars shall be fillet-welded along the edge of both lapped parts . . . to prevent opening of the joint under maximum loading.

**Sample Problem 6**   A lap-welded joint of A36 steel is to be designed for a total load of 50 000 lb. The plates are $\frac{3}{8}$ in thick and $\frac{5}{16}$-in fillet welds

**TABLE 9-4   ALLOWABLE LOADS FOR FILLET WELDS**

| Weld Size (Leg), in | Allowable Load (lb) per Inch of Effective Weld Length | |
| --- | --- | --- |
| | For $s_s$ = 18 000 psi | For $s_s$ = 21 000 psi |
| 3/16 (minimum size, AISC) | 2 400 | 2 800 |
| 1/4 | 3 200 | 3 700 |
| 5/16 | 4 000 | 4 600 |
| 3/8 | 4 800 | 5 600 |
| 7/16 | 5 600 | 6 500 |
| 1/2 | 6 400 | 7 400 |
| 9/16 | 7 200 | 8 400 |
| 5/8 | 8 000 | 9 300 |
| 11/16 | 8 800 | 10 200 |
| 3/4 | 9 600 | 11 100 |

will be used. Find the total length of welding required to carry this load if Class E 60 series electrodes are used.

**Solution:**   From Table 9-3, $s_s$ = 18 000 psi for A36 steel with Class E 60 electrodes. The allowable load per inch of weld, $F'$ = 4000 lb for $\frac{5}{16}$-in welding at $s_s$ = 18 000 psi from Table 9-4. Using Eq. (9-6),

$$F = F'(L)$$
$$L = \frac{F}{F'} = \frac{50\ 000}{4000} = 12.5 \text{ in of weld required (minimum)}$$

*Note:* This welding would be placed symmetrically on the joint.

**Sample Problem 7**   If the plates in the previous problem are overlapped by 3 in and the narrower plate is 5 in wide, design two possible arrangements for the fillet welding.

**Solution:**   (*a*) Using full side fillets, turned corners, and partial end fillets, the weld pattern would include 3 in along each side (6 in total) + end fillets both top and bottom extending for 6.5/4 = 1.625 = $1\frac{5}{8}$ in from each corner toward the centerline (6.5 in total end fillet welding). This would leave a gap of 5 − $3\frac{1}{4}$ = $1\frac{3}{4}$ in on each end which would not be welded. See Fig. 9-23*a*.

(*b*) Using full end fillets, turned corners, and partial side fillets, the weld pattern would include 5 in along each end (10 in total) + side fillets extending for $\frac{5}{8}$ in from each corner along the sides ($2\frac{1}{2}$ in total side fillet welding). This would leave an unwelded gap of 3 − $1\frac{1}{4}$ = $1\frac{3}{4}$ in on each side (see Fig. 9-23*b*). Note that minimum corner returns of twice the weld size are provided with this arrangement.

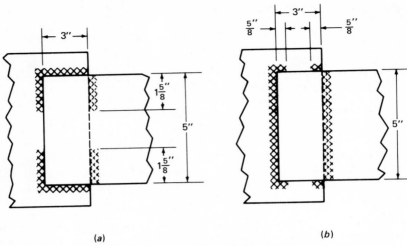

(a)                                    (b)

FIGURE 9-23   (a) Diagram for Sample Problem 7, solution a.
(b) Diagram for Sample Problem 7, solution b.

## 9-10 THIN-WALLED PRESSURE VESSELS

A thin-walled pressure vessel has a diameter at least 10 times its wall thickness. When a thin shell, such as a boiler drum, is subjected to steam pressure, or water pipes are subjected to the pressure of the water, the forces tending to rupture the vessel act normally (perpendicular) to the surface of the shell. Internal stresses are developed in the metal of the walls which must not exceed the safe working stress of the metal. These internal pressures tend to rupture the vessel in two ways: first, along seams parallel to the elements of the shell, and second, along a seam corresponding to a circumference of the shell. Figure 9-24 shows a longitudinal section of a cylindrical shell. The internal pressure $p$ acts normal to the curved surface of the cylinder.

*Longitudinal Section:* Figure 9-25a shows a free-body diagram with the pressure acting normal to the curved surface and held in equilibrium by

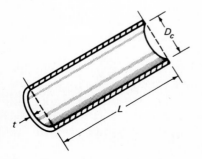

FIGURE 9-24   Longitudinal section of a thin-walled cylinder.

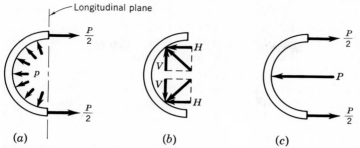

FIGURE 9-25   (a) Free-body diagram of half of cylinder.
(b) Pressure vectors and components. (c) Resultant pressure
force.

the resistance of the material in the longitudinal section. The pressure
distributes uniformly on the curved surface, but only components which
are perpendicular to the longitudinal plane tend to cause failure. Figure
9-25b shows two pressure vectors and their components. Note that the
two vertical components $V$ act in opposite directions, thus canceling each
other, while the horizontal components both act in the same direction.
Therefore, the total pressure force on the curved surface may be
represented by a single force $P$, as shown in Fig. 9-25c, which is the sum
of all $H$ components. The force $P$ is calculated by multiplying the pressure
$p$ and the area of the curved surface *projected on the longitudinal plane*. The
projected area in Fig. 9-24 is $D_c \times L$. Therefore,

$$P = pD_cL$$

The resistance of the material in the longitudinal section holds the
pressure force in equilibrium. Assuming the resistance is shared equally
by both sections in the longitudinal plane (see Fig. 9-25c), from force
equals area times stress, the total resistance is

$$2\left(\frac{P}{2}\right) = (2Lt)s_t \qquad \text{or} \qquad P = (2Lt)s_t$$

where           $2Lt$ = total resistive area, in²; m²
                $s_t$ = tensile stress in material, psi; Pa

Equating the resistance to the pressure force (for equilibrium) gives

$$2Lts_t = pD_cL \qquad \text{or} \qquad s_t = \frac{pD_c}{2t} \qquad (9\text{-}7)$$

where $s_t$ = average tensile stress in material on longitudinal section, psi; Pa
 $p$ = internal pressure in cylinder, psi; Pa
 $D_c$ = inside cylinder diameter, in; mm
 $t$ = thickness of plate material, in; mm

Note that the pressure $p$ is actually the difference between the absolute internal pressure and the absolute external pressure. Usually, the external pressure is atmospheric. In such cases, the correct value for $p$ is the internal *gage* pressure.

It is also of interest to determine the force which a given length of longitudinal joint must resist. This relation has application in designing bolted joints and lap welds for longitudinal seams on cylindrical pressure vessels. The total force which the material in the longitudinal section must withstand is

$$P = pD_c l$$

Since each of the two cross-hatched areas in Fig. 9-26 is assumed to carry one-half the total force $P$, the force $F$ on *each* seam is

$$F = \frac{pD_c l}{2} \tag{9-8}$$

where   $F$ = force on one longitudinal seam of length $l$, lb; N
 $p$ = internal pressure in cylinder, psi; Pa
 $D_c$ = inside cylinder diameter, in; mm
 $l$ = length of seam, in; mm (not necessarily entire length)

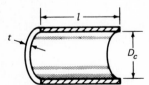

FIGURE 9-26   Dimensions of longitudinal section of cylinder.

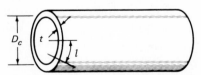

FIGURE 9-27   Transverse section of a thin-walled cylinder.

***Transverse Section:***   The stresses which exist in the material on a transverse section (the transverse plane is perpendicular to the axis of the cylinder) may be analyzed from Figs. 9-27 and 9-28. In Fig. 9-28 the internal pressure $p$ is shown acting against the closed circular end of the cylinder. The total pressure force $P$, which acts to the right, is

$$P = pA \quad \text{or} \quad P = p\left(\frac{\pi D_c^2}{4}\right)$$

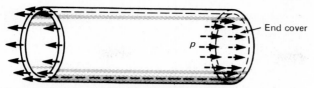

FIGURE 9-28 Free-body diagram of transverse section.

This force must be balanced by resistive forces in the thin ring of metal on the transverse section. The area of metal in this ring is approximately equal to its circumference $\pi D_c$ times its thickness $t$. Therefore, the total resistive force will equal $P$, and this will also equal the tensile stress in the transverse section $s_t$ times the metal area, or

$$P = (\pi D_c t)s_t$$

Equating the resistive force to the pressure force (for equilibrium) gives

$$\pi D_c t s_t = p\left(\frac{\pi D_c^2}{4}\right)$$

$$s_t = \frac{pD_c}{4t} \qquad (9\text{-}9)$$

where $s_t$ = average tensile stress in material on transverse section, psi; Pa
$\quad\ p$ = internal pressure in cylinder, psi; Pa
$\quad D_c$ = inside cylinder diameter, in; mm
$\quad\ t$ = thickness of plate, in; mm

Again, it may be of interest to find the force acting on a given length $l$ of transverse joint (Fig. 9-27). The force on this length of joint is given by

$$F = \frac{pD_c l}{4} \qquad (9\text{-}10)$$

where $F$ = force on a transverse seam of length $l$, lb; N
$\quad\ l$ = length of seam, in; mm (not necessarily entire circumference)

The derivation of Eq. (9-10) is left as an exercise for the student. It should be noted that Eqs. (9-9) and (9-10) apply to any cross section through the center of a spherical pressure vessel.

**\*Sample Problem 8** What water pressure will burst a 400-mm-diameter ASTM Class 60 cast-iron water pipe, if the wall thickness is 16 mm?

**Solution:** From the Table of Properties of Metals, (App. B, Table 1), the ultimate tensile strength of Class 60 cast iron is $410(10^6)$ Pa. In a pipe, failure due to internal pressure tends to occur on longitudinal sections rather than on transverse sections. This is evident from a comparison of Eqs. (9-7) and (9-9). Therefore, by Eq. (9-7) for a longitudinal section,

$$s_t = \frac{pD_c}{2t} \quad \text{or} \quad p = \frac{2ts_t}{D_c}$$

$$p = \frac{2(16)(410)(10^6)}{400} = 32.8(10^6)\ \text{Pa} = 32.8\ \text{MPa}$$

**\*Sample Problem 9** A 3.5-m-diameter sphere of 6061-T6 aluminum alloy is to contain nitrogen gas under a pressure of 6.0 MPa. The wall thickness is 20 mm. A minimum safety factor of 5 (based on ultimate) is specified. Will this spherical tank meet specifications?

**Solution:** From the discussion in Sec. 9-10, Eq. (9-9) applies to the sphere. The stress in the sphere on any transverse section is given by

$$s_t = \frac{pD_c}{4t}$$

$$= \frac{6.0(10^6)(3500)}{4(20)} = 263(10^6)\ \text{Pa}$$

Factor of safety is

$$N_u = \frac{\text{ultimate stress}}{\text{actual stress}}$$

The ultimate stress is $310(10^6)$ Pa from the Table of Properties of Metals (App. B, Table 1).

$$N_u = \frac{310(10^6)}{263(10^6)} = 1.18$$

This tank does not meet specifications. There are several alternatives which could produce a satisfactory solution. These include: (*a*) increasing the wall thickness; (*b*) reducing the tank diameter; (*c*) changing to a stronger material.

**Sample Problem 10** A cylindrical tank 3 ft in diameter and 10 ft long is to contain compressed air. The tank has a double-bolted longitudinal butt joint with 1-in-diameter SA325 steel bolts, arranged as shown in Fig. 9-29. The tank is made of SA36 steel and its wall thickness is $\frac{3}{4}$ in. Use the ASME code. Find the maximum safe pressure in the tank.

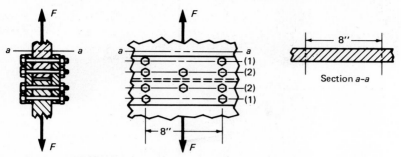

FIGURE 9-29    Diagram for Sample Problem 10

**Solution:**    The strength of the joint will be the maximum load which can be placed on the repeated section of the tank plate.

*Shear:*

$$F_s = \left[ n\left( \frac{\pi d^2}{4} \right) \right] s_s$$

where $n = 6$
$\quad s_s = 15\ 400$ psi
$\quad d = 1$ in (Table 9-2)

$$F_s = 6(0.785)(1^2)(15\ 400)$$
$$= 6(0.785)(15\ 400) = 72\ 500\ \text{lb}$$

*Bearing:*

$$F_c = (ndt)s_c$$

where $n = 3$
$\quad s_c = 20\ 200$ psi

$$F_c = 3(1)(0.75)(20\ 200)$$
$$= 45\ 400\ \text{lb}$$

*Tension*    (*Row* 1):

$$F_t = [(l - nD)t]s_t$$

where $n = 1$
$\quad s_t = 12\ 600$ psi
$\quad l = 8$ in

$$F_t = [8 - 1(1)](0.75)(12\ 600)$$
$$= 66\ 200\ \text{lb}$$

**Tension**   (Row 2):

$$\tfrac{2}{3}F_t = [(l - nD)t]s_t$$

where $n = 2$
$\quad s_t = 12\,600$ psi
$\quad l = 8$ in

$$\tfrac{2}{3}F_t = [8 - 2(1)](0.75)(12\,600)$$
$$= 56\,700 \text{ lb}$$
$$F_t = 85\,000 \text{ lb}$$

The strength of the repeating joint section is 45 400 lb. The maximum safe internal tank pressure is calculated by Eq. (9-8).

$$F = \frac{pD_c l}{2}$$

$$p = \frac{2F}{D_c l}$$

$$= \frac{2(45\,400)}{36(8)} = 315 \text{ psi (maximum safe tank pressure)}$$

It may be of interest to calculate the stress induced (by this pressure) in the *plate material* of the tank on a longitudinal section. This may be done in either of two ways. Equation (9-7) may be used to give

$$s_t = \frac{pD_c}{2t} = \frac{315(36)}{2(0.75)} = 7560 \text{ psi}$$

or the stress can be determined by considering a section of thickness $t$ whose length equals that of the repeating section (section $aa$, Fig. 9-29) subjected to a tensile force $F$.

$$s_t = \frac{F}{A} = \frac{F}{tl} = \frac{45\,400}{0.75(8)} = 7560 \text{ psi}$$

Note that, although the code permits a tensile stress of 12 600 psi, the plate carries only 7560 psi because of the weakness of the bolted joint in bearing.

**Sample Problem 11**   What is the maximum safe pressure in the tank of the previous problem if the bolted joint is replaced by the following?

(a)   A single bead of lap (fillet) welding (Fig. 9-30)—allowable shear strength = 18 000 psi
(b)   A complete-penetration butt weld

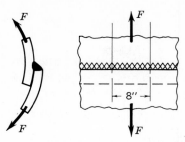

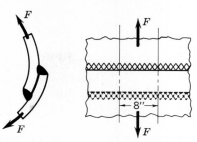

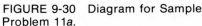

FIGURE 9-30   Diagram for Sample
Problem 11a.

FIGURE 9-31   Diagram for Sample
Problem 11c.

(c)   A double bead of fillet welding (Fig. 9-31)—allowable shear strength = 18 000 psi

**Solution a:**   The tank plates are $\frac{3}{4}$ in thick; hence, a fillet weld with a $\frac{11}{16}$-in leg may be placed (see Table 9-3). To allow comparison with the previous problem, an 8-in length of joint will be considered. (Note that if such a comparison were not being made, a 1-in length of joint would be used.) Equation (9-6) may be adapted to give

$$F = F'(L) = 8800(8) = 70\ 400\ \text{lb (maximum safe force on 8 in of weld)}$$

The stress in the plate material due to this tensile force is

$$s_t = \frac{F}{A} = \frac{F}{tl} = \frac{70\ 400}{0.75(8)} = 11\ 700\ \text{psi}$$

The maximum safe pressure in the tank is determined from Eq. (9-7).

$$s_t = \frac{pD_c}{2t}$$

$$p = \frac{2ts_t}{D_c}$$

$$= \frac{2(0.75)(11\ 700)}{36} = 488\ \text{psi (maximum safe pressure in tank)}$$

or, from Eq. (9-8),

$$p = \frac{2F}{D_c l} = \frac{2(70\ 400)}{36(8)} = 488\ \text{psi}$$

**Solution b:**   The butt-welded joint is allowed to be stressed in tension equal to the allowable plate stress. Thus, $s_t = 12\ 600$ psi. The pressure, from Eq. (9-7) is

$$p = \frac{2ts_t}{D_c} = \frac{2(0.75)(12\ 600)}{36} = 525 \text{ psi (maximum safe pressure in tank)}$$

**Solution c:**   The double-bead weld will withstand twice the force which the single-bead weld of part *a* resists. Thus, $F = 140\ 800$ lb is the maximum safe force on two 8-in lengths of weld. However, the stress in the plate material is

$$s_t = \frac{F}{tl} = \frac{140\ 800}{0.75(8)} = 23\ 500 \text{ psi}$$

which exceeds the ASME code allowable tensile stress of 12 600 psi. Consequently, the value of 12 600 psi must be used as the plate stress. This gives a safe tank pressure of 525 psi, as in part *b*. In this case, the weld is not being used to its full load-carrying capacity.

**Sample Problem 12**   An open cylindrical water tank is 25 ft high and 40 ft in diameter. What plate thickness must be used if the longitudinal joints are to be fillet-welded with a single bead? The AISC code applies. Plate material is A242 steel welded with Class E 70 series electrodes.

**Solution:**   Assume the tank to be filled with water to the full 25-ft height. The pressure at a given depth in a fluid is equal to the depth times its density, or

$$p = \frac{h\rho}{144} \tag{9-11}$$

where $p$ = fluid pressure, psi
   $h$ = depth from top surface, ft
   $\rho$ = fluid density, lb /ft$^3$

The density of water may be taken as 62.4 lb/ft$^3$ and considered to remain constant. From Eq. (9-11) it is evident that the greatest pressure occurs at the greatest depth of water. Thus, the lower end of the longitudinal (vertical) weld is subjected to the largest force. Consider the last inch of longitudinal weld at the bottom of the tank. The water pressure at the bottom is

$$p = \frac{h\rho}{144} = \frac{25(62.4)}{144} = 10.83 \text{ psi}$$

The tensile force on 1 in of longitudinal section near the bottom is

$$F = \frac{pD_c l}{2} = \frac{10.83(40)(12)(1)}{2} = 2600 \text{ lb}$$

The required height of leg of a 1-in-long fillet weld can be found from Table 9-4. To resist 2600 lb, a $\frac{3}{16}$-in weld leg is required; thus, $\frac{3}{16}$-in-thick plates are needed. (The AISC code specifies that weld size be equal to the plate thickness for plate thickness less than $\frac{1}{4}$ in.) Check the stress in the plate material against the AISC allowable tensile stress of 30 000 psi (Table 9-1).

$$s_t = \frac{F}{A} = \frac{F}{tl} = \frac{2600}{(\frac{3}{16})(1)} = 13\,900 \text{ psi}$$

Use $\frac{3}{16}$-in plates.

## PROBLEMS

**9-1.** A double-bolted lap joint is made by connecting two 4- by $\frac{1}{2}$-in A36 steel plates with two $\frac{3}{4}$-in A325 high-strength bolts (threads do not occur at the shear plane). What safe load will the joint carry? Use the AISC code, bearing-type connection.

**9-2.** Two SA285 Grade B steel plates 5 in wide and $\frac{1}{2}$ in thick are connected as a lap joint by three $\frac{5}{8}$-in SA325 steel bolts arranged in one row. Find the strength of the joint using the ASME code.

**9-3.** Two SA36 steel plates, each 10 in wide and $\frac{1}{2}$ in thick, are connected by $\frac{3}{4}$-in steel bolts to form a lap joint. Based on shear, how many bolts will be needed if the pull on the joints is 15 000 lb? Use the ASME code.

**9-4.** Two plates forming a lap joint carrying 5000 lb in tension are connected by two A307 steel bolts. What should be the diameter of the bolts based on the AISC code? Consider shear only.

**9-5.** Two 12-in-wide by $\frac{3}{4}$-in-thick A440 steel plates are joined by a single-bolted butt joint with $\frac{1}{2}$-in cover plates. The plates are connected by four $\frac{7}{8}$-in A490 steel bolts on each side. Find the strength of the joint using the AISC code. The bolt threads do not occur at the shear plane. Assume friction-type connection.

**9-6.** Two A36 steel plates 2 in wide and $\frac{3}{8}$ in thick are formed into a double-bolted butt joint with two $\frac{5}{16}$-in cover plates. There are two $\frac{1}{2}$-in A307 steel bolts on each side. If the load is 15 600 lb, find the shear and bearing stresses on the bolts and the maximum tensile stress in the plate. Use the AISC code to determine the hole size, and compare the calculated stresses with the code allowable stresses.

**9-7.** Figure Problem 9-7 shows the lower joint of a roof truss. Eleven $\frac{5}{8}$-in A307 steel bolts are used to fasten the member $A$ to the $\frac{1}{2}$-in gusset plate. The tensile force in member $A$ is 56 000 lb. Is the bolting safe? What may be the compressive force on member $B$, on the assumption that the bolts are used to their full capacity? Use the AISC code. The angles and plate are of A36 steel.

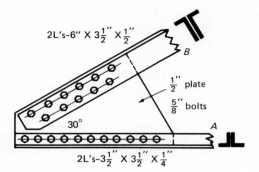

2L's-6" X $3\frac{1}{2}$" X $\frac{1}{2}$"

B

$\frac{1}{2}$" plate

$\frac{5}{8}$" bolts

30°

A

2L's-$3\frac{1}{2}$" X $3\frac{1}{2}$" X $\frac{1}{4}$"

**FIGURE PROBLEM 9-7**

**9-8.** In a butt joint connecting $\frac{3}{8}$-in SA612 Grade A plates, how many $\frac{3}{4}$-in SA325 steel bolts are needed to carry 80 000 lb safely? What would be the number of bolts for a lap joint? Use the ASME code.

**9-9.** The joint at a point in the lower chord of a roof truss is shown in Fig. Prob. 9-9. If the compression in member $A$ is 18 000 lb and the tension in member $B$ is 14 000 lb, how many $\frac{5}{8}$-in bolts (bearing-type connection) are needed to fasten these members to the $\frac{3}{8}$-in gusset plate? Use the AISC code. The angles and gusset plate are of A36 steel. The bolts are of A325 steel with the stress developing at the threaded portion of the shank.

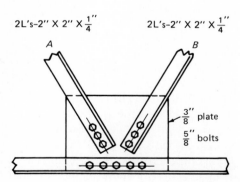

2L's-2" X 2" X $\frac{1}{4}$"          2L's-2" X 2" X $\frac{1}{4}$"

A                                      B

$\frac{3}{8}$" plate

$\frac{5}{8}$" bolts

**FIGURE PROBLEM 9-9**

**9-10.** Find the strength and efficiency for a repeating section of a single-bolted lap joint (friction-type connection), using the AISC code, if the A242 steel plates are $\frac{1}{2}$ in thick and the $\frac{5}{8}$-in A490 steel bolts are placed four bolt diameters apart.

**9-11.** Determine the strength and efficiency of a double-bolted two-cover-plate butt joint. The SA36 plates are 10 in wide and $\frac{3}{4}$ in thick. The SA499 steel bolts are 1 in in diameter. There is one bolt in the first row and three bolts in the second row. Use ASME allowable values.

**9-12.** Two $\frac{3}{4}$-in thick plates are joined by a lap joint. The SA285, Grade A steel plates are 12 in wide. The joint contains a total of 12 SA325 steel bolts, arranged in four rows of three bolts each. The bolts are $\frac{5}{8}$ in in diameter. Use the ASME code. Find the strength and efficiency of the joint.

**9-13.** A triple-bolted butt joint with two cover plates has one, two, and four bolts in the first, second, and third rows, respectively. The pitch of the repeating section is 10 in. The SA36 steel plates are $\frac{5}{8}$ in thick and the SA325 steel bolts are $\frac{3}{4}$ in in diameter. Find the strength and efficiency of the joint using the ASME code.

**9-14.** Two A36 steel plates $\frac{1}{2}$ in thick are connected by a triple-bolted two-cover plate butt joint (bearing-type connection). The bolts are $\frac{5}{8}$ in in diameter and the length of the repeating section is 12 in. There is one bolt in the first row, three bolts in the second row, and six bolts in the third row. Find the strength and efficiency of the joint. Use the AISC code. The bolts are A325 steel with the stress developing at the unthreaded portion of the shank.

**9-15.** In Fig. 9-19, p. 186, the narrow plate is 6 by $\frac{1}{2}$ in and resists a pull of 16 000 lb. The plate material is A36 steel, the electrodes are Class E 60 series, and the weld size is $\frac{7}{16}$ in (see Sec. 9-9).
 *a.* What must be the length of each side weld?
 *b.* Could a single end weld resist the pull?

**9-16.** In a lap weld, the narrow plate is 6 by $\frac{1}{2}$ in and is subjected to a pull of 60 000 lb. What must be the length of each $\frac{7}{16}$-in side weld if a total of 4 in of end weld is also to be used? The plates are A242 steel and Class E 70 series electrodes are used.

**9-17.** *a.* Two 5- by $\frac{1}{4}$-in plates are joined by top and bottom end fillet welding. What allowable load will the $\frac{3}{16}$-in weld carry? Check the plate strength in tension. The plate material is A36 steel and Class E 60 series electrodes will be used.
 *b.* What safe load would the joint carry if the plates were butt-welded?

**9-18.** *a.* The narrower plate in Fig. 9-20 is 6 by $\frac{1}{2}$ in and the plates are to be top and bottom end fillet-welded. Find the safe tensile load this joint can carry if the plates are A441 steel and Class E 70 series electrodes develop $\frac{7}{16}$-in welds.
 *b.* What length of plate overlap is required to carry the same load if only side fillet welds are used?

**9-19.** The plates in Fig. Prob. 9-19 are to be lap-welded. The thickness of each plate is $\frac{5}{16}$ in and the weld size is $\frac{1}{4}$ in. The plates are A36 steel and Class E 70 series electrodes will be used.
 *a.* What overlap is required to carry a load of $P = 32\,000$ lb if fillet welding is used on sides $A$ and $C$ only?

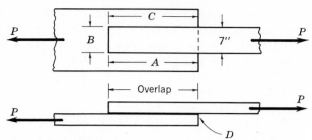

FIGURE PROBLEM 9-19

        *b.*   What maximum safe load may be carried if an overlap of 12 in is used and sides $A$, $B$, $C$, and $D$ are fillet-welded for the full length?

**9-20.**  *a.*   Two A441 steel plates $3\frac{1}{2}$ in wide are to carry a load of 25 000 lb. They are to be top and bottom end fillet-welded with Class E 70 series electrodes. What is the required thickness of the plate?
        *b.*   Assuming a complete-penetration butt weld, what would be the required thickness?
        *c.*   Assuming 1-in A307 steel bolts are to be used, find how many would be required in shear.

**9-21.**  A cylindrical pressure vessel is subjected to an internal pressure of 70 psi. It is 7 ft in diameter and the plates are $\frac{3}{8}$ in thick.
        *a.*   Find the maximum stress.
        *b.*   If it is made of Class 40 cast iron, is it safe? Assume steady stress conditions.

**\*9-22.**  What should be the thickness of the plate for a 254-mm steam pipe to resist a pressure of 550 kPa? The stress in the pipe should not exceed 62 MPa.

**9-23.**  What safe internal pressure should be used for a steel water pipe 6 in in diameter with a $\frac{1}{2}$-in shell if the allowable stress is 11 000 psi?

**\*9-24.**  Nitrogen gas is to be placed in a sphere 4.5 m in diameter under 800-kPa pressure. If the material is aluminum alloy 6061-T6, what thickness of plate will be required? Assume varying load conditions.

**9-25.**  If an 18-in Class 20 cast-iron vertical pipe is subjected to a head of 140 ft of water, what thickness of plate will be required? Assume varying stress conditions.

**\*9-26.**  A tank 2.4 m in diameter and 1.8 m deep is hooped with 12-mm round wrought-iron rods. What must be the spacing of the rods if the tank contains water? (*Note:* Mass density of water = 1000 kg/m³.)

**9-27.**  Find the necessary spacing of wrought-iron hoops $\frac{3}{4}$ in in diameter for a wood-stave water tank 20 ft in diameter and 16 ft deep. Assume $s_t = 10\ 000$ psi for wrought iron.

**9-28.** A boiler 6 ft in diameter carries steam at a pressure of 80 psi. The SA285, Grade B steel plates are $\frac{3}{8}$ in thick, the SA325 steel bolts are $\frac{3}{4}$ in in diameter, and the pitch is 3 in. If the longitudinal joint is a double-bolted lap, what are the stresses and efficiency? Use the ASME code.

**9-29.** A bolted steel water pipe is 2 ft in diameter and is made of $\frac{3}{8}$-in SA36 steel plates. The single-bolted longitudinal lap joint has $\frac{3}{4}$-in SA325 steel bolts with a pitch of $2\frac{1}{2}$ in. What is the allowable water pressure if the ASME code is used?

**9-30.** A cylindrical storage tank for compressed air has an inside diameter of 28 in and a SA36 steel plate thickness of $\frac{3}{8}$ in. The longitudinal double-bolted lap joint has a pitch of 3 in in each row. The ends are held by single-bolted lap joints, each joint having 30 steel bolts. All bolts are SA499 steel, $\frac{7}{8}$ in in diameter. What is the allowable air pressure in the tank if the ASME code is specified?

**9-31.** A spherical gas container made of $\frac{3}{4}$-in SA285, Grade B steel plate must maintain gas at a pressure of 50 psi. The great-circle joint is a double-bolted butt joint using 1-in-diameter SA325 steel bolts, with a pitch of 3 in in each row. Using the ASME code, determine the maximum diameter of the sphere.

**9-32.** A cylindrical oil tank is 100 ft in diameter and 28 ft high. Using $\frac{7}{8}$-in SA36 steel plates and $\frac{3}{4}$-in SA499 steel bolts, design a double-bolted butt joint within the ASME code. (Density for oil is 56 lb/ft³.)

**\*9-33.** A cylindrical water tank (axis vertical) has a longitudinal double-V butt weld. The tank diameter is 6.5 m. The plates are 10 mm thick. The allowable tensile stress on the weld is 110 MPa. What maximum height of water can be safely maintained in this tank? (*Note:* Mass density of water = 1000 kg/m³.)

**9-34.** A pipe 2 ft in diameter is constructed from $\frac{1}{4}$-in A242 steel plates. What is the allowable working pressure on the assumption that longitudinal complete-penetration butt-welded joints are used?

**9-35.** *a.* A gas tank 3 ft in diameter is to be operated under a pressure of 250 psi. What will be the thickness of A36 steel plate if the longitudinal seam is lap-welded with Class E 70 series electrodes?

*b.* If the seams are butt-welded, what will be the thickness?

**9-36.** *a.* A boiler 3 ft in diameter with SA36 steel plates $\frac{1}{4}$ in thick has a longitudinal seam which is a double-bolted lap joint. If the SA499 steel bolts are $\frac{3}{8}$ in in diameter and pitch = 2 in, find the safe internal pressure. Use the ASME code.

*b.* Replacing the bolts by a single-bead lap weld, find the safe internal pressure if the $\frac{3}{16}$-in weld has an allowable shear stress of 18 000 psi.

*c.* Replacing the bolts with a butt weld, find the safe internal pressure.

**9-37.**  *a.*   A water tank 18 ft in diameter is made of $\frac{5}{8}$-in SA36 steel plates. The longitudinal joint is a double-bolted lap joint with $\frac{3}{4}$-in SA499 steel bolts and a 3-in pitch. What would be the safe height of water in the tank? Use the ASME code.

 *b.*   What would be the safe height if a single-bead $\frac{9}{16}$-in. lap-welded joint with an allowable shear stress of 18 000 psi is used?

 *c.*   What would be the safe height if a butt-welded joint is used?

**\*9-38.**  A seamless pipe 600 mm in diameter is to be used to manufacture a pressure vessel designed to carry steam under a pressure of 2 MPa. The vessel is made up of a continuous length of pipe with end covers that are butt-welded to the pipe on transverse sections. The pipe material has an ultimate tensile strength of 900 MPa, and a safety factor (based on ultimate) of 5 is specified. The allowable stress for the butt weld is 110 MPa. What should the wall thickness of the pipe be?

# CHAPTER
# 10
# Center of Gravity, Centroids, and Moment of Inertia

## 10-1 CENTER OF GRAVITY OF A BODY

Every part of a body possesses weight. Weight is the force of attraction between a body and the Earth and is proportional to mass of the body. The weights of all parts of a body can be considered as parallel forces directed toward the center of the Earth. Therefore, they may be combined into a resultant force whose magnitude is equal to their algebraic sum. If a supporting force, equal and opposite to the resultant, is applied to the body along the line of action of the resultant, the body will be in equilibrium. This line of action will pass through the center of gravity of the body.

The center of gravity of some objects may be found by balancing the object on a point. Take a thin plate of the thickness $t$, shown in Fig. 10-1. Draw the diagonals of the upper and lower faces to intersect at $J$ and $K$, respectively. If the plate is placed on a pivot at $K$, the plate will not fall. That is, it is balanced. If suspended from $J$, the plate will hang horizontally. The center of gravity of the plate is at the center of the line $JK$.

If we suspend a uniform rod by a string (Fig. 10-2) and move the position of the string until the rod hangs horizontally, we can determine that the center of gravity of the rod lies at its center. Through the use of similar procedures it can be established that a body which has an axis, or line, of symmetry has its center of gravity located on that line, or axis. Of course, if a body has more than one axis of symmetry, the center of gravity must lie at the intersection of the axes.

205

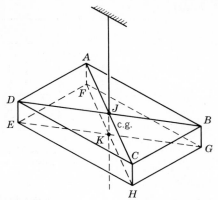

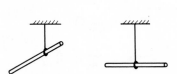

FIGURE 10-1   Locating the center of gravity of a plate.

FIGURE 10-2   Locating the center of gravity of a rod.

## 10-2 CENTER OF GRAVITY OF AREA—CENTROID

Strictly speaking, there is no center of gravity of an area, for an area does not have weight. The point $J$, at the intersection of the diagonals of the face $ABCD$ of Fig. 10-1, is the center of area of a rectangle. Because of its relation to the center of gravity of the plate, it is spoken of as the center of gravity, or *centroid*, of the rectangle.

The center of gravity of the area of the cross section of irregular bodies may be found by a simple experiment. Cut a piece of cardboard to the shape of the cross section and find the balancing point. This will give the position of the center of gravity for all practical purposes.

Another method for determining the center of gravity is by suspension. Take an object, the section of which is shown in Fig. 10-3. Suspend it from some point $A$. The body will not come to rest until its resultant weight is vertically downward from $A$. Through $A$, draw a vertical line $AC$. Then suspend the body from a point $B$, and let it come to rest.

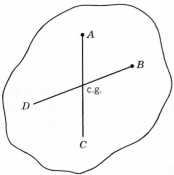

FIGURE 10-3   Center of gravity by suspension.

**TABLE 10-1**  CENTROIDS OF SIMPLE AREAS

| Shape | Area | $\bar{x}$ | $\bar{y}$ | |
|---|---|---|---|---|
| Rectangle | $bh$ | $\dfrac{b}{2}$ | $\dfrac{h}{2}$ | |
| Triangle | $\dfrac{bh}{2}$ | $\dfrac{b}{3}$ | $\dfrac{h}{3}$ | |
| Circle | $\dfrac{\pi d^2}{4}$ | $\dfrac{d}{2}$ | $\dfrac{d}{2}$ | |
| Semicircle | $\dfrac{\pi d^2}{8}$ | $\dfrac{d}{2}$ | $\dfrac{4r}{3\pi}$ (0.425r) | |
| Quadrant | $\dfrac{\pi d^2}{16}$ | 0.425r | 0.425r | |

Through $B$, draw a vertical line $BD$. The point of intersection of $AC$ and $BD$ is the position of the center of gravity, or centroid.

The centroids, or centers of gravity, for some simple areas are given in Table 10-1. The horizontal and vertical distances to the centroid are indicated by $\bar{x}$ (read as $x$ bar) and $\bar{y}$ ($y$ bar).

## 10-3 MOMENT OF AN AREA

In order to determine the location of the centroid of a plane figure, the term *moment of an area* must be understood. Imagine a force (perpendicular to the page) of as many pounds as there are square inches in the area applied at the centroid of the rectangle in Fig. 10-4. Then the moment of the rectangular area about the $y$ axis will be the product of the area and the distance from the centroid to the $y$ axis. That is,

$$M_y = A(\bar{x})$$

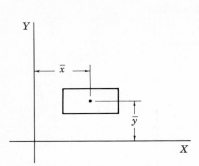

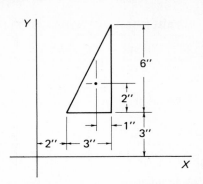

FIGURE 10-4 Determining the
moment of an area.

FIGURE 10-5 Diagram for Sample
Problem 1.

In a like manner, the moment of the area about the $x$ axis will be the
product of the area and the distance from the centroid to the $x$ axis. That
is,

$$M_x = A(y)$$

**Sample Problem 1** Determine the moment of the triangular area shown
about the $x$ axis and the $y$ axis, Fig. 10-5.

**Solution:**

$$\text{Area} = \frac{3(6)}{2} = 9 \text{ in}^2$$

The distance from the centroid to the horizontal base of the triangle
is $h/3 = \frac{6}{3} = 2$ in. Therefore, the distance from the centroid to the $x$ axis
is $2 + 3 = 5$ in. Then

$$M_x = A(\bar{y}) = 9(5) = 45 \text{ in}^3$$

The distance from the centroid to the vertical edge of the triangle
is $b/3 = \frac{3}{3} = 1$ in. Therefore, the distance from the centoid to the $y$ axis
is $2 + 2 = 4$. Then

$$M_x = A(y) = 9(5) = 45 \text{ in}^3$$

## 10-4 CENTROIDS OF COMPOSITE AREAS

The location of the centroid of a plane figure can be thought of as the
average distance of the area to an axis. Usually, the axes involved will be
the $x$ and $y$ axes.

In determining the location of the centroid, the reader will find it

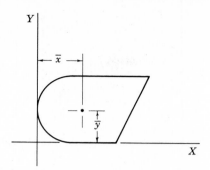

FIGURE 10-6   Centroid of a composite area.

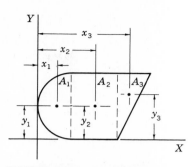

FIGURE 10-7   Composite area divided into simple areas.

advantageous to place the $x$ axis through the lowest point and the $y$ axis through the left edge of the figure. This places the plane area entirely within the first quadrant where $x$ and $y$ distances are positive (Fig. 10-6). Then divide the area into simple areas, such as rectangles, triangles, etc. (Fig. 10-7). Take the moment of each simple area about the $x$ axis. Sum up the moments about the $x$ axis. Since the centroid of the composite figure is the point at which the entire area is assumed to be concentrated, the moment of the entire area about the $x$ axis must be equal to the sum of the moments of its component parts about the $x$ axis. That is,

$$(A_1 + A_2 + \cdots + A_n)\bar{y} = A_1 y_1 + A_2 y_2 + \cdots + A_n y_n$$

Therefore,

$$\bar{y} = \frac{A_1 y_1 + A_2 y_2 + \cdots + A_n y_n}{A_1 + A_2 + \cdots + A_n}$$

or

$$\bar{y} = \frac{\Sigma A y}{\Sigma A} \tag{10-1}$$

Following the same procedure for moments about the $y$ axis,

$$\bar{x} = \frac{A_1 x_1 + A_2 x_2 + \cdots + A_n x_n}{A_1 + A_2 + \cdots + A_n}$$

or

$$\bar{x} = \frac{\Sigma A x}{\Sigma A} \tag{10-2}$$

If a hole exists in the plane figure, treat it as a negative area. The moment of a negative area will be negative, provided that the entire figure lies in the first quadrant.

**\*Sample Problem 2**   Locate the centroid of the cross section of an angle 150 by 100 by 12 mm (Fig. 10-8).

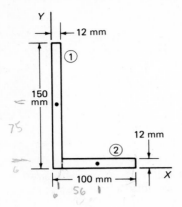

FIGURE 10-8  Diagram for Sample Problem 2.

**Solution:**  Divide the figure into two rectangles, 1 and 2.

$$A_1 = 150(12) = 1800 \text{ mm}^2$$
$$A_2 = \phantom{0}88(12) = \underline{1056} \text{ mm}^2$$
$$\Sigma A = 2856 \text{ mm}^2$$

Using Eqs. (10-1) and (10-2),

$$\bar{x} = \frac{(1800)(6) + (1056)(56)}{2856} = \frac{10\,800 + 59\,140}{2856} = \frac{69\,940}{2856} = 24.5 \text{ mm}$$

$$\bar{y} = \frac{(1800)(75) + (1056)(6)}{2856} = \frac{135\,000 + 6340}{2856} = \frac{141\,340}{2856} = 49.5 \text{ mm}$$

Some readers may prefer to systematize the procedure as follows:

| Area | Dimen. | A | x | Ax | y | Ay |
|------|--------|---|---|-----|---|-----|
| 1 | 150 × 12 | 1 800 | 6 | 10 800 | 75 | 135 000 |
| 2 | 88 × 12 | 1 056 | 56 | 59 140 | 6 | 6 340 |
| | | $\Sigma A$ = 2 856 | | $\Sigma Ax$ = 69 940 | | $\Sigma Ay$ = 141 340 |

$$\bar{x} = \frac{\Sigma Ax}{\Sigma A} = \frac{69\,940}{2856} = 24.5 \text{ mm}$$

$$\bar{y} = \frac{\Sigma Ay}{\Sigma A} = \frac{141\,340}{2856} = 49.5 \text{ mm}$$

**Sample Problem 3**  Locate the centroid of the piece of sheet metal shown in Fig. 10-9.

**Solution:**  Divide the figure into three simple areas.

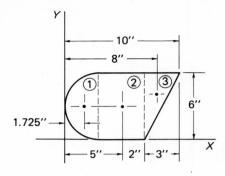

FIGURE 10-9   Diagram for Sample Problem 3, $\bar{x}$ calculation.

1. Semicircle:

$$A_1 = \frac{\pi d^2}{8} = \frac{\pi (6)^2}{8} = 14.13 \text{ in}^2$$

2. Rectangle:

$$A_2 = bh = 4(6) = 24.00 \text{ in}^2$$

3. Triangle:

$$A_3 = \frac{bh}{2} = \frac{3(6)}{2} = 9.0 \text{ in}^2$$

$$\Sigma A = 47.13 \text{ in}^2$$

Taking moments of areas about the $y$ axis (Fig. 10-9),

$$\bar{x} = \frac{(14.13)(1.725) + (24)(5) + (9)(8)}{47.13}$$
$$= \frac{24.4 + 120 + 72}{47.13}$$
$$= \frac{216.4}{47.13} = 4.59 \text{ in} \quad \text{Say, 4.6 in}$$

Taking moments about the $x$ axis (Fig. 10-10),

$$\bar{y} = \frac{(14.13)(3) + (24)(3) + (9)(4)}{47.13}$$
$$= \frac{42.39 + 72 + 36}{47.13}$$
$$= \frac{150.39}{47.13} = 3.19 \text{ in} \quad \text{Say, 3.2 in}$$

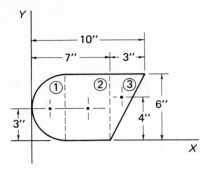

FIGURE 10-10   Diagram for Sample Problem 3, $\bar{y}$ calculation.

| Area | Dimen. | $A$ | $x$ | $Ax$ | $y$ | $Ay$ |
|------|--------|-----|-----|------|-----|------|
| 1 | $d = 6$ | 14.13 | 1.73 | 24.4 | 3.0 | 42.39 |
| 2 | $6 \times 4$ | 24.0 | 5.0 | 120.0 | 3.0 | 72.0 |
| 3 | $b = 3$ | 9.0 | 8.0 | 72.0 | 4.0 | 36.0 |
|   | $h = 6$ |     |     |      |     |      |
|   |         | $\Sigma A = 47.13$ |     | $\Sigma Ax = 216.4$ |     | $\Sigma Ay = 150.39$ |

$$\bar{x} = \frac{\Sigma Ax}{\Sigma A} = \frac{216.4}{47.13} = 4.59 \text{ in}  \text{ Say, 4.6 in}$$

$$\bar{y} = \frac{\Sigma Ay}{\Sigma A} = \frac{150.39}{47.13} = 3.19 \text{ in}  \text{ Say, 3.2 in}$$

\***Sample Problem 4**   Determine the location of the centroid of the plane figure shown in Fig. 10-11.

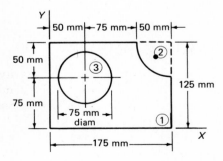

FIGURE 10-11   Diagram for Sample Problem 4.

**Solution:**   Divide the composite figure into three simple areas, a rectangle 175 by 125 mm, a quadrant with a 50-mm radius, and a circle with a 75-mm diameter. The rectangle is a positive area. The quadrant and hole are treated as negative areas.

$$A_1 = bh = 175(125) = 21\ 880 \text{ mm}^2$$

$$A_2 = \frac{-\pi d^2}{16} = \frac{-\pi(100)^2}{16} = -1963 \text{ mm}^2$$

$$A_3 = \frac{-\pi d^2}{4} = \frac{-\pi(75)^2}{4} = -4417 \text{ mm}^2$$

$$\Sigma A = 15\ 500 \text{ mm}^2$$

$$\bar{x} = \frac{\Sigma Ax}{\Sigma A} = \frac{(21\ 880)(87.5) - 1963(153.75) - 4417(50)}{15\ 500}$$

$$\bar{x} = \frac{1915(10^3) - 302(10^3) - 221(10^3)}{15\ 500} = \frac{1392(10^3)}{15.5(10^3)} = 89.8 \text{ mm}$$

$$\bar{y} = \frac{\Sigma Ay}{\Sigma A} = \frac{(21\ 880)(62.5) - 1963(103.75) - 4417(75)}{15\ 500}$$

$$= \frac{1368(10^3) - 204(10^3) - 331(10^3)}{15\ 500} = \frac{833(10^3)}{15.5(10^3)} = 53.7 \text{ mm}$$

or:

| Area | Dimen. | $A$ | $x$ | $Ax$ | $y$ | $Ay$ |
|------|--------|-----|-----|------|-----|------|
| 1 | $175 \times 125$ | 21 880 | 87.5 | $1\ 915(10^3)$ | 62.5 | $1\ 368(10^3)$ |
| 2 | $r = 50$ | $-1\ 963$ | 153.75 | $-302(10^3)$ | 103.75 | $-204(10^3)$ |
| 3 | $d = 75$ | $-4\ 417$ | 50 | $-221(10^3)$ | 75 | $-331(10^3)$ |
| | | $\Sigma A = 15\ 500$ | | $\Sigma Ax = 1\ 392(10^3)$ | | $\Sigma Ay = 833(10^3)$ |

$$\bar{x} = \frac{\Sigma Ax}{\Sigma A} = \frac{1392(10^3)}{15.5(10^3)} = 89.8 \text{ mm}$$

$$\bar{y} = \frac{\Sigma Ay}{\Sigma A} = \frac{833(10^3)}{15.5(10^3)} = 53.7 \text{ mm}$$

## 10-5  CENTER OF GRAVITY OF SIMPLE SOLIDS

The weight of a body is a force acting at its own center of gravity and directed toward the center of the Earth. The position of the center of bodies weighing $W_1$, $W_2$, $W_3$, etc., is found in the same manner as the resultant of parallel forces.

$$\bar{x} = \frac{\Sigma Wx}{\Sigma W}$$

$$\bar{y} = \frac{\Sigma Wy}{\Sigma W}$$

$$\bar{z} = \frac{\Sigma Wz}{\Sigma W}$$

(10-3)

If the bodies are all the same material weighing $\rho$ lb/ft$^3$, then

$$W_1 = \rho V_1 \qquad W_2 = \rho V_2 \qquad W_3 = \rho V_3 \qquad \text{etc.}$$

Substituting in Eqs. (10-3), we have

$$
\begin{aligned}
\bar{x} &= \frac{\Sigma \rho V x}{\Sigma \rho V} = \frac{\Sigma V x}{\Sigma V} \\
\bar{y} &= \frac{\Sigma \rho V y}{\Sigma \rho V} = \frac{\Sigma V y}{\Sigma V} \\
\bar{z} &= \frac{\Sigma \rho V z}{\Sigma \rho V} = \frac{\Sigma V z}{\Sigma V}
\end{aligned}
\qquad (10\text{-}4)
$$

That is, if the bodies are made of the same material and are of the same density throughout, the center of gravity of the bodies is their center of volume. If the bodies are of the same cross section but perhaps of different lengths,

$$V_1 = al_1 \qquad V_2 = al_2 \qquad V_3 = al_3 \qquad \text{etc.}$$

Substituting in Eqs. (10-4) produces,

$$
\begin{aligned}
\bar{x} &= \frac{\Sigma a l x}{\Sigma a l} = \frac{\Sigma l x}{\Sigma l} \\
\bar{y} &= \frac{\Sigma a l y}{\Sigma a l} = \frac{\Sigma l y}{\Sigma l} \\
\bar{z} &= \frac{\Sigma a l z}{\Sigma a l} = \frac{\Sigma l z}{\Sigma l}
\end{aligned}
\qquad (10\text{-}5)
$$

If the bodies are parts of a wire, pipe, or rod of constant cross section, then their center of gravity may be found from the center of their lengths.

The center of gravity of a right circular cylinder is at the center of its geometric axis, Fig. 10-12.

The center of gravity of a right circular cone is on the geometric axis, $\frac{1}{4} h$ from the base or $\frac{3}{4} h$ from the vertex, where $h$ is the altitude of the cone, Fig. 10-13.

The center of gravity of a hemisphere of radius $r$ is $\frac{3}{8} r$ from the center of the sphere, Fig. 10-14.

**\*Sample Problem 5**  A thin wire is bent into a shape shown in Fig. 10-15. Find the coordinates of its center of gravity.

**Solution:**  Since lengths are involved, Eqs. (10-5) will be used. It will be seen that for $A$, the center of the 200-mm length,

$$x = 0 \qquad y = 0 \qquad z = 100$$

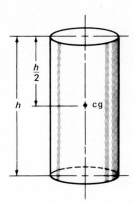

FIGURE 10-12   Center of gravity of cylinder.

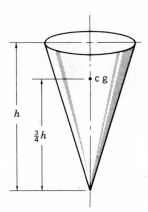

FIGURE 10-13   Center of gravity of cone.

For $B$,           $x = 125$     $y = 0$     $z = 0$

For $C$,   $x = 250 + 150 \cos 45° = 356$     $y = 150 \sin 45° = 106$     $z = 0$

| $l$ | $x$ | $y$ | $z$ | $lx$ | $ly$ | $lz$ |
|---|---|---|---|---|---|---|
| 200 | 0 | 0 | 100 | 0 | 0 | 20 000 |
| 250 | 125 | 0 | 0 | 31 250 | 0 | 0 |
| 300 | 356 | 106 | 0 | 106 800 | 31 800 | 0 |
| 750 | | | | 138 050 | 31 800 | 20 000 |

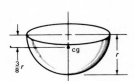

FIGURE 10-14   Center of gravity of hemisphere.

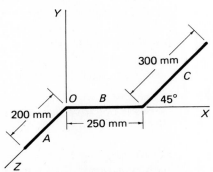

FIGURE 10-15   Diagram for Sample Problem 5.

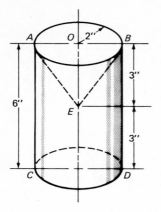

FIGURE 10-16   Diagram for Sample Problems 6 and 7.

Then

$$\bar{x} = \frac{138\,050}{750} = 184.1 \text{ mm}$$

$$\bar{y} = \frac{31\,800}{750} = 42.4 \text{ mm}$$

$$\bar{z} = \frac{20\,000}{750} = 26.7 \text{ mm}$$

Note that the center of gravity does not lie on the wire.

**Sample Problem 6** A cylinder for which $r = 2$ in and $h = 6$ in has an inverted cone cut from it (Fig. 10-16). The cone, for which $r = 2$ in and $h = 3$ in, has its base in the upper base of the cylinder. Find the center of gravity of the part remaining.

**Solution:**   Choose $O$ as the origin. Since the figure is symmetrical with respect to the axis $OE$, the center is on the axis and we need only find $\bar{y}$. The volume of the cylinder is

$$V_1 = \frac{\pi d^2 h}{4} = \frac{\pi(4)^2}{4}(6) = 24\pi \text{ in}^3$$

and for the cone,

$$V_2 = \frac{1}{3}\left(\frac{\pi d^2}{4}\right)h = \frac{-\pi(4^2)}{12}(3) = -4\pi \text{ in}^3$$

$V_2$ is negative because it is volume removed. Then, from Eqs. (10-4),

$$\bar{y} = \frac{\Sigma Vy}{\Sigma V} = \frac{(24\pi)(3) - 4\pi(\frac{1}{4})(3)}{24\pi - 4\pi}$$

$$= \frac{72 - 3}{20} = 3.45 \text{ in}$$

**Sample Problem 7**  Suppose the cylinder in Fig. 10-16 is made of steel ($\rho = 490$ lb/ft³) and, after the conical hole is bored out, it is filled with concrete ($\rho = 150$ lb/ft³). Find the center of gravity of the solid.

**Solution:**   The weight of the steel cylinder after the conical hole is bored, but before concrete is added, is

$$W_s = V_s\rho_s = (24\pi - 4\pi)\left(\frac{490}{1728}\right) = 17.81 \text{ lb}$$

The center of gravity for the steel is 3.45 in from point $O$ (see previous problem).

The weight of the concrete cone is

$$W_c = V_c\rho_c = 4\pi\left(\frac{150}{1728}\right) = 1.09 \text{ lb}$$

The center of gravity for the concrete is 0.75 in from point $O$. From Eqs. (10-3),

$$\bar{y} = \frac{\Sigma Wy}{\Sigma W} = \frac{17.81(3.45) + 1.09(0.75)}{17.81 + 1.09} = 3.3 \text{ in}$$

# 10-6  AREAS AND VOLUMES—CENTROID METHOD

Since the center of gravity of an area or a body is the point at which the area or mass of the body may be assumed to be concentrated, it can be said that the distance through which an area or a body moves is the same as the distance described by its center of gravity. This relation is used in finding areas and volumes.

Thus, a line moving parallel to its original position is said to generate an area that is equal to the length of the line multiplied by the distance through which its centroid moves. That is, area is equal to length times width.

Also, an area moving parallel to its original position is said to develop the volume of a prism that is equal to the area multiplied by the distance through which the centroid moves. That is, a volume is equal to the area of the base times the altitude.

Similarly, a line rotating about one end will develop the area of a circle. A right triangle rotating about either leg will develop the volume of a cone. In each case, a line or an area moves through a distance equal to the length of a path described by the centroid of either the line or the area. Many determinations of areas or volumes are simplified by the use of this method.

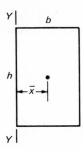

FIGURE 10-17   Diagram for Sample Problem 8.

**Sample Problem 8**   Find the volume of a vertical cylinder by the method of revolving an area (Fig. 10-17).

**Solution:**   The vertical section of the cylinder is a rectangle. It may be seen that the volume of this cylinder can be developed by revolving a rectangle about a vertical edge $YY$.

$$A_{\text{rectangle}} = bh$$

where                     $b$ = radius of cylinder
                          $h$ = height of cylinder

The distance from edge $YY$ to the centroid is $\bar{x} = b/2$. In one revolution, the centroid moves a distance

$$l = 2\pi \frac{b}{2} = \pi b$$

Then,                 $V_{\text{cylinder}} = A_{\text{rectangle}}(l) = bh(\pi b) = \pi b^2 h$

This result is the same as that for volume determined by the ordinary methods, where $\pi b^2$ is the area of the base and $h$ is the altitude.

## 10-7  MOMENT OF INERTIA

In determining the strength of members which are loaded in bending, such as beams, a term appears in the strength equations, which is called the *moment of inertia*. It is necessary to have an understanding of the moment of inertia of an area before these members can be analyzed.

Figure 10-18 indicates a rectangle $ABCD$, with an axis parallel to the base, passing through the centroid. The figure is divided into a number of small rectangles, as shown. The areas are represented by $a$ with a subscript. The distance from the axis $XX$ to the centroid of each rectangle is represented by $y$ with a subscript corresponding to the area. Thus, for

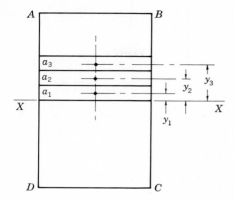

FIGURE 10-18   Determining the moment of inertia.

area $a_1$, there is a moment arm $y_1$, etc. If each area is multiplied by the square of its moment arm and the sum of all such terms taken, because the number of small areas into which the rectangle $ABCD$ is divided becomes very great, the result is called the moment of inertia $I_x$ of the rectangle about the centroidal axis $XX$. Thus,

$$I_x = \sum_{i=1}^{n} a_i y_i^2 \tag{10-6}$$

The derivation of formulas for $I_x$ is a problem of calculus and is beyond the scope of this text. Fortunately, the areas involved in engineering problems are often rectangles, triangles, circles, or a combination of one or more of these figures. Thus, computation of moment of inertia is not always a difficult problem.

The formulas for centroidal moments of inertia for some common areas (Table 10-2) have been developed from the calculus.

**Sample Problem 9** For the rectangle shown in Fig. 10-19, find the moment of inertia about its centroidal axes.

**Solution:** The dimension parallel to the axis is $b$, and $h$ is the dimension perpendicular to the axis

$$I_x = \frac{bh^3}{12} \quad b = 6 \text{ in} \qquad h = 10 \text{ in}$$

$$I_x = \frac{6(10)^3}{12} = 500 \text{ in}^4$$

$$I_y = \frac{bh^3}{12} \quad b = 10 \text{ in} \qquad h = 6 \text{ in}$$

$$I_y = \frac{10(6)^3}{12} = 180 \text{ in}^4$$

**TABLE 10-2**   CENTROIDAL MOMENTS OF INERTIA FOR SIMPLE AREAS

| Shape | Moment of Inertia | |
|-------|-------------------|---|
| Rectangle | $I_x = \dfrac{bh^3}{12}$ | |
| Triangle | $I_x = \dfrac{bh^3}{36}$ | |
| Circle | $I_x = \dfrac{\pi d^4}{64}$ | |
| Semicircle | $I_x = 0.11r^4$  $I_y = \dfrac{\pi d^4}{128}$ | |
| Quadrant | $I_x = 0.055r^4$ | |

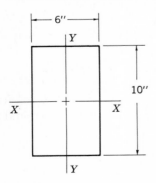

FIGURE 10-19   Diagram for Sample Problem 9.

Moments of inertia are expressed in inches to the fourth power or meters to the fourth power, units which have no apparent physical significance.

The results of this problem show that the moment of inertia of this rectangle is greater about the axis $XX$ than about axis $YY$. The moment of inertia is then seen to depend on the arrangement of the area with reference to the axis.

**\*Sample Problem 10**  Find the moment of inertia of the triangle (shown in Fig. 10-20) with reference to centroidal axis $XX$.

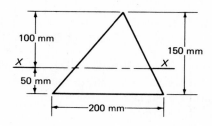

100 mm

150 mm

$X$                    $X$

50 mm

200 mm

FIGURE 10-20   Diagram for Sample Problem 10.

**Solution:**

$$I_x = \frac{bh^3}{36} \quad b = 200 \text{ mm} \qquad h = 150 \text{ mm}$$

$$I_x = \frac{(200)(150)^3}{36} = 18.8(10^6) \text{ mm}^4 = 18.8(10^{-6}) \text{ m}^4$$

**Sample Problem 11**  Find the moment of inertia for an 8-in-diameter circle about any centroidal axis.

**Solution:**

$$I_x = \frac{\pi d^4}{64} \quad d = 8 \text{ in}$$

$$I_x = \frac{\pi(8)^4}{64} = 201 \text{ in}^4$$

Note that the circle is symmetrical about any centroidal axis.

**\*Sample Problem 12**  Find the moment of inertia of the semicircle, indicated in Fig. 10-21, about the horizontal and vertical centroidal axes.

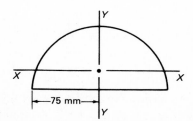

$Y$

$X$                    $X$

75 mm

$Y$

FIGURE 10-21   Diagram for Sample Problem 12.

**Solution:**

$$I_x = 0.11r^4 = 0.11(75)^4 = 3.48(10^6) \text{ mm}^4 = 3.48(10^{-6}) \text{ m}^4$$
$$I_y = \frac{\pi d^4}{128} = \frac{\pi(150)^4}{128} = 12.4(10^6) \text{ mm}^4 = 12.4(10^{-6}) \text{ m}^4$$

## 10-8 TRANSFER FORMULA

It is often desirable to obtain the moment of inertia of an area about some axis parallel to a centroidal axis, but not passing through the centroid of the area. This can be accomplished by calculating the moment of inertia about the gravity axis of the figure and adding to this value the product of the area of the figure and the square of the distance between the gravity axis and the other axis. Referring to Fig. 10-22,

$$I_{a-a} = I_x + Ad^2 \qquad (10\text{-}7)$$

Equation (10-7) is referred to as the *transfer formula*. Transfer may be made only between parallel axes.

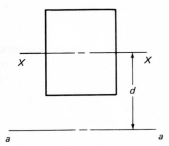

FIGURE 10-22  Transfer of moment of inertia.

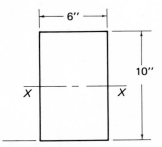

FIGURE 10-23  Diagram for Sample Problem 13.

**Sample Problem 13**  Find the moment of inertia of a 6- by 10-in rectangle about its 6-in base (Fig. 10-23).

**Solution:**  Let $XX$ be a gravity axis parallel to the base.

$$I_x = \frac{bh^3}{12} = \frac{6(10)^3}{12} = 500 \text{ in}^4$$
$$d = 5 \text{ in} \qquad \text{(the perpendicular distance between the axes)}$$

Then, by Eq. (10-7),

$$I_{a-a} = I_x + Ad^2 = 500 + 60(5)^2 = 2000 \text{ in}^4$$

## 10-9  MOMENTS OF INERTIA OF COMPOSITE AREAS

The transfer principle discussed above can now be used to determine the moment of inertia of a composite area; that is, an area which can be broken into simple areas. The T section shown in Fig. 10-24 is such an area.

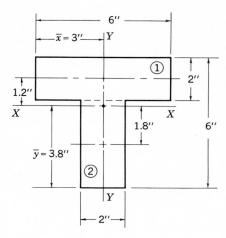

FIGURE 10-24   Composite area.

To calculate the moment of inertia of this area about its horizontal and vertical gravity axes, the centroid must be located. The composite section can be divided into two rectangles which will be identified by 1 and 2. Because of the symmetry of the figure, it can be seen that

$$\bar{x} = 3 \text{ in (from left edge)}$$

Calculating $\bar{y}$,

$$\bar{y} = \frac{\Sigma Ay}{\Sigma A} = \frac{(12)(5) + (8)(2)}{12 + 8} = \frac{60 + 16}{20} = \frac{76}{20} = 3.80 \text{ in} \qquad \text{(from base)}$$

The moment of inertia of this area about a gravity axis will be found by computing the moment of inertia of each rectangle about its own centroidal axis, transferring the moment of inertia of each rectangle to the parallel gravity axis of the entire figure, and then summing the transformed moments of inertia.

Moment of inertia about horizontal gravity axis $XX$:

**For Area 1:**

$$I_{x,1} = I_1 + A_1 d^2$$

$$I_1 = \frac{bh^3}{12} = \frac{6(2)^3}{12} = 4 \text{ in}^4$$

$$I_{x,1} = 4 + 12(1.2)^2 = 4 + 12(1.44)$$
$$= 4 + 17.28 = 21.28 \text{ in}^4$$

**For Area 2:**

$$I_{x,2} = I_2 + A_2 d^2$$
$$I_2 = \frac{bh^3}{12} = \frac{2(4)^3}{12} = 10.67 \text{ in}^4$$
$$I_{x,2} = 10.67 + 8(1.8)^2 = 10.67 + 8(3.24)$$
$$= 10.67 + 25.92 = 36.59 \text{ in}^4$$
$$I_x = I_{x,1} + I_{x,2} = 21.28 + 36.59 = 57.87 \text{ in}^4$$

Moment of inertia about vertical gravity axis $YY$:

**For Area 1:**

$$I_{y,1} = I_1 + A_1 d^2$$
$$I_1 = \frac{bh^3}{12} = \frac{2(6)^3}{12} = 36 \text{ in}^4$$
$$I_{y,1} = 36 + 12(0)^2 = 36 + 0 = 36 \text{ in}^4$$

**For Area 2:**

$$I_{y,2} = I_2 + A_2 d^2$$
$$I_2 = \frac{bh^3}{12} = \frac{4(2)^3}{12} = 2.67 \text{ in}^4$$
$$I_{y,2} = 2.67 + 8(0)^2 = 2.67 + 0 = 2.67 \text{ in}^4$$
$$I_y = I_{y,1} + I_{y,2} = 36.00 + 2.67 = 38.67 \text{ in}^4$$

The calculations performed above can be systematized through use of the following tabular form.

$I_x$:

| Area | $I$ | $A$ | $d$ | $d^2$ | $Ad^2$ | $I + Ad^2$ |
|------|-----|-----|-----|-------|--------|------------|
| 1 | 4 | 12 | 1.2 | 1.44 | 17.28 | 21.28 |
| 2 | 10.67 | 8 | 1.8 | 3.24 | 25.92 | 36.59 |
| | | | | | | $I_x = 57.87 \text{ in}^4$ |

$I_y$:

| Area | $I$ | $A$ | $d$ | $d^2$ | $Ad^2$ | $I + Ad^2$ |
|------|-----|-----|-----|-------|--------|------------|
| 1 | 36 | 12 | 0 | 0 | 0 | 36 |
| 2 | 2.67 | 8 | 0 | 0 | 0 | 2.67 |
| | | | | | | $I_y = 38.67 \text{ in}^4$ |

In the preceding problem, it can be seen that the vertical gravity axis of the entire figure coincides with the vertical centroidal axis of each rectangle. The transfer distance is therefore zero.

*Sample Problem 14 Determine the moment of inertia of the area shown in Fig. 10-25:

  (a)  About the vertical gravity axis $YY$
  (b)  About a horizontal axis $a$-$a$ 50 mm below the base

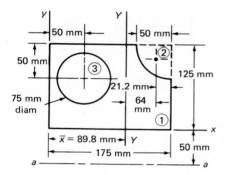

FIGURE 10-25  Diagram for Sample Problem 14a.

In calculating the moment of inertia of an area that has holes or cutouts in it, treat the areas and moments of inertia of these holes and cutouts as negative values.

Solution a:  From Sample Problem 4, $x = 89.8$ mm from the left edge. About vertical axes,

$$I_1 = \frac{bh^3}{12} = \frac{125(175)^3}{12} = 55.83(10^6) \text{ mm}^4$$

$$I_2 = -0.055r^4 = -0.055(50)^4 = -0.34(10^6) \text{ mm}^4$$

$$I_3 = \frac{-\pi(d)^4}{64} = \frac{-\pi(75)^4}{64} = 1.55(10^6) \text{ mm}^4$$

| Area | $I$ | $A$ | $d$ | $d^2$ | $Ad^2$ | $I + Ad^2$ |
|------|-----|-----|-----|-------|--------|------------|
| 1 | $55.83(10^6)$ | 21 880 | 2.3 | 5.29 | $0.12(10^6)$ | $55.95(10^6)$ |
| 2 | $-0.34(10^6)$ | $-1$ 963 | 64 | 4 096 | $-8.04(10^6)$ | $-8.38(10^6)$ |
| 3 | $-1.55(10^6)$ | $-4$ 417 | 39.8 | 1 584 | $-7.00(10^6)$ | $-8.55(10^6)$ |

$$I_y = 39.02(10^6) \text{ mm}^4$$
$$\text{say, } I_y = 39(10^6) \text{ mm}^4$$
$$= 39(10^{-6}) \text{ m}^4$$

**Solution b:** About axis $aa$, from Fig. 10-26.

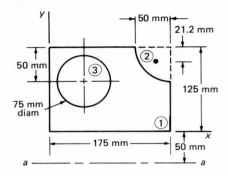

FIGURE 10-26   Diagram for Sample Problem 14b.

$$I_1 = \frac{bh^3}{12} = \frac{175(125)^3}{12} = 28.48(10^6) \text{ mm}^4$$

$$I_2 = -0.055r^4 = -0.055(50)^4 = -0.34(10^6) \text{ mm}^4$$

$$I_3 = \frac{-\pi d^4}{64} = \frac{-\pi(75)^4}{64} = -1.55(10^6) \text{ mm}^4$$

| Area | $I$ | $A$ | $d$ | $d^2$ | $Ad^2$ | $I + Ad^2$ |
|------|-----|-----|-----|-------|--------|-----------|
| 1 | $28.48(10^6)$ | $21\,880$ | $112.5$ | $12\,660$ | $277(10^6)$ | $305.48(10^6)$ |
| 2 | $-0.34(10^6)$ | $-1\,963$ | $153.8$ | $23\,650$ | $-46.42(10^6)$ | $-46.76(10^6)$ |
| 3 | $-1.55(10^6)$ | $-4\,417$ | $125$ | $15\,620$ | $-68.99(10^6)$ | $-70.54(10^6)$ |

$$I_{a-a} = 188.18(10^6) \text{ mm}^4$$
$$\text{say, } I_{a-a} = 188(10^6) \text{ mm}^4$$
$$= 188(10^{-6}) \text{ m}^4$$

# PROBLEMS

**10-1.** Locate the centroid of the T section in Fig. Prob. 10-1 from point $O$.

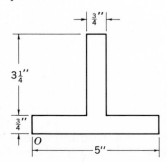

FIGURE PROBLEM 10-1 and 10-21

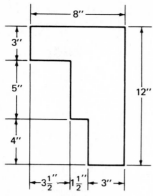

FIGURE PROBLEM 10-2

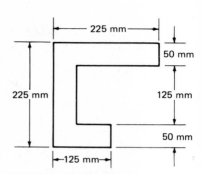

FIGURE PROBLEM 10-3

**10-2.** Locate the centroid of Fig. Prob. 10-2.

**\*10-3.** Locate the centroid of Fig. Prob. 10-3 with reference to axes through the lower left-hand corner of the figure.

**10-4.** Three unfinished planks are nailed to an unfinished plank, as shown in Fig. Prob. 10-4. How far is the center of gravity above the base?

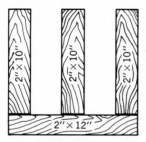

FIGURE PROBLEM 10-4
and 10-23

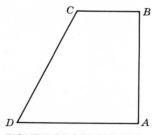

FIGURE PROBLEM 10-5

**10-5.** In Fig. Prob. 10-5, $AB = 16$ ft, $BC = 6$ ft, $AD = 12$ ft. Find the coordinates of the centroid, using $D$ as a reference point.

**\*10-6.** Find the centroid of the area shown in Fig. Prob. 10-6.

**10-7.** Locate the centroid of the channel shown in Fig. Prob. 10-7.

**10-8.** Find the centroid of each of three angle irons having the following dimensions, respectively: 4 by 3 by $\frac{1}{4}$; 6 by 6 by $\frac{3}{4}$; 8 by 6 by 1. Verify the results from App. B, Tables 7 and 8.

**\*10-9.** Find the center of gravity of the piece of sheet iron shown in Fig. Prob. 10-9.

**\*10-10.** Find the center of gravity of a plate 150 by 250 mm from which two circular holes have been cut, as shown in Fig. Prob. 10-10.

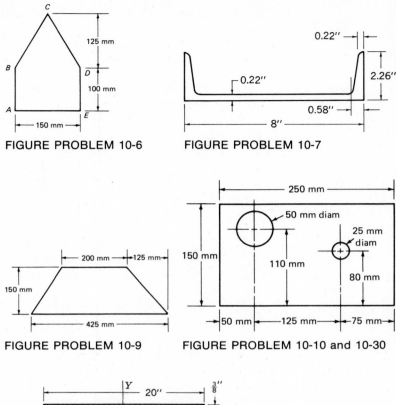

FIGURE PROBLEM 10-6

FIGURE PROBLEM 10-7

FIGURE PROBLEM 10-9

FIGURE PROBLEM 10-10 and 10-30

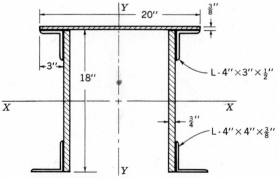

FIGURE PROBLEM 10-11 and 10-28

**10-11.** Find the centroid of the section of a bridge member made of two 18- by $\frac{3}{4}$-in web plates, a 20- by $\frac{3}{8}$-in top plate, two top angles 4 by 3 by $\frac{1}{2}$ in, and two bottom angles 4 by 4 by $\frac{3}{8}$ in (Fig. Prob. 10-11).

**10-12.** A 6-in cube of wood weighing 40 lb/ft³ has a 1-in-diameter hole drilled through it parallel to one edge. The long axis of the hole is 2 in from the upper surface and 2 in from the left side of the cube.

*a.* Locate the centroid.

*b.* If this hole is filled with lead weighing 710 lb/ft³, where is the centroid?

**\*10-13.** In Fig. Prob. 10-13, the dam has a mass of 450 kg and $F = 3.6$ kN. If $AB = 4.8$ m, $BC = 1.8$ m, and $AD = 4.8$ m, find the maximum height above $A$ where the force $F$ may be applied without overturning the dam about point $D$. (*Note:* $W$ acts at the centroid of $ABCD$.)

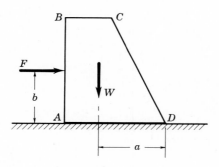

FIGURE PROBLEM 10-13

**10-14.** A dam stands 12 ft high. It is 4 ft wide at the top and 10 ft wide at the base. The upstream face is vertical. The coefficient of friction between the dam and its foundation is 0.6. What are the factors of safety against sliding and overturning when the water has reached the top of the dam? The dam is made of concrete (density = 150 lb/ft³).

**\*10-15.** Find the volume of a cone by the method of revolving a right triangle with a 75-mm base and 100-mm altitude about the vertical edge.

**10-16.** Find the volume of a sphere by revolving a semicircle about a diameter.

**10-17.** Find the volume of the rim of a flywheel 6 ft in diameter if the rim is 12 in wide and 8 in thick.

**\*10-18.** Find the moment of inertia about the horizontal and vertical gravity axes of the following rectangles:

*a.* 100 by 250 mm

*b.* 200 by 300 mm

*c.* 250 by 250 mm

**10-19.** Find the moment of inertia about the horizontal and vertical centroidal axes for the following figures:

*a.* Right triangle with horizontal base of 5 in and altitude of 7 in

    *b.*   $3\frac{1}{2}$-in-diameter circle

    *c.*   $1\frac{3}{4}$-in-radius semicircle with horizontal base

    *d.*   $1\frac{3}{4}$-in-radius quadrant (90° segment) with one side horizontal

**10-20.**  Find *I* about the gravity axis parallel to the base of the area shown in Fig. Prob. 10-20.

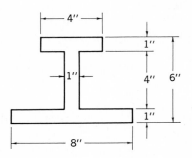

FIGURE PROBLEM 10-20

**10-21.**  Find the moments of inertia with respect to both principal axes (Fig. Prob. 10-21, page 226).

**\*10-22.**  In Fig. 10-8, p. 210, find the moment of inertia with respect to the horizontal gravity axis and also with respect to the left edge of the figure.

**10-23.**  What are the moments of inertia with respect to the principal axes in Fig. Prob. 10-23, page 227?

**10-24.**  Find $I_x$ and $I_y$ for each of two angle irons, 4 by 4 by $\frac{1}{2}$ in and 4 by 3 by $\frac{1}{2}$ in. Compare results with the values given in App. B, Tables 7 and 8.

**10-25.**  Find $I_x$ and $I_y$ for a girder made with four 7- by 4- by $\frac{3}{8}$-in angles and one 12- by $\frac{3}{8}$-in web plate (Fig. Prob. 10-25).

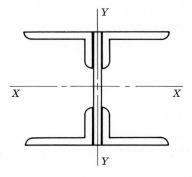

FIGURE PROBLEM 10-25 and 10-26

**10-26.**  Find $I_x$ and $I_y$ for a girder built up of four 5- by $3\frac{1}{2}$- by $\frac{1}{2}$-in angles and a 10- by $\frac{1}{2}$-in web plate (Fig. Prob. 10-26).

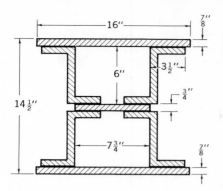

FIGURE PROBLEM 10-27

**10-27.** What is the moment of inertia with respect to the horizontal gravity axis of the column section shown in Fig. Prob. 10-27? The area of a Z bar is 8.63 in², and the moment of inertia of a Z bar with respect to its gravity axis parallel to the shorter leg is 42.12 in⁴.

**10-28.** Find the moment of inertia of the section shown in Fig. Prob. 10-28 (page 228) with respect to both principal axes.

**10-29.** A box girder is shown in Fig. Prob. 10-29. What is the moment of inertia with respect to the horizontal gravity axis if the bolt holes are deducted? Bolts are located in the center of the angle legs.

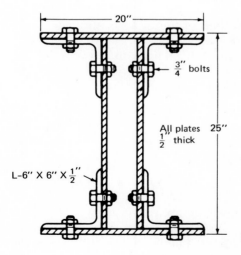

FIGURE PROBLEM 10-29

**10-30.** Find the moment of inertia about the horizontal and vertical gravity axes for the plate shown in Fig. Prob. 10-30 (page 228).

# CHAPTER
# 11

# Beams — Shear Forces and Bending Moments

## 11-1 TYPES OF BEAMS

A beam is a member which resists transverse loads and forces by bending elastically. The beams discussed in this study will be confined to those which meet the following limitations.

1. All forces acting on the beam lie in the same plane; this (longitudinal) plane passes through the centroids of all the cross sections of the beam.
2. The elastic limit of the beam material is not exceeded.
3. The cross section of the beam is uniform in size and shape for the entire length.
4. The beam is thick enough to prevent localized buckling or wrinkling.
5. Vibration, shock, and impact loading do not occur.

Beams may be generally categorized as either statically determinate or statically indeterminate. A statically determinate beam is one whose unknown forces or moments can be evaluated by applying the conditions for static equilibrium, Eqs. (11-1).

$$\Sigma F_x = 0 \qquad \Sigma F_y = 0 \qquad \Sigma M = 0 \tag{11-1}$$

Common beams of this type are shown in Fig. 11-1. Among these are a *simply supported beam* (Fig. 11-1a), an *overhanging beam* (Fig. 11-1b), and a *cantilever beam* (Fig. 11-1c). The loads, whether concentrated ($F$) or uniformly distributed ($W$), are assumed to be known, while the reactions ($R$) of the supports and the wall moment ($M$) are unknown. This chapter deals only with statically determinate beams.

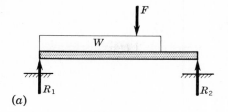

(a)

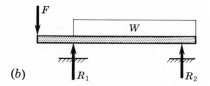

(b)

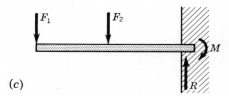

(c)

FIGURE 11-1   Common statically determinate beams: (a) simply supported beam, $R_1$ and $R_2$ unknown; (b) overhanging beam, $R_1$ and $R_2$ unknown; (c) cantilever beam, $R$ and $M$ unknown.

The unknown forces acting on statically indeterminate beams cannot be found by Eqs. (11-1) only. Special procedures are used to solve problems of this type. Chapter 16 deals with statically indeterminate beams.

## 11-2  BEAM THEORY

Let Fig. 11-2 represent a simply supported beam carrying a load $F$ at its center. The load produces a bending in the beam at all points. Any sections such as $AB$ and $CD$ are rotated into new positions by the bending of the beam. The length $BD$ is shortened or compressed, and length $AC$ is lengthened or stretched. The resistance offered to this shortening and lengthening of the fibers is called *internal fiber stress* or simply *stress*. The upper fibers are in compression and the lower ones in tension.

It was shown in Chap. 8 that the deformation of a body, if within the elastic limit, is proportional to the applied force. The diagram shows that the maximum internal stresses occur at the outer fibers $AC$ and $BD$, for the deformation is greatest at these points and decreases to zero at the center $G$ or $H$.

Axes $GG$ and $HH$ (Fig. 11-3) are called the *neutral axes of the sections*, or *axes of zero stress*. The neutral axis of a section of a beam passes through the centroid of the section.

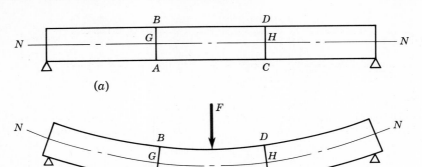

(a)

(b)

FIGURE 11-2  Simply supported beam: (a) before load is applied.
(b) after load is applied.

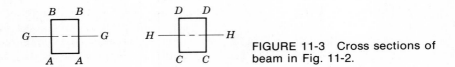

FIGURE 11-3  Cross sections of beam in Fig. 11-2.

In addition to the bending of a beam, there is a tendency of one section of a beam to slip past the adjacent section. This tendency is called *shear*, and the shear forces must be resisted by the fibers of the beam. Shearing forces are parallel to the plane of the section (Fig. 11-4).

Figure 11-5 shows the beam at section *AB*. The fibers resisting bending above the neutral axis *G* are subjected to compression forces, and those below *G* are subjected to tension forces. The compressive and tensile forces due to bending are indicated. Note that these forces increase as the distance from the neutral axis *G* increases. Consequently, we usually expect the most severe stresses to occur at the extreme fibers. The vertical shear force *V* at section *AB* is also shown. In Fig. 11-6 the same situation is represented, except that the compressive forces above the neutral axis (between *B* and *G* in Fig. 11-5) are replaced by their resultant *C* and,

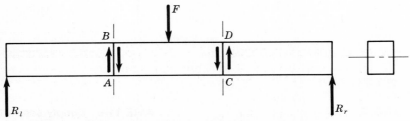

FIGURE 11-4  Shear forces in simply supported beam.

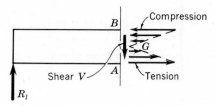

FIGURE 11-5    Shear and bending forces at a beam section.

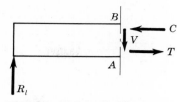

FIGURE 11-6    Shear and resultant bending forces at a beam section.

similarly, the tensile forces are replaced by resultant $T$. For the portion of the beam shown in Fig. 11-6, the only horizontal forces acting are $C$ and $T$. Applying the condition for equilibrium that $\Sigma F_x = 0$, it is evident that $T = C$ in magnitude although their directions are opposite. These equal, opposite, parallel forces form a *couple*, the effect of which is to produce a *resisting moment*. This moment must be equal and opposite (direction of rotation) to the moment produced by external forces acting on the left portion of the beam so that the condition $\Sigma M = 0$ is satisfied. The reaction $R_l$ at the left support is the only external force which produces a moment in Fig. 11-6 (the weight of the beam has been neglected). Its moment arm is the horizontal distance from the support to section $AB$.

It is customary to use the portion of the beam *to the left* of a section in solving problems. This practice will be followed in almost all cases.

For problems with statically determinate beams, it is usual to first calculate the unknown reactions. Types of loading which are symmetrically placed on simply supported beams enable the reactions to be determined by inspection. In Fig. 11-7, a simply supported beam with a concentrated load $F$ at the center of the span is shown. Neglecting the weight of the beam, apply the condition $\Sigma F_y = 0$. It is seen that $R_l + R_r - F = 0$, and since the load $F$ is centrally located on the beam, then each reaction equals half the load, or $R_l = F/2$ and $R_r = F/2$. A similar analysis applies to the beam in Fig. 11-8 which carries a uniformly distributed load whose total weight is $W$. Again it is found that each reaction equals half the load, or $R_l = W/2$ and $R_r = W/2$. Uniform loads are often specified by the load they apply either per linear foot or per linear meter of beam. The symbol

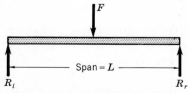

FIGURE 11-7    Simply supported beam with load at mid-span.

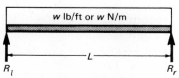

FIGURE 11-8    Simply supported beam with uniformly distributed load.

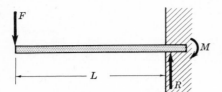
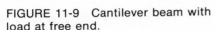
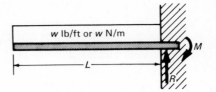

FIGURE 11-9 Cantilever beam with load at free end.

FIGURE 11-10 Cantilever beam with uniformly distributed load.

$w$ will be used for this purpose. In the U.S. customary system, $w$ will be in terms of pounds per foot; in the SI metric system, $w$ will be in terms of newtons per meter. Note that the total uniform load ($W$) equals the weight per foot ($w$) times the length of the load. For the load in Fig. 11-8, the length of the load equals the span length ($L$) of the beam; thus, $W = wL$.

For cantilever beams with vertical loading only, the unknown reaction $R$ may be evaluated by applying the condition $\Sigma F_y = 0$. The beam shown in Fig. 11-9 is a cantilever with a concentrated load $F$ at the unsupported end. The wall must provide a force (reaction) $R$ to counteract the load $F$ so that $\Sigma F_y = 0$ is satisfied. Therefore, $R - F = 0$, or $R = F$. Similarly, the cantilever beam in Fig. 11-10, carrying a uniform load $W = wL$, has a wall reaction $R$ which balances the total load $W$. Thus, $R - W = 0$, or $R = W$. The weight of the beam has been neglected in both of the previous cases.

The unknown moment $M$ at the wall for a cantilever beam may be calculated from $\Sigma M = 0$. Since there can be no unbalanced moment in static equilibrium, the moment due to the load on the beam must be counteracted by $M$ at the wall. The moment arm of the load is always measured to the wall for this calculation. For the beam in Fig. 11-9, $M = FL = 0$, or $M = FL$. For the purpose of taking moments, the total weight of a uniform load is considered to act at the center of the load. Then, for the beam in Fig. 11-10, $M - W(L/2) = 0$, or $M = W(L/2)$.

**Sample Problem 1** Calculate the reactions for a simply supported beam with a 12-ft span which carries a concentrated load of 8000 lb placed 3 ft from the left support, if:

    (a)   The weight of the beam is neglected
    (b)   The beam weighs 50 lb/ft

**Solution a:** A sketch of the beam showing the loading and the reactions should be drawn as in Fig. 11-11.

To ensure static equilibrium, Eqs. (11-1) must be satisfied. Since there are no horizontal forces nor forces with horizontal components, the condition $\Sigma F_x = 0$ is satisfied, but it gives no useful equation to help solve for the unknowns. The other conditions for equilibrium are $\Sigma F_y = 0$ and

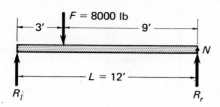

FIGURE 11-11   Diagram for Sample Problem 1a.

$\Sigma M = 0$. Each gives a useful equation which enables the solution of the unknowns $R_l$ and $R_r$.

$\Sigma F_y = 0$

Up forces will be considered positive and down forces negative.

$$R_l + R_r - F = 0$$
$$R_l + R_r = 8000$$

$\Sigma M = 0$

This relation states that there can be no unbalanced moment about *any axis of rotation*. Hence, any point on or off the beam may be selected as an axis about which to take moments. The usual procedure, however, is to select an axis so that one (or more) of the unknowns is eliminated. Such an axis would be located on the line of action of one of the unknown reactions, since, if its line of action passes through the axis of rotation, the reaction will have a zero moment arm and thus a zero moment. Point $N$ in Fig. 11-11 represents an axis passing through the centroid of the right end section of the beam in a direction perpendicular to the plane of the paper. Reaction $R_r$ will have no moment about axis $N$. The common notation for this is $\Sigma M_r = 0$; i.e., the sum of moments about $R_r$ (actually axis $N$) is zero. Clockwise moments will be called positive, while counterclockwise moments will be taken as negative.

$\Sigma M_r = 0$

$$R_l(12) - 8000(9) = 0$$
$$R_l = \frac{8000(9)}{12} = 6000 \text{ lb}$$

Substituting this value in the equation from $\Sigma F_y = 0$ gives

$$6000 + R_r = 8000$$
$$R_r = 2000 \text{ lb}$$

Note that the load $F$ is supported by each reaction in proportion to the distance of $F$ from the other reaction. The load is 3 ft from the left end of the 12-ft span; thus, $R_r = \frac{3}{12} F = 3(8000)/12 = 2000$ lb. $F$ is 9 ft from

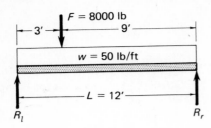

FIGURE 11-12   Diagram for Sample Problem 1b.

the right end; thus, $R_l = \frac{9}{12} F = 9(8000)/12 = 6000$ lb. This method of division of loading between the reactions is called the *lever rule*.

**Solution b:**   Figure 11-12 shows the beam with the loading and reactions. The weight of the beam is treated as a uniformly distributed load of $w = 50$ lb/ft. The total weight of the beam is $W = wL = 50(12) = 600$ lb. The procedure is similar to part $a$ of this problem.

$\Sigma F_y = 0$

$$R_l + R_r - F - W = 0$$
$$R_l + R_r = 8000 + 600$$
$$R_l + R_r = 8600$$

$\Sigma M_r = 0$

$$R_l(12) - 8000(9) - 600(6) = 0$$
$$R_l = \frac{8000(9) + 600(6)}{12} = \frac{72\,000 + 3600}{12}$$
$$= \frac{75\,600}{12} = 6300 \text{ lb}$$

Substituting,

$$6300 + R_r = 8600$$
$$R_r = 2300 \text{ lb}$$

Comparison of these results with those of part $a$ indicates that the effect of the uniform load is divided equally between $R_l$ and $R_r$.

**\*Sample Problem 2**   A 6-m-long beam is supported 0.5 m from the right end and 1.5 m from the left end. The beam carries a concentrated load of 4.5 kN at the left end of the beam and a uniform load of 3000 N/m extending from the left support to the right end of the beam. Calculate the reactions at the supports. Neglect the weight of the beam.

**Solution:**   Sketch the beam as in Fig. 11-13.

$$W = 3000(4.5) = 13\,500 \text{ N} = 13.5 \text{ kN}$$

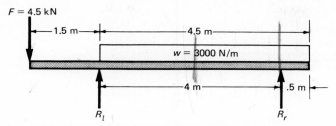

FIGURE 11-13   Diagram for Sample Problem 2.

$\Sigma F_y = 0$

$$R_l + R_r - F - W = 0$$
$$R_l + R_r - 4.5 - 13.5 = 0$$
$$R_l + R_r = 18$$

$\Sigma M_r = 0$

$$R_l(4) - 4.5(5.5) - 13.5(1.75) = 0$$
$$4R_l = 24.75 + 23.62 = 48.37$$
$$R_l = 12.1 \text{ kN}$$

Substituting,
$$12.1 + R_r = 18$$
$$R_r = 5.9 \text{ kN}$$

## 11-3  SHEAR-FORCE DIAGRAM

It was indicated in Sec. 11-2 that a vertical force exists at a given section in a beam. It is useful to know what shear force a beam must resist at *every section*. The shear force at any section of a beam is the algebraic sum of all vertical forces acting on the beam to the left of that section. This information is conveniently represented in a shear-force diagram which is drawn in projection with the sketch of the beam it represents. The shear-force diagram is a plot of the *net external shearing forces* which act at each beam cross section. These forces are caused by the loading on the beam. The fibers of the beam material must resist these forces to maintain static equilibrium.

Several examples will demonstrate the procedure for obtaining a shear-force diagram. In these examples, the shear force is calculated at sections 1 ft apart (0.5 m apart when using metric units), starting from the left end. External up forces are considered positive, and external down forces are considered negative. All reaction forces must be known before sketching the shear diagram.

**Sample Problem 3**  Find the shear forces on a simply supported beam, 10 ft long, due to a concentrated load of 1000 lb at the center of the span.

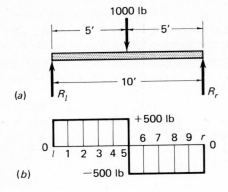

(a)

(b)

FIGURE 11-14 (a) Beam diagram for Sample Problem 3. (b) Shear-force diagram.

Neglect the weight of the beam. Sketch the shear-force diagram for this beam.

**Solution:** From the beam shown in Fig. 11-14a it is apparent that the reactions are 500 lb each due to the symmetrical loading.

At the left end of the beam the only force acting is the reaction (500 lb up); therefore, the net shear force at the left end (actually, just slightly in from the end) is $V_l = +500$ lb. At a point 1 ft in from the reaction no new forces are encountered; thus $V_1 = +500$ lb. As we move from left to right along the beam and examine the net shear force at every foot, we find that

$$V_2 = +500 \text{ lb}$$
$$V_3 = +500 \text{ lb}$$
$$V_4 = +500 \text{ lb}$$

At a section just to the left of the center (say at 4.99 ft from the left end), no new load has been encountered; thus $V_5 = +500$ lb. However, at the center of the beam a down load of 1000 lb appears. The net shear force is now $V_5' = +500 - 1000 = -500$ lb. No new loads are encountered for the remainder of the beam until the right reaction is reached; therefore,

$$V_6 = -500 \text{ lb}$$
$$V_7 = -500 \text{ lb}$$
$$V_8 = -500 \text{ lb}$$
$$V_9 = -500 \text{ lb}$$

Just before reaching the right reaction, the shear force is $V_r = -500$ lb. When the reaction is reached, the net shear force becomes $V_r' = -500 + 500 = 0$ lb.

These shear forces may be plotted (to a suitable scale) in projection with the beam. Such a plot is called the *shear-force diagram*, as shown in

Fig. 11-14*b*. Establish a horizontal line *OO* whose length is the same as the beam. Plot values of shear force *V* at the location along line *OO* corresponding to the beam section where *V* acts. Plot positive *V* values above line *OO* and negative values below line *OO*. Connect the plotted points with straight lines. The resulting diagram gives the net shear force at *all* sections across the beam.

The shear diagram shows a constant value of shear force between the left end of the beam and the concentrated load. Therefore, it is un-necessary to calculate shear forces for intermediate sections if no new forces are present in this range.

**\*Sample Problem 4** Determine the shear diagram for a simply sup-ported beam, 5 m long, carrying a uniformly distributed load of 1400 N/m, including its own weight.

**Solution:**    The total load is 1400 N/m × 5 m = 7000 N. Each reaction is one-half the total load, or 3500 N. Therefore, $V_0 = 3500$ N. Between the 0.5-m section and the left end of the beam there are 700 N of load. Hence, $V_{0.5} = 3500 - 700 = 2800$ N. The forces to the left of the 1-m section are 3500 N up and 1400 N of load down.

$$V_1 = 3500 - 1400 = 2100 \text{ N}$$

Similarly,

$$V_{1.5} = 1400 \text{ N}$$
$$V_2 = 700 \text{ N}$$
$$V_{2.5} = 0$$
$$V_3 = -700 \text{ N}$$
$$V_{3.5} = -1400 \text{ N}$$
$$V_4 = -2100 \text{ N}$$
$$V_{4.5} = -2800 \text{ N}$$
$$V_5 = -3500 \text{ N}$$

These values are plotted in projection with the beam, as shown in Fig. 11-15.

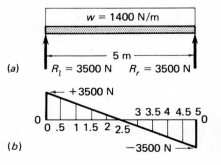

(a)

(b)

FIGURE 11-15   (a) Beam diagram for Sample Problem 4. (b) Shear-force diagram.

A uniform load on a beam results in a sloped straight line on the shear diagram. To construct the shear-force diagram in this problem, it is necessary only to calculate and plot the shear forces at each end of the uniform load $V_0$ and $V_5$ and to join these with a straight line.

***Sample Problem 5**  Determine the shear diagram for a cantilever beam, 3 m long, with a concentrated load of 70 kN at the free end. Neglect the weight of the beam.

**Solution:**  Cantilever beams should always be sketched with the free end at the left so that the signs (+ or −) given to forces and moments remain consistent.

At the left end of the beam, the shear force is 70 kN down due to the concentrated load. No other forces are encountered until the fixed end of the beam, where a reaction of $R = 70$ kN (up) must occur to ensure that $\Sigma F_y = 0$. Thus, the shear diagram can be sketched (Fig. 11-16) without intermediate calculations at other sections.

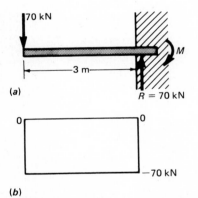

FIGURE 11-16   (a) Beam diagram for Sample Problem 5. (b) Shear-force diagram.

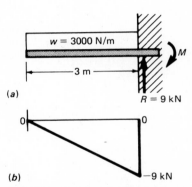

FIGURE 11-17   (a) Beam diagram for Sample Problem 6. (b) Shear-force diagram.

***Sample Problem 6**  Using the same cantilever beam as in Sample Problem 5 and replacing the concentrated load with a uniform load of 3000 N/m, determine the shear diagram.

**Solution:**  $W = 3000$ N/m × 3 m = 9000 N = 9 kN. The shear diagram for this beam is shown in Fig. 11-17.

**Sample Problem 7**  Determine the shear-force diagram for the overhanging beam shown in Fig. 11-18. The beam weighs 35 lb/ft.

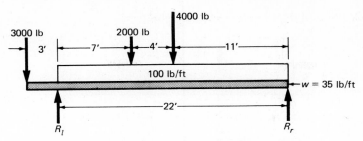

FIGURE 11-18   Beam diagram for Sample Problem 7.

**Solution:**   Reactions $R_l$ and $R_r$ must be determined before the shear diagram can be sketched.

$\Sigma F_y = 0$

$$R_l + R_r - 3000 - 2000 - 4000 - 100(22) - 35(25) = 0$$
$$R_l + R_r = 12\,075 \text{ lb}$$

$\Sigma M_r = 0$

$$R_l(22) - 3000(25) - 2000(15) - 4000(11)$$
$$- 100(22)(11) - 35(25)(12.5) = 0$$
$$R_l = \frac{75\,000 + 30\,000 + 44\,000 + 24\,200 + 10\,940}{22} = \frac{184\,140}{22}$$

$$= 8370 \text{ lb}$$
$$R_r = 12\,075 - 8370 = 3705 \text{ lb}$$

Start the shear diagram from the left end of the beam.

At the left end of the beam there is a down force of 3000 lb. The 3-ft section between the left end and $R_l$ has a uniform load of 35 lb/ft due to the weight of the beam.

$$V_3 = -[3000 + 35(3)] = -[3000 + 105] = -3105 \text{ lb}   \text{(down)}$$

The reaction force, $R_l = 8370$ lb, acts up.

$$V_3' = 8370 - 3105 = 5265 \text{ lb}   \text{(up)}$$

The 7-ft section between $R_l$ and the 2000-lb concentrated load has a uniform load of 135 lb/ft.

$$V_{10} = 5265 - 135(7) = 5265 - 945 = 4320 \text{ lb}   \text{(up)}$$

The effect of the 2000-lb load is considered next.

$$V_{10}' = 4320 - 2000 = 2320 \text{ lb}   \text{(up)}$$

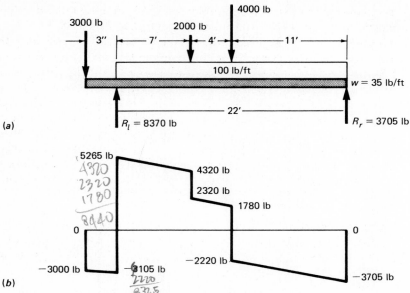

FIGURE 11-19 (a) Beam diagram for Sample Problem 7. (b) Shear-force diagram.

The next 4-ft section has a uniform load of 135 lb/ft.

$$V_{14} = 2320 - 135(4) = 2320 - 540 = 1780 \text{ lb} \quad \text{(up)}$$

The 4000-lb concentrated load will cause the net shear force to become negative.

$$V'_{14} = 1780 - 4000 = -2220 \text{ lb} \quad \text{(down)}$$

The remaining 11-ft section increases the net down force by 135 lb/ft.

$$V_{25} = -2220 - (135)11 = -2220 - 1485 = -3705 \text{ lb} \quad \text{(down)}$$

But the right reaction $R_r = 3705$ lb (up); thus the shear diagram returns to zero.

The shear diagram for this beam is shown in Fig. 11-19.

## 11-4 MOMENT DIAGRAM

In order that the designer may select a beam to carry a given load, he must know how the bending moment changes at sections from one end to the beam of the other. For this purpose, the bending moment is calculated about axes through sections 1 ft apart (0.5 m apart when using metric units) and is plotted just as in the case of shear. The figure thus obtained is called the *bending-moment diagram*. Bending-moment diagrams are also used in determining the patterns of deflection in beams. The bending moment at any section of the beam is the algebraic sum of the moments due to forces acting to the left of that section. The bending-moment diagram is a plot of the *net external moments* which act at each beam cross section. These external moments, which are caused by external forces, must be resisted by the beam material to maintain equilibrium. As a result of resisting these moments, the beam material is stressed and bending occurs. For purposes of calculation, external *clockwise moments* are taken as *positive*, while external *counterclockwise moments* are considered *negative*.

**Sample Problem 8** Determine the bending-moment diagram for a simply supported beam, 10 ft long, with a concentrated load of 1000 lb at the center of the span. Neglect the weight of the beam. *Note:* Refer to Sample Problem 3 and Fig 11-14. Also, refer to Fig. 11-20.

**Solution:** Calculate the bending moment at sections 1 ft apart along the length of the beam, starting from the left end.

   At the left end of this beam the only external effect is the 500-lb reaction. The line of action of this force passes through the end section. Therefore, this force produces no moment at the end section. $M_0 = 0$.

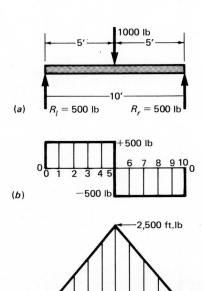

FIGURE 11-20 (*a*) Beam diagram for Sample Problem 8. (*b*) Shear–force diagram. (*c*) Moment diagram.

To calculate the moment about an axis through the section 1 ft from the left end, all external effects *to the left of that section* are considered. The reaction force of 500 lb has a moment arm of 1 ft. Therefore, $M_1 = 500(1) = 500$ ft·lb (clockwise). Note that $M_1$ is the net external moment at the 1-ft section. Equilibrium is maintained when the beam material successfully counterbalances $M_1$. By following a similar procedure, the moments at other sections are

$$M_2 = 500(2) = 1000 \text{ ft·lb}$$
$$M_3 = 500(3) = 1500 \text{ ft·lb}$$
$$M_4 = 500(4) = 2000 \text{ ft·lb}$$
$$M_5 = 500(5) = 2500 \text{ ft·lb}$$

At the 6-ft section, the counterclockwise moment effect of the 1000-lb force must be considered. Hence,

$$M_6 = 500(6) - 1000(1) = 3000 - 1000 = 2000 \text{ ft·lb}$$

Similarly,

$$M_7 = 500(7) - 1000(2) = 1500 \text{ ft·lb}$$
$$M_8 = 500(8) - 1000(3) = 1000 \text{ ft·lb}$$
$$M_9 = 500(9) - 1000(4) = 500 \text{ ft·lb}$$
$$M_{10} = 500(10) - 1000(5) = 0 \text{ ft·lb}$$

The moment diagram shown in Fig. 11-20 is obtained by plotting the values of $M$ calculated above.

*Sample Problem 9** Determine the bending-moment diagram for a simply supported beam, 5 m long, carrying a uniformly distributed load of 1400 N/m, including its own weight. *Note:* Refer to Sample Problem 4, p. 241.

**Solution:** Calculate the bending moment at sections 0.5 m apart along the length of the beam, starting from the left end.

At the left end of this beam the only external effect is the 3500-N reaction. Since the line of action of this force passes through the end section, the moment $M_0 = 0$. For the section 0.5 m from the left end, consider only those external effects *to the left of that section*. There are two such effects: the 3500-N reaction force with a 0.5-m moment arm, and the portion of the uniform load which lies to the left of the section with its moment arm. The portion of the uniform load in question is 0.5 m long; therefore, its *weight* is 1400 N/m × 0.5m = 700 N. Since the weight of a uniform load can be considered concentrated at the center of gravity of that load for purposes of taking moments, the moment arm for this

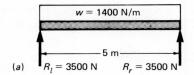

(a)

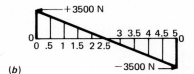

(b)

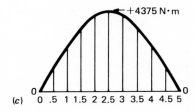

(c)

FIGURE 11-21  (a) Beam diagram for Sample Problem 9. (b) Shear-force diagram. (c) Moment diagram.

portion of the uniform load is 0.25 m. Thus, the moment at the 0.5-m section is

$$M_{0.5} = 3500(0.5) - 700(0.25) = 1750 - 175 = 1575 \text{ N·m}$$

Similarly,

$$M_1 = 3500(1) - 1400(0.5) = 3500 - 700 = 2800 \text{ N·m}$$
$$M_{1.5} = 3500(1.5) - 2100(0.75) = 5250 - 1575 = 3675 \text{ N·m}$$
$$M_2 = 3500(2) - 2800(1) = 7000 - 2800 = 4200 \text{ N·m}$$
$$M_{2.5} = 3500(2.5) - 3500(1.25) = 8750 - 4375 = 4375 \text{ N·m}$$
$$M_3 = 3500(3) - 4200(1.5) = 10\,500 - 6300 = 4200 \text{ N·m}$$
$$M_{3.5} = 3500(3.5) - 4900(1.75) = 12\,250 - 8575 = 3675 \text{ N·m}$$
$$M_4 = 3500(4) - 5600(2) = 14\,000 - 11\,200 = 2800 \text{ N·m}$$
$$M_{4.5} = 3500(4.5) - 6300(2.25) = 15\,750 - 14\,175 = 1575 \text{ N·m}$$
$$M_5 = 3500(5) - 7000(2.5) = 17\,500 - 17\,500 = 0 \text{ N·m}$$

Plotting these values to a suitable scale gives the moment diagram in Fig. 11-21.

$$W = 1400 \text{ N/m} \times 5 \text{ m} = 7000 \text{ N}$$

**\*Sample Problem 10**  Determine the moment diagram for a cantilever beam, 3 m long, with a concentrated load of 70 kN at the free end. Neglect the weight of the beam. *Note:* Refer to Sample Problem 5, p. 242.

**Solution:** With the free end at the left, calculate the bending moment at sections 0.5 m apart along the length of the beam, starting from the free left end.

The load at the free end has no moment arm about the end section; thus, $M_0 = 0$ kN·m. At the section 0.5 m from the left end, the 70-kN load has a 0.5-m arm; thus, the counterclockwise moment

$$M_{0.5} = -70 \text{ kN}(0.5 \text{ m}) = -35 \text{ kN·m}$$

Similarly,
$$M_1 = -70 \text{ kN}(1\text{m}) = -70 \text{ kN·m}$$
$$M_{1.5} = -105 \text{ kN·m}$$
$$M_2 = -140 \text{ kN·m}$$
$$M_{2.5} = -175 \text{ kN·m}$$
$$M_3 = -210 \text{ kN·m}$$

When these values are plotted to scale, a moment diagram as shown in Fig. 11-22 is obtained.

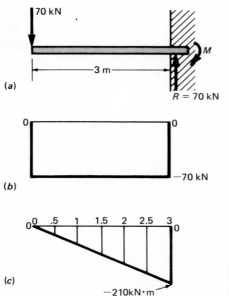

(a)

(b)

(c)

FIGURE 11-22 (a) Beam diagram for Sample Problem 10. (b) Shear-force diagram. (c) Moment diagram.

**\*Sample Problem 11** Determine the moment diagram for a cantilever beam, 3 m long, with a uniform load of 3000 N/m (including the weight of the beam) for the entire length. *Note:* Refer to Sample Problem 6, p. 242.

**Solution** $W = 3000 \text{ N/m} \times 3 \text{ m} = 9000 \text{ N} = 9 \text{ kN}$

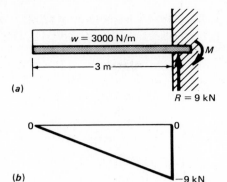

(a)

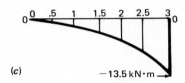

(b)

(c)

FIGURE 11-23 (a) Beam diagram for Sample Problem 11. (b) Shear force diagram. (c) Moment diagram.

With the free end at the left, calculate the bending moment at sections 0.5 m apart along the length of the beam, starting from the free left end.

Since the load at the free end is zero, $M_0 = 0$. The portion of the uniform load to the left of the 0.5-m section weighs 1500 N. The weight of a uniform load can be considered as acting at its center of gravity; therefore, the moment arm about the 0.5-m section is 0.25 m. Thus,

$$M_{0.5} = -1.5(0.25) = -0.375 \text{ kN} \cdot \text{m} \quad \text{(counterclockwise)}$$

Similarly,
$$M_1 = -3(0.5) = -1.5 \text{ kN} \cdot \text{m}$$
$$M_{1.5} = -4.5(0.75) = -3.4 \text{ kN} \cdot \text{m}$$
$$M_2 = -6(1) = -6 \text{ kN} \cdot \text{m}$$
$$M_{2.5} = -7.5(1.25) = -9.4 \text{ kN} \cdot \text{m}$$
$$M_3 = -9(1.5) = -13.5 \text{ kN} \cdot \text{m}$$

This beam, with its shear and moment diagrams, is shown in Fig. 11-23.

# 11-5 RELATION BETWEEN BEAM LOADING, SHEAR DIAGRAM, AND MOMENT DIAGRAM

Examination of the previous examples (Sample Problems 8 to 11), and especially the sketches in these examples (Figs. 11-20 to 11-23), reveals that certain relationships exist among beam loading, the shape of the shear-force diagram, and the shape of the bending-moment diagram. These relationships can be summarized as follows:

| Beam load | Shear diagram | Moment diagram |
|---|---|---|
| 1. Concentrated load | Horizontal line | Sloped straight line |
| 2. Uniform load | Sloped straight line | Curved line |

Item 1 above is demonstrated in Figs. 11-20 and 11-22. Item 2 above is shown in Figs. 11-21 and 11-23. The mathematical expressions which explain these relations will be discussed in Sec. 11-6.

In general, the *vertical height* (ordinate) at any point on a diagram will equal the *slope* of the next higher degree diagram at the same point. Thus, the vertical height at any point on the shear diagram equals the slope of the bending-moment diagram at that same point. One important application of this relationship is revealed by close inspection of how the shear-force diagrams relate to the corresponding bending-moment diagrams in Figs. 11-20 to 11-23. The relationship may be stated as follows.

*When the shear diagram passes through zero or equals zero for a particular beam section, then the value on the moment diagram for that section will be either:*

(a)  A maximum (see Figs. 11-20 to 11-23)
(b)  Zero (see Figs. 11-20 to 11-23)
(c)  A relative maximum (see Fig. 11-25)

**Sample Problem 12**  Determine the moment diagram for the overhanging beam shown in Fig. 11-24. Neglect the weight of the beam.

**Solution:**  Application of the principles stated above can facilitate the determination of the moment diagram. Thus, the moment diagram will be curved for the portion of the beam between the reactions due to the presence of a uniform load. The overhanging portion, which has no uniform load, will have a straight sloped line for its moment diagram.

Furthermore, the shear diagram indicates two possible locations (10 and 20 ft from the left end) for maximum bending moment where the shear diagram goes through zero. Although the shear diagram is zero at both ends of the beam, the moments at these end sections are zero for a *free* or *simply supported end*.

If an accurate moment diagram is required, then moments should be calculated for (1) each section where the shear diagram changes direction or jogs, (2) sections where the shear diagram crosses through or equals zero, and (3) several intermediate sections in regions where the moment diagram will curve.

Rather than calculate moments for each 1-ft section, this solution will

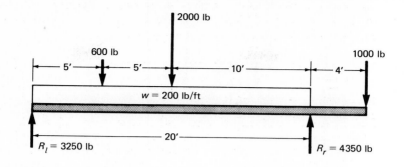

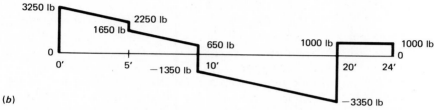

FIGURE 11-24 (a) Beam diagram for Sample Problem 12. (b) Shear-force diagram.

show only two or three intermediate calculations between important sections so that the shape of the diagram may be determined.

$M_o = 0$ (by inspection)

$M_2 = 3250(2) - 200(2)(1) = 6500 - 400 = 6100\ \text{ft·lb}$

$M_4 = 3250(4) - 200(4)(2) = 13\,000 - 1600 = 11\,400\ \text{ft·lb}$

$M_5 = 3250(5) - 200(5)(2.5) = 16\,250 - 2500 = 13\,750\ \text{ft·lb}$

$M_7 = 3250(7) - 600(2) - 200(7)(3.5) = 22\,750 - 1200 - 4900$
$$= 16\,650\ \text{ft·lb}$$

$M_9 = 3250(9) - 600(4) - 200(9)(4.5) = 29\,250 - 2400 - 8100$
$$= 18\,750\ \text{ft·lb}$$

$M_{10} = 3250(10) - 600(5) - 200(10)(5) = 32\,500 - 3000 - 10\,000$
$$= 19\,500\ \text{ft·lb}$$

$M_{12} = 3250(12) - 600(7) - 2000(2) - 200(12)(6) = 39\,000 - 4200$
$$- 4000 - 14\,400 = 16\,400\ \text{ft·lb}$$

$M_{15} = 3250(15) - 600(10) - 2000(5) - 200(15)(7.5) = 48\,750$
$$- 6000 - 10\,000 - 22\,500 = 10\,250\ \text{ft·lb}$$

$M_{18} = 3250(18) - 600(13) - 2000(8) - 200(18)(9) = 58\,500$
$$- 7800 - 16\,000 - 32\,400 = 2300\ \text{ft·lb}$$

$M_{20} = 3250(20) - 600(15) - 2000(10) - 200(20)(10) = 65\,000$
$$- 9000 - 20\,000 - 40\,000 = -4000\ \text{ft·lb}$$

$M_{24} = 0$ (by inspection)

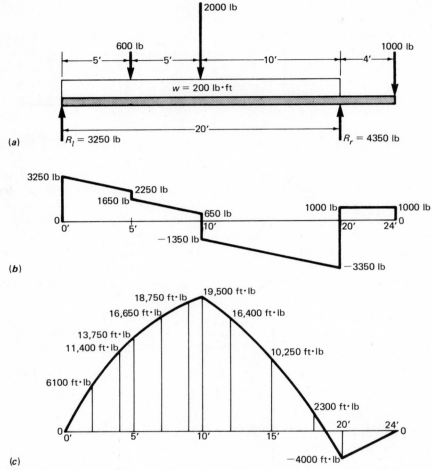

FIGURE 11-25  (a) Beam diagram for Sample Problem 12. (b) Shear-force diagram. (c) Moment diagram.

A plot of these moment values results in the moment diagram shown in Fig. 11-25. The calculations and the diagram reveal that the maximum bending moment occurs at the section 10 ft from the left end (note that the shear diagram passes through zero) and a relative maximum moment occurs at the 20-ft section (shear diagram passes through zero).

## 11-6 BENDING MOMENT FROM SHEAR-DIAGRAM AREA

A mathematical relation exists between a bending moment at a given section in a statically determinate beam and the area of the shear diagram from the end of the beam to that section. The expression is

$$M_x = A_{0-x} \qquad\qquad (11\text{-}2)$$

where $A_{0-x}$ = shear-diagram area between left end of beam and section
$x$. Since the shear diagram is a plot of shear force $V$ in
pounds (or newtons), and beam length in feet (or meters),
areas on this diagram have units of foot pounds (or newton-
meters)

$\qquad M_x$ = bending moment (ft·lb or N·m) at section $x$

Equation (11-2) is a specific application of a more general relationship
between successive diagrams. In general, the *area* between any two vertical
heights (ordinates) on a diagram will equal the *difference in length* of the
two corresponding vertical heights (ordinates) on the next higher degree
diagram.

To demonstrate the use of Eq. (11-2), consider a simply supported
weightless beam with a central concentrated load of 900 lb on a 12-ft
span. Suppose the moment is required at the section 4 ft from the left
end. Then the shaded area on the shear diagram of Fig. 11-26 will equal
the value of the moment $M_4$.

$$M_4 = +450(4) = +1800 \text{ ft·lb}$$

Note that shear-diagram areas above the zero line are positive, while areas
below the zero line are negative.

By the same method, the maximum bending moment at the center
section (shear diagram crosses zero) is

$$M_6 = M_{\max} = +450(6) = +2700 \text{ ft·lb}$$

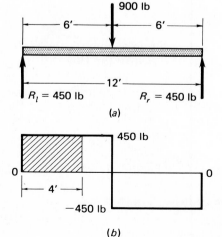

FIGURE 11-26   (a) Beam diagram.
(b) Shear-force diagram. Cross-
hatched area equals bending moment
at 4-ft section.

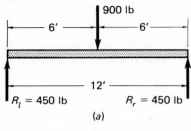

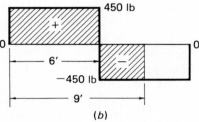

FIGURE 11-27 (a) Beam diagram. (b) Shear-force diagram. Algebraic sum of cross-hatched areas equals bending moment at 9-ft section.

Suppose it is required to find the moment for the section 9 ft from the left end of this beam. From Eq. (11-2),

$$M_9 = A_{0-9}$$

Thus, the shear-diagram area from the left end to the 9-ft section equals $M_9$. Figure 11-27 shows that some of the shaded area $A_{0-9}$ is plus while some of it is minus. In totaling the areas, plus and minus signs must be considered; therefore,

$$M_9 = [+450(6)] + [-450(3)] = +2700 - 1350 = +1350 \text{ ft} \cdot \text{lb}$$

Although these examples are fairly simple, the method is similar regardless of the shape and complexity of the shear diagram. Bending moments determined by the shear-diagram-area method correctly represent the effect of the beam loading from which the shear diagram was sketched. However, if the beam is subjected to an external moment (for example, by a couple acting at the end of the beam), then the true bending moment is determined by algebraically adding the external moment to the shear-diagram area.

**Sample Problem 13** Determine the shear-force and bending-moment diagrams for the beam shown in (a) Fig. 11-28 and in (b) Fig. 11-29.

**Solution a:** This simply supported beam is solved in the same way as in Sample Problems 8, 9, and 12.

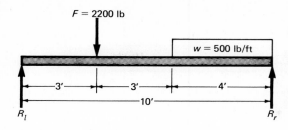

FIGURE 11-28   Beam diagram for Sample Problem 13a.

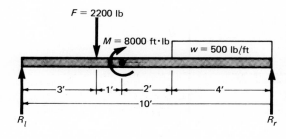

FIGURE 11-29   Beam diagram for Sample Problem 13b.

$\Sigma M_r = 0$

$$R_l(10) - F(7) - w(4)(2) = 0$$

$$R_l = \frac{2200(7) + 500(4)2}{10} = 1940 \text{ lb (up)}$$

$\Sigma F_y = 0$

$$R_l + R_r - F - w(4) = 0$$

$$R_r = 2200 + 500(4) - 1940 = 2260 \text{ lb (up)}$$

**Solution b:**   The beam shown in Fig. 11-29 is identical to the one in Fig. 11-28 except that a couple is applied which adds an external moment at a section 4 ft from the left support. Note that both the shear and moment diagrams are affected by the addition of the couple.

$\Sigma M_r = 0$

$$R_l(10) - F(7) + M - w(4)(2) = 0$$
$$R_l(10) - 2200(7) + 8000 - 2000(2) = 0$$
$$R_l = \frac{2200(7) - 8000 + 2000(2)}{10} = 1140 \text{ lb (up)}$$

$\Sigma F_y = 0$

$$R_l + R_r - 2200 - 2000 = 0$$

$$R_r = 4200 - 1140 = 3060 \text{ lb (up)}$$

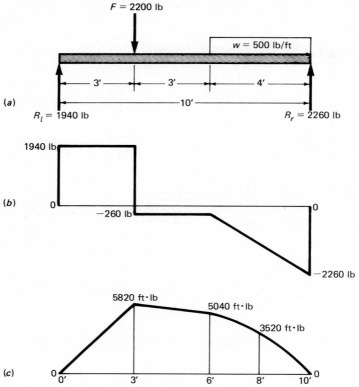

FIGURE 11-30   (a) Beam diagram for Sample Problem 13a.
(b) Shear-force diagram. (c) Moment diagram.

**\*Sample Problem 14**   Determine the shear-force and bending-moment diagrams for the beam shown in (a) Fig. 11-32 and in (b) Fig. 11-33.

**Solution a:**   This cantilever beam is solved in the same way as in Sample Problems 10 and 11.

$$\Sigma M_r = 0$$

$$-F(8) - w(4)(2) + M = 0$$

$$M = 5(8) + 1(4)(2) = 48 \text{ kN} \cdot \text{m}   \text{(clockwise)}$$

$$\Sigma F_y = 0$$

$$-F - w(4) + R_r = 0$$

$$R_r = 5 + 1(4) = 9 \text{ kN}   \text{(up)}$$

**Solution b:**   The beam shown in Fig. 11-33 is identical to the one in Fig. 11-32, except that a couple is applied which adds an external moment at

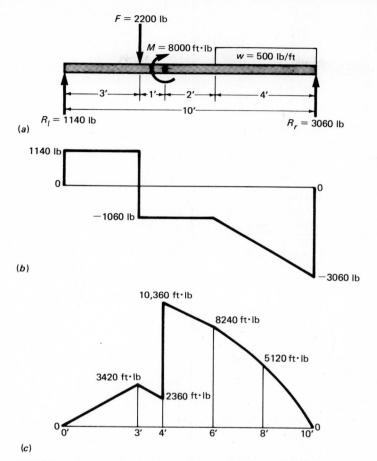

FIGURE 11-31   (a) Beam diagram for Sample Problem 13b.
(b) Shear-force diagram. (c) Moment diagram.

a section 2 m from the left end. Note that since there is only one reaction $R_r$, its value remains the same in both cases and therefore the shear-force diagram is unchanged. The moment diagram, however, is directly affected by the addition of the couple.

$$\Sigma M_r = 0$$

$$-F(8) + M_1 - w(4)(2) + M_2 = 0$$

$$M_2 = 5(8) - 15 + 1(4)(2)$$

$$M_2 = 40 - 15 + 8 = 33 \text{ kN} \cdot \text{m} \quad \text{(clockwise)}$$

$$\Sigma F_y = 0$$

$$-F - w(4) + R_r = 0$$

$$R_r = 5 + 1(4) = 9 \text{ kN} \quad \text{(up)}$$

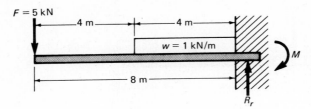

FIGURE 11-32   Beam diagram for Sample
Problem 14a.

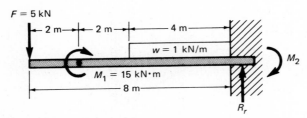

FIGURE 11-33   Beam diagram for Sample
Problem 14b.

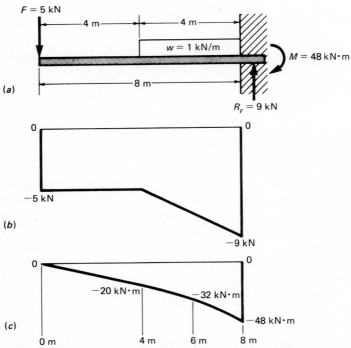

FIGURE 11-34   (a) Beam diagram for Sample Problem 14a.
(b) Shear-force diagram. (c) Moment diagram.

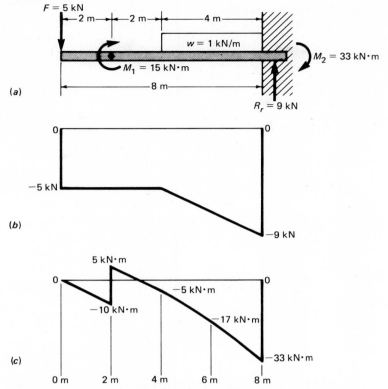

FIGURE 11-35   (a) Beam diagram for Sample Problem 14b.
(b) Shear-force diagram. (c) Moment diagram.

## 11-7 MOVING LOADS

In calculating shear forces and bending moments, the assumption has
been made that the loads on the beam were in a fixed position. There are
many cases where this is not true. A loaded truck or train passing over
a bridge and a heavy machine being moved over a floor are examples of
moving loads transmitted through wheels which are a fixed distance from
each other. It is evident that the magnitude of the shear forces and
bending moments will change as the load system moves across a beam.
The problem then is to find the magnitude and location of the maximum
shear force and bending moment. The following discussion will deal with
simple beams only.

(a) *Maximum Shear:*   For simple beams, the maximum shear load
($V_{max}$) occurs at and is equal to the maximum reaction support force.
Since the magnitude of the reactions will change as the loads move across
the beam, it will be necessary to consider the loads at different locations

in order to establish $V_{max}$. This is accomplished by alternately placing each load over a reaction and calculating the magnitude of that reaction. It should be kept in mind that the distance between the loads must not change and that the loads can move in either direction.

**(b) Maximum Bending Moment:** To establish the maximum bending moment ($M_{max}$) which develops due to the moving loads, the maximum moment under each load is calculated and the greatest of all these values is $M_{max}$. The following discussion will develop the theory of this problem, and, although only three loads will be considered, the conclusions drawn will be general and can be applied to any number of loads.

Fig. 11-36 shows a simple beam subjected to three moving loads ($F_1$, $F_2$, $F_3$). $F$ is the resultant of the loads.

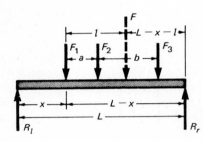

FIGURE 11-36 Moving loads on a simply supported beam.

First, determine the reaction $R_l$.

$$\Sigma M_r = 0$$

$$R_l(L) - F(L - x - l) = 0$$

$$R_l = \frac{F(L - x - l)}{L}$$

For bending moment under $F_1$, consider the section of the beam to the left of $F_1$.

$$M_x = R_l x$$

$$= \frac{F(Lx - x^2 - lx)}{L}$$

$$= -\frac{F}{L}[x^2 - (L - l)x]$$

But bending moment is a maximum when the expression in the brackets is greatest, since $F$ is constant. Now, in algebra, it is shown that the quadratic expression $Ax^2 + Bx + C$ is greatest or least when

$$x = -\frac{B}{2A}$$

In $x^2 - (L - l)\,x$, $A = 1$, $B = -(L - l)$, and $C = 0$

Then $x = (L - l)/2$ gives a maximum or a minimum. Since the minimum moment at the supports is zero, then the preceding value of $x$ is for the maximum moment. The distance from $R_l$ to $F$ is

$$x + l = \frac{L - l}{2} + l = \frac{L + l}{2}$$

It is now evident that the distance from $R_l$ to the resultant $F$ is $(L + l)/2$. That is, the center of the beam is midway between the load $F_1$ and the resultant $F$ of all the loads.

The rule may be stated as follows: *The maximum bending moment under any load of a set of moving concentrated loads occurs when that load is as far on one side from the center of the beam as the resultant of all the loads on the beam is on the other side.* This rule applies to simple beams only.

To determine the maximum bending moment for any possible placement of loads, the above calculation should be made for each of the moving loads placed in its maximum bending-moment position. The bending moment used for beam design will be the largest of the maximum bending moments.

**\*Sample Problem 15**  Two loads of 40 kN and 20 kN, 3 m apart, roll over a beam 10 m long. Find (*a*) the maximum shear load ($V_{max}$) and (*b*) the maximum bending moment ($M_{max}$).

**Solution:**  (*a*) Figs. 11-37 and 11-38 show the position of the loads in order to develop maximum shear at each of the reactions. $V_{max}$ will occur at one of the two reactions.

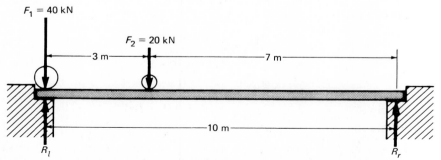

FIGURE 11-37  Beam diagram for Sample Problem 15. Load positions arranged for maximum shear at $R_l$.

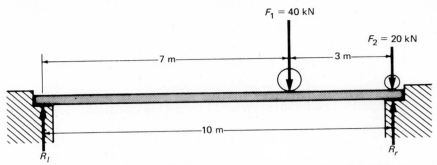

FIGURE 11-38  Beam diagram for Sample Problem 15. Load positions arranged for maximum shear at $R_r$.

From Fig. 11-37:  $R_l = 40 + \dfrac{7}{10}(20) = 40 + 14 = 54 \text{ kN}$

From Fig. 11-38:  $R_r = 20 + \dfrac{7}{10}(40) = 20 + 28 = 48 \text{ kN}$

Therefore,  $V_{\max} = 54 \text{ kN}$

(b) Referring to Fig. 11-39, find the position of the resultant with reference to $F_1$ by taking moments about $F_1$.

$$F_2(3) - F(l) = 0$$
$$20(3) - 60(l) = 0$$
$$60 = 60l$$
$$l = 1 \text{ m}$$

Then, for maximum moment,

$$x = \frac{L - l}{2} = \frac{10 - 1}{2} = 4.5 \text{ m}$$

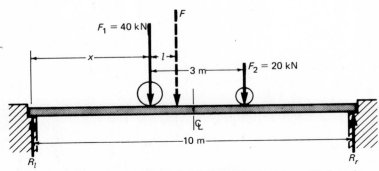

FIGURE 11-39  Beam diagram for Sample Problem 15. Loads and resultant arranged in a general position.

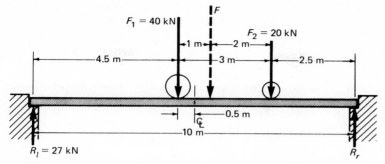

FIGURE 11-40   Beam diagram for Sample Problem 15. Load positions arranged for maximum moment under $F_1$.

or $F_1$ is 0.5 m to the left of the center of the beam, and the resultant $F$ is 0.5 m to the right.

Figure 11-40 shows the location of the loads with $F_1$ in its maximum bending-moment position.

$$\Sigma M_r = 0$$

$$R_l(10) - 40(5.5) - 20(2.5) = 0$$
$$10R_l = 270$$
$$R_l = 27 \text{ kN}$$

Find the moment at the section under $F_1$ (40-kN load).

$$M_1 = 27(4.5) = 121.5 \text{ kN} \cdot \text{m}$$

$M_1$ is the maximum moment that is developed under the 40-kN load. Since $F$ is 2 m from $F_2$, the maximum moment under $F_2$ load occurs when $F$ and $F_2$ are each 1 m from the center of the beam, as shown in Fig. 11-41.

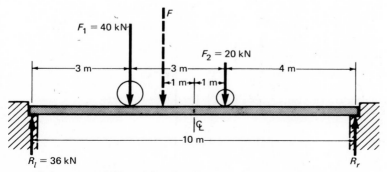

FIGURE 11-41   Beam diagram for Sample Problem 15. Load positions arranged for maximum moment under $F_2$.

$$\Sigma M_r = 0$$

$$R_l(10) - 40(7) - 20(4) = 0$$
$$10R_l = 360$$
$$R_l = 36 \text{ kN}$$

Next, take a section of the beam at $F_2$ (20-kN load), and find moments of the load to the left of this section.

$$M_2 = 36(6) - 40(3) = 216 - 120 = 96 \text{ kN} \cdot \text{m}$$

This is the maximum moment developed under the 20-kN load. Since the maximum moment under $F_1$ is greater than that under $F_2$, 121.5 kN·m is the maximum for the beam.

**Sample Problem 16**   Three loads of 6000, 9000, and 5000 lb are placed 6 and 10 ft apart, respectively (Fig. 11-42). They are moved across a bridge of span 40 ft. Find the maximum bending moment.

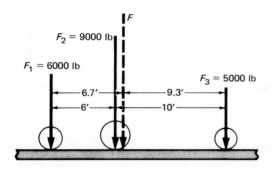

FIGURE 11-42   Load diagram for Sample Problem 16.

**Solution:**   First, locate the resultant $F$ with reference to $F_1$.

$$20\,000l = 6(9000) + 16(5000)$$
$$l = 6.7 \text{ ft}$$

The free-body diagrams are shown for each load in the position for maximum moments (Figs. 11-43, 11-44, and 11-45), and results given. The work of verifying results is left to the reader.

$$M_1 = 16.65(8320) = 138\,500 \text{ ft} \cdot \text{lb}$$
$$M_2 = 19.65(9820) - 6(6000) = 157\,000 \text{ ft} \cdot \text{lb}$$
$$M_3 = 24.65(12\,320) - 16(6000) - 10(9000) = 117\,700 \text{ ft} \cdot \text{lb}$$

Therefore, $M = 157\,000$ ft·lb is the maximum bending moment.

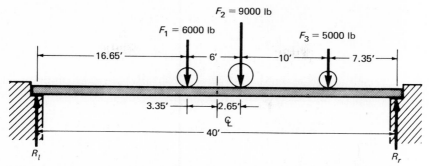

FIGURE 11-43 Beam diagram for Sample Problem 16. Maximum moment at $F_1$.

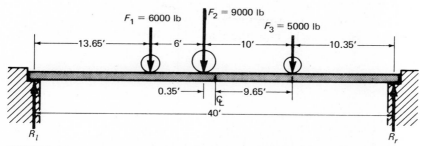

FIGURE 11-44 Beam diagram for Sample Problem 16. Maximum moment at $F_2$.

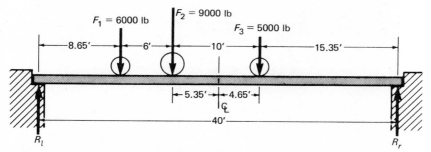

FIGURE 11-45 Beam diagram for Sample Problem 16. Maximum moment at $F_3$.

## PROBLEMS

Calculate the unknown reactions. Sketch and label the shear-force diagram for Probs. 11-1 to 11-10.

**11-1.** A simply supported beam with a 10-ft span carries a concentrated load of 2000 lb acting 4 ft from the left support. Neglect the weight of the beam.

**\*11-2.** The beam in Fig. Prob. 11-2.

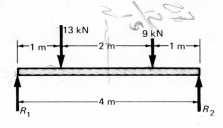

FIGURE PROBLEM 11-2 and 11-12

**\*11-3.** A simply supported beam, 5 m long, carries a uniformly distributed load of 1.5 kN/m, including the weight of the beam and a concentrated load of 15 kN, 1.5 m from the right end.

**11-4.** The beam in Fig. Prob. 11-4.

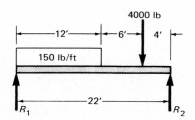

FIGURE PROBLEM 11-4 and 11-14

**\*11-5.** A 3.5-m-long cantilever beam of negligible weight carries concentrated loads of 10 and 8 kN, acting at 1.5 and 3.5 m, respectively, from the wall.

**11-6.** The beam in Fig. Prob. 11-6.

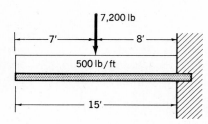

FIGURE PROBLEM 11-6 and 11-16

**11-7.** The beam in Fig. Prob. 11-7.
**\*11-8.** The beam in Fig. Prob. 11-8.
**\*11-9.** The beam in Fig. Prob. 11-9.
**11-10.** The beam in Fig. Prob. 11-10.

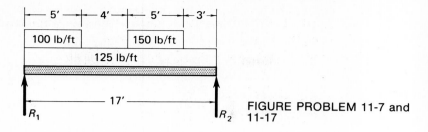

FIGURE PROBLEM 11-7 and 11-17

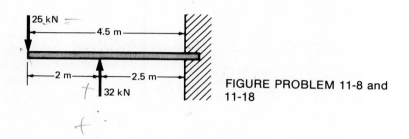

FIGURE PROBLEM 11-8 and 11-18

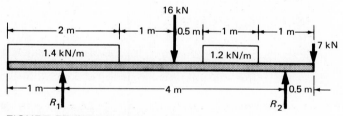

FIGURE PROBLEM 11-9 and 11-19

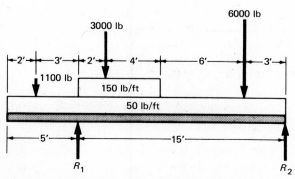

FIGURE PROBLEM 11-10 and 11-20

Determine the location and magnitude of the maximum bending moment and sketch the bending-moment diagram for the beams in Probs. 11-11 to 11-20.

**11-11.** The beam in Fig. Prob. 11-1.
**11-12.** The beam in Fig. Prob. 11-12.
**11-13.** The beam in Fig. Prob. 11-3.
**11-14.** The beam in Fig. Prob. 11-14.
**11-15.** The beam in Fig. Prob. 11-5.
**11-16.** The beam in Fig. Prob. 11-16.
**11-17.** The beam in Fig. Prob. 11-17.
**11-18.** The beam in Fig. Prob. 11-18.
**11-19.** The beam in Fig. Prob. 11-19.
**11-20.** The beam in Fig. Prob. 11-20.
**11-21.** For the beam in Fig. Prob. 11-21, determine the following.
    *a.* The reactions in terms of $P$ and $L$
    *b.* The location and magnitude of the maximum shear force
    *c.* The magnitude of the maximum bending moment
    *d.* The magnitude of the bending moment at section $x$ ft from $R_1$

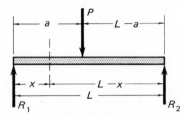

FIGURE PROBLEM 11-21

**\*11-22.** Determine the unknown reactions, sketch, and label the shear-force diagram and the bending-moment diagram for the beam in Prob. 11-12, except that there is an external clockwise couple of 20 kN·m applied at midspan.

**11-23.** Determine the unknown reactions, sketch, and label the shear-force diagram and the bending-moment diagram for the beam in Prob. 11-16, except that there is an external counterclockwise couple of 25 000 ft·lb applied at a point 4 ft from the wall.

**11-24.** A two-axle steam roller with axles 10 ft apart crosses a 25-ft simply supported beam bridge. If the load on each axle is 16 000 lb, determine the position of the wheels to produce (*a*) the maximum shear load, and (*b*) the maximum bending moment. Calculate the value and location of $V_{max}$ and $M_{max}$.

**\*11-25.** A three-axle truck crosses a 15-m simply supported span. The front and center axles are spaced 3 m apart and the center and rear axles are 4.2 m apart. The front axle carries 36 kN, the center 81 kN, and the rear 153 kN. Calculate the maximum bending moment as this truck rolls across the span from left to right and specify the axle positions for the maximum moment.

# CHAPTER
# 12
# Beams—Design

## 12-1 STRESS DUE TO BENDING

In Sec. 11-2, it was shown that the fiber stresses over any cross section of a beam were a maximum at the extreme outer fibers, decreasing to zero at the netural axis.

The formula for the stress in a beam due to bending will be derived so that it will be applicable to the cross section of beams of any shape. For simplicity, a rectangular section will be used. The result will be true, in general, within the elastic range, as proved in more advanced works on strength of materials.

Let Fig. 12-1 be the section of a rectangular beam. $XX$ is the centroidal axis. Take a thin rectangle of area $a$ whose centroid is at the distance $y_a$ from $XX$. Let $s_a$ be the stress on area $a$. Then the force $F_a$ on area $a$ is $F_a = s_a a$. The moment of $F_a$ about $XX$ is

$$F_a y_a = s_a a y_a$$

But the fiber stress is 0 at the centroidal axis and increases in proportion to the distance $y$ to a maximum value $s$ at 1-2, the extreme fiber where $y = c$. Then,

$$\frac{s_a}{y_a} = \frac{s}{c}$$

or

$$s_a = \frac{s y_a}{c}$$

Substituting in the equation for the moment, above, we obtain

$$F_a y_a = \frac{s y_a}{c} a y_a$$

$$= \frac{s}{c} y_a{}^2 a$$

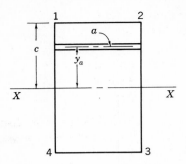

FIGURE 12-1   Section of a rectangular beam.

If we divide the remaining cross-sectional area into a large number of very thin areas similar to $a$, the bending moment for each may be represented by a formula as above. Since these moments are referred to a common axis $XX$, they may be added. Thus the total moment $M$ is

$$M = \Sigma\left(\frac{s}{c}y^2a\right)$$

But $s/c$ is the same for every term in the summation; therefore, $s/c$ may be factored, giving

$$M = \frac{s}{c}\Sigma y^2a$$

From Sec. 10-7, $\Sigma y^2a = I$, the moment of inertia about axis $XX$. In Sec. 11-2, the moment of the resisting couple was shown to be equal to the bending moment of the external loads to the left of the section. Therefore,

$$M = \frac{s}{c}I$$

or
$$s = \frac{Mc}{I} \tag{12-1}$$

The expression $I/c$ is called the *section modulus S*. Equation (12-1) may then be written

$$s = \frac{Mc}{I} = \frac{M}{I/c}$$

$$s = \frac{M}{S} \tag{12-2}$$

where $s$ = maximum fiber stress due to bending, psi; Pa
    (usually expressed as $10^6$ Pa or MPa)
$M$ = bending moment, in·lb; N·m
$c$ = distance from neutral axis to the extreme fiber, in; m
$I$ = moment of inertia about neutral axis, in⁴; m⁴
$S$ = section modulus, in³; m³

For a given cross section, the moment of inertia $I$ and the distance $c$ depend only on the size, shape, and axis location. Therefore, the section modulus $S = I/c$ also depends on these factors.

Expressions for $I$ and $S$ for a variety of common cross sections and structural shapes are given in App. B.

The equation for stress due to bending [Eq. (12-1)] applies only within the elastic range of the material, since proportionality of stress and strain was used in its derivation. An index of the rupture strength of the material in bending can be found by using the moment at the breaking point $M_r$ in the flexure equation Eq. (12-1). Thus,

$$s_r = \frac{M_r c}{I} \qquad (12\text{-}3)$$

where $s_r$ is a numerical index of rupture strength called *modulus of rupture*. While $s_r$ is useful for comparative purposes, it is not the true ultimate bending stress.

**Sample Problem 1** An 8- by 10-in dressed-size timber beam carries a uniformly distributed load on a simply supported 14-ft span. Assume an allowable bending stress of 1100 psi. Find the maximum safe distributed load (pounds per foot).

**Solution:** The American Standard dressed size for a nominal 8 by 10 in is $7\frac{1}{2}$ by $9\frac{1}{2}$ in (see App. B, Table 11, for American Standard Timber Sizes).

Assume the cross section to be placed for maximum strength; that is, maximum moment of inertia about the neutral axis. Figure 12-2a shows the cross section placed for maximum strength, while Fig. 12-2b shows the same cross section in a weaker position. Refer to Fig. 11-21, which shows that the maximum bending moment occurs at the center of the beam. The maximum stress occurs where the bending moment is maximum. Therefore, the maximum bending moment for a simply supported beam with a uniformly distributed load across the entire span, as shown in Fig. 12-3, is

$$M_{\max} = R_l\left(\frac{L}{2}\right) - \frac{W}{2}\left(\frac{L}{4}\right) = \frac{WL}{4} - \frac{WL}{8} = \frac{WL}{8} = \frac{wL^2}{8}$$

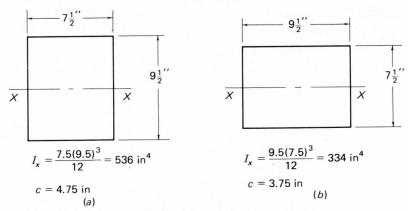

$$I_x = \frac{7.5(9.5)^3}{12} = 536 \text{ in}^4$$

$$c = 4.75 \text{ in}$$

(a)

$$I_x = \frac{9.5(7.5)^3}{12} = 334 \text{ in}^4$$

$$c = 3.75 \text{ in}$$

(b)

FIGURE 12-2 Diagram for Sample Problem 1: (a) cross section placed for maximum bending strength; (b) cross section placed for minimum bending strength.

For the particular beam in Fig. 12-3,

$$M_{max} = \frac{W(14)}{8} = 1.75W \text{ ft} \cdot \text{lb}$$

or

$$M_{max} = 1.75(12)W = 21W \text{ in} \cdot \text{lb}$$

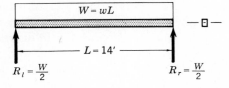

FIGURE 12-3 Beam diagram for Sample Problem 1.

From Eq. (12-1),

$$s = \frac{Mc}{I} \qquad 1100 = \frac{21W(4.75)}{536}$$

$$W = \frac{1100(536)}{21(4.75)} = 5910 \text{ lb}$$

$$w = \frac{W}{L} = \frac{5910}{14} = 422 \text{ lb/ft}$$

(maximum safe load, including beam weight)

**Sample Problem 2** A run of 4-in schedule 40 seamless steel pipe (4.50 in OD, 0.237 in wall thickness) is to carry a ¼-ton-capacity chain hoist attached midway between pipe support hangers. The ultimate tensile

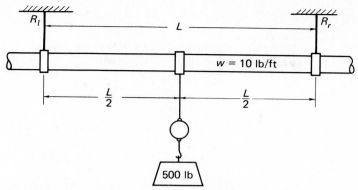

FIGURE 12-4   Diagram for Sample Problem 2.

strength of the steel pipe is 48 000 psi. A safety factor of 4 is specified. The pipe weighs 10 lb/ft. Assume no additional stress or load due to internal pressure. Treat the length of pipe between hangers as a simply supported beam. Find the maximum safe spacing of pipe support hangers.

**Solution:**   Sketch the system as in Fig. 12-4.

$\Sigma F_y = 0$

$$R_l + R_r - 10L - 500 = 0$$
$$R_l = R_r = 5L + 250 \text{ (due to symmetry)}$$

Find $M_{max}$ at center of span.

$$M_{max} = R_l\left(\frac{L}{2}\right) - \frac{10L}{2}\left(\frac{L}{4}\right)$$
$$= (5L + 250)\frac{L}{2} - \frac{5L^2}{4}$$
$$= \frac{5L^2}{2} + 125L - \frac{5L^2}{4} = \frac{5L^2}{4} + 125L \text{ ft·lb}$$
$$= 15L^2 + 1500L \text{ in·lb}$$

For a hollow circular section (Table 12, App. B),

$$I = \frac{\pi}{64}(d_0{}^4 - d_i{}^4) = \frac{\pi}{64}(4.50^4 - 4.03^4)$$
$$= \frac{\pi}{64}(410 - 262) = \frac{\pi}{64}(148) = 7.26 \text{ in}^4$$
$$c = \frac{4.50}{2} = 2.25 \text{ in}$$

Allowable stress = 48 000/4 = 12 000 psi. From Eq. (12-1),

$$s = \frac{Mc}{I} \qquad M = \frac{sI}{c}$$

$$M = \frac{12\,000\,(7.26)}{2.25} = 38\,700 \text{ in} \cdot \text{lb}$$

$$15L^2 + 1500L = 38\,700$$
$$L^2 + 100L - 2580 = 0$$

This quadratic equation may be solved by quadratic formula.

$$L = \frac{-100 \pm \sqrt{100^2 - 4(1)(-2580)}}{2} = \frac{-100 \pm \sqrt{10\,000 + 10\,320}}{2}$$

$$= \frac{-100 \pm \sqrt{20\,320}}{2} = \frac{-100 \pm 143}{2}$$

Selecting the positive result, we obtain

$$L = \frac{-100 + 143}{2} = \frac{43}{2} = 21.5 \text{ ft (maximum safe spacing of supports)}$$

**Sample Problem 3**   For the beam shown in Fig. 12-5, find the following.

(a)   The maximum compressive stress in the beam
(b)   The maximum tensile stress in the beam
(c)   The factors of safety in tension and compression based on the ultimate stresses if the material is AISI 1020 steel

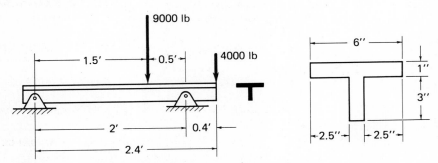

FIGURE 12-5   Beam diagram for Sample Problem 3.

*Note:* In this problem, the weight of the beam will be neglected. As a rule of thumb, if the total beam weight is less than 10 percent of the total load, the weight of the beam may be neglected without serious error. The density of steel is about 0.30 lb/in³. The volume of this beam is

$$V = [(1)(6) + (1)(3)](2.4)(12) = 9(2.4)(12)$$
$$V = 259.2 \text{ in}^3 \qquad \text{say } V = 260 \text{ in}^3$$
$$W = 260(0.30) = 78 \text{ lb (weight of beam)}$$

But
$$9000 + 4000 = 13\,000 \text{ lb (total load)}$$

Therefore, neglect weight of beam.

**Solution a:**    To determine stress, the centroid and the moment of inertia of the T section must be found.

By symmetry, the centroid is located somewhere on the vertical centerline (see Fig. 12-6). The distance $\bar{y}$ from the base of the T is found.

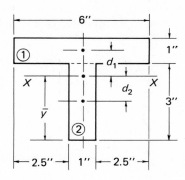

FIGURE 12-6    Diagram for Sample Problem 3. Location of centroid of cross section.

| Code | Dimen. | $A$ | $y$ | $Ay$ |
|------|--------|-----|-----|------|
| 1 | $1 \times 6$ | 6 | 3.5 | 21 |
| 2 | $1 \times 3$ | 3 | 1.5 | 4.5 |
|   |   | $\Sigma A = 9$ |   | $25.5 = \Sigma Ay$ |

$$\bar{y} = \frac{\Sigma Ay}{\Sigma A} = \frac{25.5}{9} = 2.83 \text{ in}$$

The moment of inertia about the horizontal centroidal axis is found as follows:

$$I_x = (I + Ad^2)_1 + (I + Ad^2)_2$$
$$d_1 = y_1 - \bar{y} = 3.5 - 2.83 = 0.67 \text{ in}$$
$$d_2 = \bar{y} - y_2 = 2.83 - 1.5 = 1.33 \text{ in}$$
$$I_x = \left[\frac{6(1)^3}{12} + 6(0.67)^2\right] + \left[\frac{1(3)^3}{12} + 3(1.33)^2\right]$$
$$= (0.5 + 2.68) + (2.25 + 5.33) = 3.18 + 7.58$$
$$= 10.76 \text{ in}^4$$

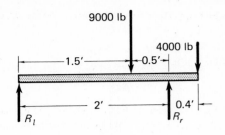

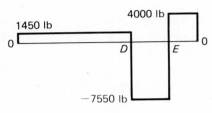

FIGURE 12-7 (a) Beam diagram for Sample Problem 3. (b) Shear-force diagram.

The distances to the extreme fibers are

$$c_1 = 4 - 2.83 = 1.17 \text{ in (to top)}$$
$$c_2 = 2.83 \text{ in (to base)}$$

To determine the maximum bending motion, calculate the reactions $R_l$ and $R_r$, and sketch the shear diagram as in Fig. 12-7.

$$\Sigma F_y = 0$$
$$R_l + R_r - 9000 - 4000 = 0$$
$$R_l + R_r = 13\,000 \text{ lb}$$

$$\Sigma M_l = 0$$
$$-R_r(2) + 9000(1.5) + 4000(2.4) = 0$$
$$R_r = \frac{23\,100}{2} = 11\,550 \text{ lb}$$
$$R_l = 1450 \text{ lb}$$

From the shear diagram, it is apparent that sections $D$ and $E$ must be investigated for maximum bending moment. Moments at $D$ and $E$ may be calculated either by taking moments of all forces to the left of the sections (see Sec. 11-4 and Sample Problem 12, Chap. 11) or by the shear-diagram-area method (see Sec. 11-6). By the shear-diagram-area method,

$$M_d = 1450(1.5) = 2180 \text{ ft·lb}$$
$$M_e = 1450(1.5) - (7550)(0.5) = 2180 - 3780$$
$$= -1600 \text{ ft·lb}$$

Therefore, maximum bending moment $= M_d = 2180$ ft·lb.

From Eq. (12-1), the maximum compressive stress in the beam occurs at the top fiber of section $D$ and is

$$s_c = \frac{Mc_1}{I} = \frac{2180(12)(1.17)}{10.76} = 2840 \text{ psi}$$

**Solution b:** The maximum tensile stress in the beam occurs at the bottom fiber of section $D$ and is

$$s_t = \frac{Mc_2}{I} = \frac{2180(12)(2.83)}{10.76} = 6880 \text{ psi}$$

**Solution c:** From App. B, Table 1, the ultimate tensile and compressive strengths for AISI 1020 steel are both equal to 65 000 psi.

$$N_c = \frac{65\,000}{2840} = 22.9$$

$$N_t = \frac{65\,000}{6880} = 9.4$$

**Sample Problem 4** A cantilever beam carries a uniform load of 400 lb/ft across the entire 9-ft length of span and a concentrated load of 3000 lb at the free end. For a maximum allowable stress of 22 000 psi, determine the most economical beam (lightest beam that satisfies the strength requirement), assuming that the section is:

    (*a*)   A W beam (a wide-flanged I-shaped beam)
    (*b*)   An S beam (a standard I-shaped beam)
    (*c*)   A standard channel

**Solution:** A cantilever beam has its maximum bending moment at the wall. Figure 12-8 shows this beam with its shear-force diagram. The solution will omit the weight of beam until a selection is made, at which time the weight of beam will be compared to the total loading of 6600 lb for significance.

The maximum bending moment is

$$M = -3000(9) - 3600(4.5) = -27\,000 - 16\,200$$
$$= -43\,200 \text{ ft} \cdot \text{lb}$$

From Eq. (12-2),

$$s = \frac{M}{S}$$

$$S = \frac{M}{s} = \frac{43\,200(12)}{22\,000} = 23.56 \text{ in}^3 \text{ (minimum required)}$$

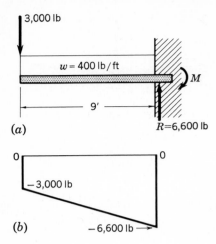

(a)

(b)

FIGURE 12-8  (a) Beam diagram for Sample Problem 4. (b) Shear-force diagram.

For strength, the beam selected must have a section modulus greater than 23.56 in³. The most economical beam of a given type is selected for its light weight (therefore, low cost).

From the Section Modulus Table (App. B, Table 9):

(a)  W 10 × 25 (actual $S$ = 26.5 in³)
(b)  S 10 × 25.4 (actual S = 24.7 in³)
(c)  C 12 × 25 (actual S = 24.1 in³)

Note that in each case the total beam weight is less than 5 percent of the load. If the beam weight were included in the calculations it would increase both $M$ and $S$ only 2 percent.

## 12-2 HORIZONTAL AND VERTICAL SHEAR STRESSES

Shear has been defined as the tendency of two adjacent portions of a body to slide by each other, like the two blades of a pair of scissors. The subject of vertical shear in beams has already been considered in Sec. 11-2. There is also a shearing tendency in a plane at right angles to the vertical shearing plane. This is called *horizontal shear*.

The student will get the best idea of horizontal shear from a simple illustration. Place several 12- by 1-in boards on two supports, as shown in Fig. 12-9a. When the boards are of the same length, the ends will be practically straight and even. If now a load $F$ is applied at the center of the pile, the boards bend or sag in the middle, and the result is as shown in Fig. 12-9b. Each board tends to slide on the one above or below it and in this way moves the ends from their original positions. The horizontal motion of one board over the other is what causes horizontal shear.

Now, if each board is glued securely to the ones next to it and a small load applied, the ends of the pile of boards remain square and even.

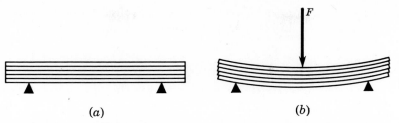

FIGURE 12-9  (a) Boards before load is applied. (b) Boards after load is applied.

There is still a tendency to slide, but the motion is prevented by the glue. This resistance measures the horizontal shear. Again, if the boards are securely bolted together and a load is applied, the boards cannot slip. The shearing resistance is offered by the bolts.

When the pile of boards is replaced by a single beam 12 by 6 in and a load $F$ is applied, the beam will bend slightly. There is still the tendency of one horizontal portion to slide past the adjacent face. However, if the beam is not overloaded, the sliding motion is prevented by the internal fibers of the beam, just as it was resisted by the glue or bolts. The resistance that the fibers are capable of offering is called their *horizontal shearing strength*.

Some fibrous materials, such as wood and wrought iron, are more likely to fail from horizontal shear than those materials that have no natural internal cleavage surfaces. Timber has a low shearing strength parallel to the grain; hence, a short beam tends to fail from horizontal shear rather than from bending.

Figure 12-10 represents a small cube cut from any part of a loaded beam.

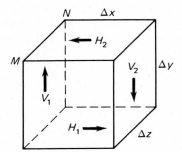

FIGURE 12-10  Small cube of material from a beam with load.

Let $V_1$ and $V_2$ be the vertical shearing forces and $H_1$ and $H_2$ the horizontal shearing forces. The dimension $\Delta x$ is very small, so that the effect of the load so far as vertical shear on the two faces is concerned may be neglected and $V_1$ may be considered equal to $V_2$. Similarly,

$$H_1 = H_2$$

Let $s_v$ be the vertical shear stress and $s_h$ the horizontal shear stress. Then

$$V_1 = \Delta y \Delta z s_v = V_2$$

and

$$H_1 = \Delta x \Delta z s_h = H_2$$

Since the block is in equilibrium under the shearing forces, moments about the line $MN$ must equal zero, or

$$\Delta y H_1 = \Delta x V_2$$

$$\Delta x \Delta y \Delta z s_h = \Delta x \Delta y \Delta z s_v$$

Therefore,     $s_h = s_v$

In Chap. 7, it was assumed that the shearing stress was of the same intensity on each square inch of the cross section. In the derivations and the solutions that follow, it will be shown that the shearing stress in a beam of symmetrical section is a maximum at the netural axis and decreases to zero at the outer fibers.

The relationship between the stresses just developed may be stated as follows: *At any point in a member subjected to shearing forces, there exists equal shearing stresses in planes mutually at right angles to each other.* The subscripts may be dropped and the equal unit shearing stresses may be designated as $s_s$.

Imagine the segment of the beam in Fig. 12-11a between planes $AB$ and $CD$ to be removed as a free body. The enlarged view of this free body is shown in Fig. 12-11b with the distribution of stresses due to bending. The stresses on section $CD$ are greater than the corresponding stresses on section $AB$ because the bending moment at $CD$ is greater than at $AB$.

In order to develop an equation for the shear stress on any horizontal plane, let us consider a plane such as $KL$ in Fig. 12-11b. In Sec. 11-4, it was seen that, in general, bending moments vary from one end of the beam to the other. Let $F_1$ be the force due to the compressive stresses on surface $AK$ (whose thickness is $b$), and $F_2$ that on surface $CL$, as in Fig. 12-11c.

Since $F_1$ and $F_2$ are different, they must be balanced by a force $H$ on the surface $KL$. This is the horizontal shear force on that surface. Since $F_2$ is greater than $F_1$, for equilibrium,

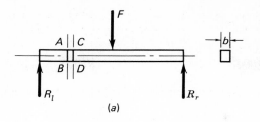

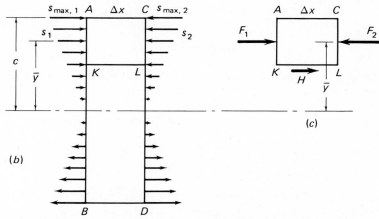

FIGURE 12-11  Horizontal shear forces in a beam: (a) beam diagram; (b) bending stresses on a beam segment; (c) bending and horizontal shear forces on a portion of a beam segment.

$$H = F_2 - F_1$$

Let $s_1$ and $s_2$ be the fiber stresses at the centroids of the faces $AK$ and $CL$, respectively, and $a$ the area of each of the faces.

$$a = b(CL) = b(AK)$$

Then
$$F_1 = s_1 a \qquad \text{and} \qquad F_2 = s_2 a$$

From Sec. 12-1,

$$\frac{s_1}{\bar{y}} = \frac{s_{\max,1}}{c} \qquad \text{and} \qquad \frac{s_2}{\bar{y}} = \frac{s_{\max,2}}{c}$$

where $\bar{y}$ is the distance from the neutral axis to the centroid of area $a$, $c$ is the distance from the neutral axis to outside fibers, and $s_{\max,1}$ and $s_{\max,2}$ are the stresses in the outside fibers. Then

$$F_1 = s_1 a = \frac{s_{\text{max},1}}{c}\, \bar{y}a$$

$$F_2 = s_2 a = \frac{s_{\text{max},2}}{c}\, \bar{y}a$$

$$H = \frac{s_{\text{max},2}}{c}\, a\bar{y} - \frac{s_{\text{max},1}}{c}\, a\bar{y} = \frac{a\bar{y}}{c}(s_{\text{max},2} - s_{\text{max},1})$$

The horizontal shear force $H$, on plane $KL$, whose area is $b\Delta x$, is also given by

$$H = (b\Delta x)s_h$$

Equating the two expressions for $H$ gives

$$b\Delta x s_h = \frac{a\bar{y}}{c}(s_{\text{max},2} - s_{\text{max},1})$$

$$s_h = \frac{a\bar{y}}{bc\Delta x}(s_{\text{max},2} - s_{\text{max},1})$$

If $M_1$ and $M_2$ are the bending moments on the respective faces, by Eq. (12-1),

$$s_{\text{max},1} = \frac{M_1 c}{I} \quad \text{and} \quad s_{\text{max},2} = \frac{M_2 c}{I}$$

Then
$$s_h = \frac{a\bar{y}}{bc\Delta x}\left(\frac{M_2 c}{I} - \frac{M_1 c}{I}\right)$$
$$= \frac{a\bar{y}}{Ib}\frac{M_2 - M_1}{\Delta x}$$

When $\Delta x$ is small, as in this discussion, then

$$\frac{M_2 - M_1}{\Delta x} = V$$

where $V$ is the vertical shear force at sections $AB$ and $CD$ (since $\Delta x$ is small). The value of $V$ is taken from the shear-force diagram at the section to be analyzed. Substituting, we obtain the final form for horizontal shear stress.

$$s_h = \frac{Va\bar{y}}{Ib} \qquad (12\text{-}4)$$

where $s_h$ = horizontal shear stress, psi; Pa
  $V$ = vertical shear force at section being considered, lb; N
  $a$ = area of cross section of beam between fiber being
     considered and nearest extreme fiber, in²; m²
  $\bar{y}$ = distance from neutral axis of entire cross section to
     centroid of area $a$, in; m
  $I$ = moment of inertia of entire cross section of beam, in⁴; m⁴
  $b$ = width of cross section at fiber being considered, in; m

**Sample Problem 5**  A 10-ft long, simply supported, 4- by 8-in rough-cut beam carries a uniform load of 100 lb/ft (Fig. 12-12a). Find the maximum horizontal shear stresses on each of the following horizontal planes.

  (a)  At the neutral axis
  (b)  1 in above the neutral axis
  (c)  2 in above the neutral axis
  (d)  3 in above the neutral axis
  (e)  4 in above the neutral axis

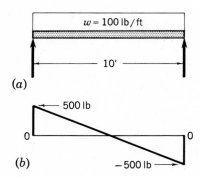

(a)

(b)

FIGURE 12-12  (a) Beam diagram for Sample Problem 5. (b) Shear-force diagram.

**Solution:**  Equation (12-4) shows that $s_h$ varies directly as $V$. Therefore, maximum $s_h$ will occur at the section where $V$ has its largest value. In Fig. 12-12b the shear-force diagram shows that $V$ is a maximum (500 lb) at both reactions

$$V = 500 \text{ lb} \qquad b = 4 \text{ in} \qquad h = 8 \text{ in}$$

$$I = \frac{bh^3}{12} = \frac{4(8)^3}{12} = 170.7 \text{ in}^4$$

  (a)  At the neutral axis $XX$ (area $a$ is shown shaded in Fig. 12-13a):

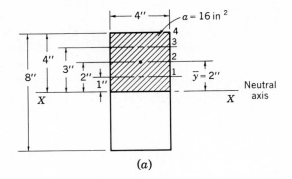

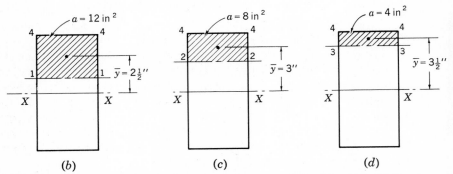

FIGURE 12-13 (a) Beam cross section for Sample Problem 5a. (b) Beam cross section for Sample Problem 5b. (c) Beam cross section for Sample Problem 5c. (d) Beam cross section for Sample Problem 5d.

$$a = 16 \text{ in}^2 \qquad \bar{y} = 2 \text{ in}$$

$$s_h = \frac{Va\bar{y}}{Ib} = \frac{500(16)(2)}{170.7(4)} = 23.4 \text{ psi}$$

(b)   At horizontal plane 1-1 (see Fig. 12-13b):

$$a = 12 \text{ in}^2 \qquad \bar{y} = 2.5 \text{ in}$$

$$s_h = \frac{500(12)(2.5)}{170.7(4)} = 21.9 \text{ psi}$$

(c)   At horizontal plane 2-2 (see Fig. 12-13c):

$$a = 8 \text{ in}^2 \qquad \bar{y} = 3 \text{ in}$$

$$s_h = \frac{500(8)(3)}{170.7(4)} = 17.5 \text{ psi}$$

(d)   At horizontal plane 3-3 (see Fig. 12-13d):

$$a = 4 \text{ in}^2 \qquad \bar{y} = 3.5 \text{ in}$$

$$s_h = \frac{500(4)(3.5)}{170.7(4)} = 10.3 \text{ psi}$$

(*e*)   At horizontal plane 4-4 (The area *a* represents the area above the horizontal plane in question. Thus, the area above plane 4-4 is zero.):

$$a = 0 \qquad \text{therefore} \qquad s_h = 0$$

The values of $s_h$ for horizontal planes below the neutral axis are determined in the same way, except that area *a* is the area below the horizontal plane. This problem illustrates the principle that $s_h$ is a maximum at the neutral axis and zero at the outer fibers.

To study the manner in which the shearing stresses vary, the stresses occurring at different distances from the neutral axis may be laid off to scale, as shown in Fig. 12-14. The vectors representing the stresses are of varying length, as can be seen from the values computed in the foregoing example. If the ends of the vectors are joined by means of a smooth curve, then any horizontal length between the vertical axis and the curve will represent the shearing stress at the corresponding point in the beam. The curve so determined is found to be a parabola.

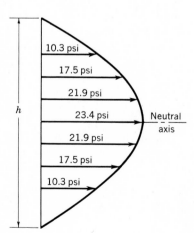

FIGURE 12-14   Distribution of horizontal shearing stresses for Sample Problem 5.

## 12-3 MAXIMUM HORIZONTAL SHEAR STRESS FOR COMMON CROSS SECTIONS

For the special case of horizontal shear stress at the neutral axis when the cross section is a rectangle, a simplified formula is available.

For a rectangular cross section *b* in [or m] wide by *h* in [or m] high,

$$a = \frac{bh}{2} \qquad \bar{y} = \frac{h}{4} \qquad I = \frac{bh^3}{12}$$

Substituting this information into Eq. (12-4),

$$s_h = \frac{Vay}{Ib} = \frac{V\left(\frac{bh}{2}\right)\left(\frac{h}{4}\right)}{\frac{bh^3}{12}} \quad (b)$$

$$s_h = \frac{3V}{2A} \qquad\qquad (12\text{-}5)$$

where $s_h$ = horizontal shear stress at the neutral axis for a rectangular
cross section, psi; Pa
$V$ = vertical shear force at section being considered, lb; N
$A = bh$, the area of the entire cross section, in$^2$; m$^2$

Similarly, for a beam of circular cross section whose diameter is $d$ in
[or m] (see Fig. 12-15),

$$a = \frac{\pi d^2}{8} \qquad \bar{y} = \frac{4r}{3\pi} = \frac{2d}{3\pi} \qquad b = d \qquad I = \frac{\pi d^4}{64}$$

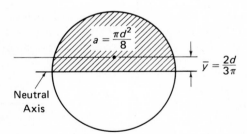

Neutral
Axis

FIGURE 12-15   Circular cross
section for determination of
maximum horizontal shear stress.

Substituting into Eq. (12-4),

$$s_h = \frac{V\left(\frac{\pi d^2}{8}\right)\left(\frac{2d}{3\pi}\right)}{(\pi d^4 / 64)d}$$

$$s_h = \frac{4V}{3A} \qquad\qquad (12\text{-}6)$$

where $s_h$ = horizontal shear stress at the neutral axis for a circular cross
section, psi; Pa
$V$ = vertical shear force at section being considered, lb; N
$A = \pi d^2 / 4$, the area of the entire cross section, in$^2$; m$^2$

The reader is cautioned that Eqs. (12-5) and (12-6) are limited to calculations at the neutral axis for rectangular and circular cross sections, respectively.

## 12-4 MAXIMUM VERTICAL SHEAR STRESS IN S-SHAPE AND W-SHAPE BEAMS

The assumption is usually made in the case of S- and W-shape beams that the flange carries all the bending stress, and the web, extending through the entire depth, takes the vertical shear. Although the assumption is not quite true, it is on the side of safety. The vertical shear stress in the webs of S- and W-shape beams is given by the equation

$$s_v = \frac{V}{td} \tag{12-7}$$

where $s_v$ = maximum vertical shear stress, psi; Pa
    $V$ = maximum vertical shear force, lb; N
    $t$ = thickness of the web, in; m
    $d$ = depth of the beam, in; m

From Sec. 12-2, $s_v = s_h$.

In the design of a beam, tests should always be made for shearing strength, as well as bending strength.

The code frequently used for the design of structural beams is the American Institute of Steel Construction (AISC) code. Table 12-1 gives the allowable stresses for the structural steels in bending and shear.

**TABLE 12-1** AISC ALLOWABLE STRESSES FOR BEAMS

| Material Specification (ASTM Designation) | Allowable Stress, psi | |
| --- | --- | --- |
| | Bending ($s_b$) | Shear ($s_v = s_h$) |
| A 36, Structural steel ($s_y = 36\ 000$ psi) | 24 000 | 14 500 |
| A 529, Structural steel ($s_y = 42\ 000$ psi) | 28 000 | 17 000 |
| A 572, Structural steel | | |
|   Grade 42 ($s_y = 42\ 000$ psi) | 28 000 | 17 000 |
|   Grade 45 ($s_y = 45\ 000$ psi) | 29 700 | 18 000 |
|   Grade 50 ($s_y = 50\ 000$ psi) | 33 000 | 20 000 |
|   Grade 55 ($s_y = 55\ 000$ psi) | 36 300 | 22 000 |
|   Grade 60 ($s_y = 60\ 000$ psi) | 39 600 | 24 000 |
|   Grade 65 ($s_y = 65\ 000$ psi) | 42 900 | 26 000 |
| A 242, Structural steel ($s_y = 50\ 000$ psi) | 33 000 | 20 000 |
| A 441, Structural steel ($s_y = 50\ 000$ psi) | 33 000 | 20 000 |
| A 588, Structural steel ($s_y = 50\ 000$ psi) | 33 000 | 20 000 |

**Sample Problem 6** An S $15\times50$ carries a total uniformly distributed load of 40 000 lb on a 20-ft simple span. Find the maximum bending stress and the shear stress in the web.

**Solution:** From App. B, Table 6, for an S $15\times50$, $S = 64.8$ in$^3$, depth $d = 15$ in, and web thickness $t = 0.550$ in. For a simply supported beam with a uniform load,

$$M_{max} = \frac{WL}{8} = \frac{40\ 000(20)}{8} = 100\ 000 \text{ ft·lb}$$

In bending, $\qquad s = \frac{M}{S} = \frac{100\ 000(12)}{64.8} = 18\ 500 \text{ psi}$

Note that the weight of beam has been neglected.
From Fig. 12-16, $V_{max} = 20\ 000$ lb. From Eq. (12-7) for web shear,

$$s_v = s_h = \frac{V}{td} = \frac{20\ 000}{(0.550)15} = 2420 \text{ psi}$$

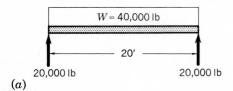

(a)

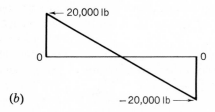

(b)

FIGURE 12-16  (a) Beam diagram for Sample Problem 6. (b) Shear-force diagram.

## 12-5 DISCUSSION OF BEAM DEFLECTION

When a load is placed on a beam, the beam tends to sag or deflect, as in Fig. 12-17.

Deflection plays a very important part in the design of structures and machines. If floor beams, or joists, deflect too far, the plaster on the ceiling under them may crack. Although no damage to the structure may result, the appearance of the ceiling may be ruined. Also, a floor supported by such beams may be so out of level that its usefulness for machinery may be impaired.

Under load, the neutral axis becomes a curved line and is called the

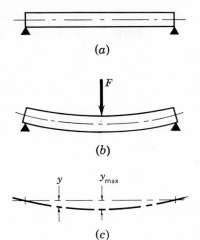

(a)

(b)

$y$    $y_{max}$

(c)

FIGURE 12-17   Beam deflection due to load: (a) beam without load; (b) beam with load; (c) comparison of loaded and unloaded neutral axes.

elastic curve. The deflection $y$ is vertical distance between a point on the elastic curve and the unloaded neutral axis.

## 12-6  RADIUS OF CURVATURE

The radius of curvature at any point on a curve is the radius of the circle which would match the shape of the curve at that point and in the immediate neighborhood of the point. In general, the elastic curve is not a circle, but, for a very short length, such as $\Delta l$, it may be considered as the arc of a circle (Fig. 12-18). Before the beam is loaded, sections $AB$ and $GH$ are parallel to each other (Fig. 12-18a). After loading, sections $AB$ and $GH$ occupy new positions $A'B'$ and $G'H'$ owing to deflection. These new positions are no longer parallel to each other but are both perpendicular to the elastic curve (neutral axis), as in Fig. 12-18b. From Fig. 12-18c it is apparent that fiber $AG$ has been shortened (compression) to length $A'G'$, and fiber $BH$ has been stretched (tension) to $B'H'$, while the portion of the neutral axis $XX$ has not changed its length. The elongation $\delta$ of fiber $BH$ is the change in its length; therefore,

$$\delta = B'B + HH' = 2B'B \quad (\text{since } B'B = HH')$$

Also,        $\Delta l = XX = BH$

The strain $\epsilon$ at the extreme fiber is

$$\epsilon = \frac{\delta}{\Delta l}$$

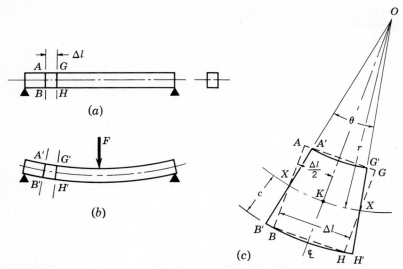

FIGURE 12-18   Radius of curvature of a deflected beam: (a) beam before load is applied; (b) beam after load is applied; (c) effect of deflection on a beam segment.

But the stress at this fiber is

$$s = \frac{Mc}{I}$$

and since modulus of elasticity $E = s/\epsilon$, then

$$E = \frac{Mc\Delta l}{\delta I}$$

By observing that triangle $OKX$ is similar to triangle $XBB'$, we find that

$$\frac{OX}{XB'} = \frac{XK}{B'B}$$

But         $OX = r$   (the radius of curvature)

$$XB' = c$$

$$XK = \frac{\Delta l}{2}$$

$$B'B = \frac{\delta}{2}$$

Substituting,       $\dfrac{r}{c} = \dfrac{\Delta l/2}{\delta/2} = \dfrac{\Delta l}{\delta}$

From which $$r = \frac{c\Delta l}{\delta}$$

Combining the equation for $r$ with the equation for $E$, we obtain

$$E = \frac{Mr}{I}$$

or $$r = \frac{EI}{M} \tag{12-8}$$

It is evident that the greater the deflection at a point in a beam, the smaller will be the radius of curvature. That is, the deflection varies inversely as $r$. From Eq. (12-8), the deflection is found to depend directly on $M$ and inversely on $EI$.

## 12-7 METHODS OF DETERMINING DEFLECTION FORMULAS

Several different methods are available for deriving beam-deflection formulas. They are as follows.

1. Moment-area method, which will be used in this book to determine maximum deflections.
2. Slope-deviation method, a more comprehensive variation of the moment-area method*
3. Double-integration method, a direct procedure involving higher mathematics*

The moment-area method and the slope-deviation method make use of relationships between various diagrams. In Secs. 11-5 and 11-6, two important principles were stated as follows.

*The area between any two vertical heights (ordinates) on a diagram will equal the difference in length of the two corresponding ordinates on the next higher degree diagram.*

*The vertical height (ordinate) at any point on a diagram will equal the slope of the next higher degree diagram at the same point.*

The first of these statements was used in determining bending moments from the shear-force diagram areas. The second statement was used to help predict the shape of successive diagrams and, most importantly, to pinpoint the possible locations of maximum bending moments from shear-diagram zero locations.

There are two more successive diagrams which can be added to those we have used thus far. They will be introduced briefly merely to emphasize the relationships. The next higher degree diagram ($EI\theta$ diagram) after

---

* For a thorough treatment of these methods, the reader may wish to refer to Robert W. Fitzgerald, *Strength of Materials*, Addison-Wesley Publishing Co., Reading, MA, 1967.

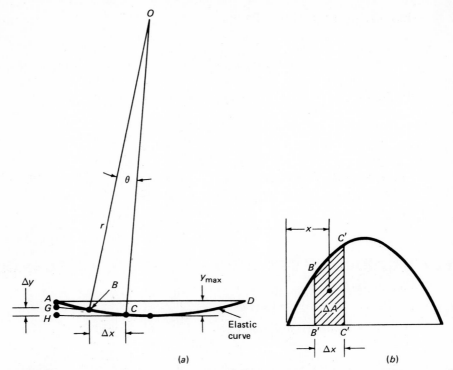

FIGURE 12-19   (a) Elastic curve of a deflected beam. (b) Moment diagram of a deflected beam.

the bending-moment diagram is one which gives the *slope*, $\theta$, of the deflected neutral axis multiplied by $EI$. The next higher degree diagram ($EIy$ diagram) gives the *deflection* $y$ of the neutral axis from the normal horizontal unloaded position, multiplied by $EI$. Thus, areas on the moment diagram relate to differences in ordinate heights on the $EI\theta$ diagram, which in turn influence the slopes of the *elastic curve* (neutral axis) in the $EIy$ diagram. These interrelationships provide the basis for the moment-area and slope-deviation methods of determining deflection.

The moment-area method for a simply supported beam is developed as follows: Consider a simply supported beam which, under load, deflects so that its neutral axis takes the shape of the curve $ABCD$ in Fig. 12-19a. A very small length along the curve between points $B$ and $C$ has a radius of curvature $r$ and subtends angle $BOC = \theta$. Now

$$\tan \theta = \frac{BC}{r}$$

But

$$BC \approx \Delta x$$

since the length is short. Thus,

$$\tan \theta = \frac{\Delta x}{r}$$

By Eq. (12-8),

$$r = \frac{EI}{M}$$

where $M$ is the bending moment at both $B'B'$ and $C'C'$ of Fig. 12-19$b$ since length $\Delta x$ is short. Combining these equations,

$$\tan \theta = \frac{M\Delta x}{EI}$$

From the moment diagram (Fig. 12-19$b$) the shaded area $\Delta A$ is approximately rectangular, such that

$$\Delta A = M \, \Delta x$$

Then 
$$\tan \theta = \frac{\Delta A}{EI}$$

If tangents to the elastic curve (Fig. 12-19$a$) are drawn at points $B$ and $C$ such as $BG$ and $CH$, then angles $OBG$ and $OCH$ are right angles and the angle between $BG$ and $CH$ is $\theta$. The vertical distance $\Delta y = GH$ along a perpendicular through point $A$ represents a *portion* of the maximum deflection in this case. Since $BG$ and $CH$ are almost horizontal, the average of their lengths may be taken as $x$, the distance from the left reaction to the centroid of area $\Delta A$.

From this discussion,

$$\tan \theta = \frac{\Delta y}{(BG + CH)/2} = \frac{\Delta y}{x}$$

Setting the two expressions for $\tan \theta$ equal,

$$\frac{\Delta y}{x} = \frac{\Delta A}{EI}$$

or 
$$\Delta y = \frac{x\Delta A}{EI}$$

This equation indicates that $\Delta y$ depends upon the *moment of an area* on the bending-moment diagram.

For several systems, $\Delta y$ is a portion of the maximum deflection $Y_{max}$. If all the $\Delta y$ values are added for all segments along the elastic curve between the left end $A$ (Fig. 12-19a) and the point of maximum deflection, they will total to $y_{max}$. This is true for simple beams with symmetrical loading and for cantilever beams. In cases where this is not true, a more general method is necessary. Derivation and detailed discussion of such methods are beyond the scope of this book.

Proceeding with the case at hand, we can find an expression for the maximum deflection by adding all the $\Delta y$ components as follows.

$$
\begin{aligned}
y_{max} &= \Sigma \, \Delta y \\
&= \Delta y_1 + \Delta y_2 + \Delta y_3 + \cdots + \Delta y_n \\
&= \frac{x_1 \Delta A_1}{EI} + \frac{x_2 \Delta A_2}{EI} + \frac{x_3 \Delta A_3}{EI} + \cdots + \frac{x_n \Delta A_n}{EI} \\
&= \frac{\displaystyle\sum_{i=1}^{n} (x_i \Delta A_i)}{EI}
\end{aligned}
$$

Equation (10-2) may be written in the following form:

$$
\bar{x} = \frac{\displaystyle\sum_{i=1}^{n} (x_i \Delta A_i)}{\displaystyle\sum_{i=1}^{n} \Delta A_i}
$$

$$
\sum_{i=1}^{n} (x_i \Delta A_i) = \bar{x} \sum_{i=1}^{n} \Delta A_i = \bar{x} A
$$

where $\bar{x}$ is the distance from the centroid of shaded area $A$ (Fig. 12-20) to the left reaction. Therefore,

$$
y_{max} = \frac{\bar{x} A}{EI} \tag{12-9}
$$

Note that deflection is a function of a moment divided by $EI$, as discussed toward the end of Sec. 12-6.

Equation (12-9) may also be used to determine the deflection at any point in a cantilever beam, which will be demonstrated later in this chapter.

To determine the deflection at any point on a simply supported beam with symmetrical loading, it is necessary to find the maximum deflection and subtract from it an amount which represents the difference between $y_{max}$ and the desired deflection.

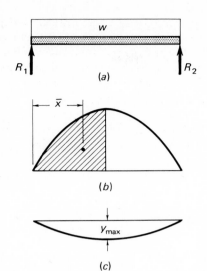

(a)

(b)

(c)

FIGURE 12-20 (a) Beam diagram of a simply supported beam with uniform load. (b) Moment diagram. (c) Elastic curve.

The equation for deflection at any point on a simply supported beam with symmetrical loading (Fig. 12-21) is

$$y = \frac{\bar{x}_1 A_1}{EI} - \frac{\bar{x}_2 A_2}{EI} \qquad (12\text{-}10)$$

where $\dfrac{\bar{x}_1 A_1}{EI} = y_{\max}$; $\quad \dfrac{\bar{x}_2 A_2}{EI} = y_{\max} - y$

and $\quad y$ = deflection at any point on a simply supported beam

$A_1$ = area of the bending-moment diagram between the left reaction and the point of maximum deflection

$\bar{x}_1$ = horizontal distance from the centroid of area $A_1$ to the left reaction

$A_2$ = area of the bending-moment diagram between the point where $y$ is to be found and the point of maximum deflection

$\bar{x}_2$ = horizontal distance from the centroid of area $A_2$ to an axis through the point where deflection is to be determined

$E$ = modulus of elasticity of the beam material

$I$ = moment of inertia of the beam cross section

Note that for the beam in Fig. 12-21,

$A_1$ = area $BCD$ of the bending-moment diagram
$\bar{x}_1$ = moment arm of centroid of $A_1$ about the left end of the beam
$A_2$ = area $GHCD$ of the bending-moment diagram
$\bar{x}_2$ = moment arm of centroid of $A_2$ about point $x$

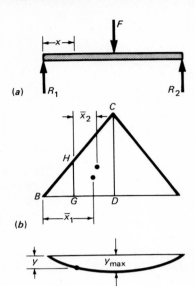

(a)

(b)

(c)

FIGURE 12-21 (a) Beam diagram of a simply supported beam with concentrated load at midspan. (b) Moment diagram. (c) Elastic curve.

## 12-8 DEFLECTION OF A SIMPLY SUPPORTED BEAM (CONCENTRATED LOAD AT CENTER)

Maximum deflection: by Eq. (12-9),

$$y_{max} = \frac{\bar{x}A}{EI}$$

From Fig. 12-22b,

$$A = \frac{1}{2}\left(\frac{FL}{4}\frac{L}{2}\right) = \frac{FL^2}{16}$$

$$\bar{x} = \frac{2}{3}\left(\frac{L}{2}\right) = \frac{L}{3}$$

Therefore,

$$y_{max} = \frac{\frac{L}{3}\left(\frac{FL^2}{16}\right)}{EI}$$

or

$$y_{max} = \frac{FL^3}{48EI} \qquad (12\text{-}11)$$

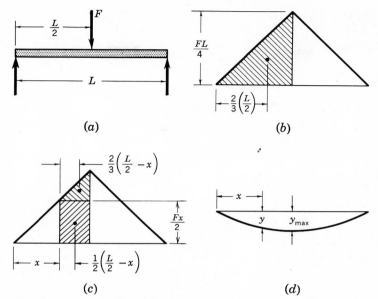

FIGURE 12-22 (a) Beam diagram of a simply supported beam with load at midspan. (b) Moment diagram for derivation of maximum deflection equation. (c) Moment diagram for derivation of equation for deflection at any point. (d) Elastic curve.

Deflection at any point: by Eq. (12-10),

$$y = \frac{\bar{x}_1 A_1}{EI} - \frac{\bar{x}_2 A_2}{EI}$$

where $\bar{x}_1 A_1 / EI = y_{\text{max}}$ (Sec. 12-7)

$\bar{x}_2 A_2$ is the moment of the entire shaded area in Fig. 12-22c. For convenience, the entire shaded area is divided into a rectangle and a triangle. The sum of the moment of the rectangular and triangular areas equals $\bar{x}_2 A_2$. Then

$$\bar{x}_2 A_2 = \left[ \frac{1}{2}\left(\frac{L}{2} - x\right) \frac{Fx}{2} \left(\frac{L}{2} - x\right) \right]_{\text{rectangle}}$$

$$+ \left[ \frac{2}{3}\left(\frac{L}{2} - x\right)\left(\frac{FL}{4} - \frac{Fx}{2}\right)\left(\frac{L}{2} - x\right)\frac{1}{2} \right]_{\text{triangle}}$$

from which

$$\frac{\bar{x}_2 A_2}{EI} = \frac{F}{48EI}\left(L^3 - 3L^2 x + 4x^3\right)$$

Substituting into Eq. (12-10),

$$y = \frac{FL^3}{48EI} - \frac{F}{48EI}(L^3 - 3L^2x + 4x^3)$$

which reduces to

$$y = \frac{Fx}{48EI}(3L^2 - 4x^2) \tag{12-12}$$

## 12-9 DEFLECTION OF A SIMPLY SUPPORTED BEAM (UNIFORM LOAD)

Maximum deflection: by Eq. (12-9),

$$y_{max} = \frac{\bar{x}A}{EI}$$

From Fig. 12-23b, $A = \frac{2}{3}$ of the enclosing rectangle.

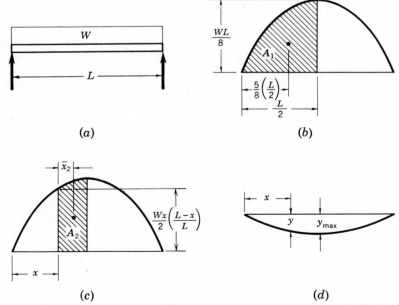

(a)

(b)

(c)

(d)

FIGURE 12-23   (a) Beam diagram of a simply supported beam with uniform load. (b) Moment diagram for derivation of maximum deflection equation. (c) Moment diagram for derivation of equation for deflection at any point. (d) Elastic curve.

$$A = \frac{2}{3}\left(\frac{WL}{8}\right)\left(\frac{L}{2}\right) = \frac{WL^2}{24}$$

$$\bar{x} = \frac{5}{8}\left(\frac{L}{2}\right) = \frac{5L}{16}$$

Substituting,
$$y_{max} = \frac{\dfrac{5L}{16}\left(\dfrac{WL^2}{24}\right)}{EI}$$

or
$$y_{max} = \frac{5WL^3}{384EI} \tag{12-13}$$

Deflection at any point: by applying a similar procedure as in Sec. 12-8 to Fig. 12-23c, we find that

$$y = \frac{Wx}{24EI}\left(\frac{L^3 - 2Lx^2 + x^3}{L}\right) \tag{12-14}$$

## 12-10 DEFLECTION OF A CANTILEVER BEAM (GENERAL)

The maximum deflection for a cantilever beam is given by

$$y_{max} = \frac{\bar{x}A}{EI} \tag{12-15}$$

where $A$ is the entire area of the moment diagram and $\bar{x}$ is the distance from its centroid to the free end, as shown in Fig. 12-24b.

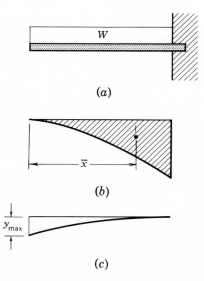

(a)

(b)

(c)

FIGURE 12-24  (a) Beam diagram of a cantilever beam with uniform load. (b) Moment diagram. (c) Elastic curve.

Equation (12-15) for a cantilever beam can be developed using a procedure similar to that in Sec. 12-7.

## 12-11 DEFLECTION OF A CANTILEVER BEAM (CONCENTRATED LOAD AT FREE END)*

Maximum deflection: by Eq. (12-15), $y_{max} = \dfrac{\bar{x}A}{EI}$

From Fig. 12-25$b$,

$$A = \frac{1}{2}(L)FL = \frac{FL^2}{2}$$

$$\bar{x} = \frac{2L}{3}$$

Substituting,

$$y_{max} = \frac{\dfrac{2L}{3}\left(\dfrac{FL^2}{2}\right)}{EI}$$

$$y_{max} = \frac{FL^3}{3EI} \tag{12-16}$$

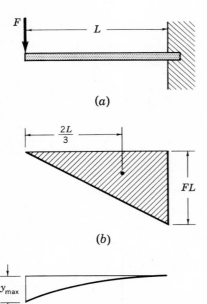

(a)

(b)

(c)

FIGURE 12-25 (a) Beam diagram of a cantilever beam with load at free end. (b) Moment diagram. (c) Elastic curve.

* For cases not shown here, refer to *Manual of Steel Construction*, American Institute of Steel Construction.

## 12-12 DEFLECTION OF A CANTILEVER BEAM (CONCENTRATED LOAD AT ANY POINT)

Maximum deflection: by Eq. (12-15),

$$y_{max} = \frac{\bar{x}A}{EI}$$

From Fig. 12-26$b$,

$$A = \frac{1}{2}(L - a)F(L - a) = \frac{F}{2}(L - a)^2$$

$$\bar{x} = \frac{2}{3}(L - a) + a = \frac{2}{3}L + \frac{a}{3}$$

Substituting,

$$y_{max} = \frac{\left(\dfrac{2L}{3} + \dfrac{a}{3}\right)\left(\dfrac{F}{2}\right)(L - a)^2}{EI}$$

which reduces to

$$y_{max} = \frac{F}{6EI}(2L^3 - 3aL^2 + a^3) \qquad (12\text{-}17)$$

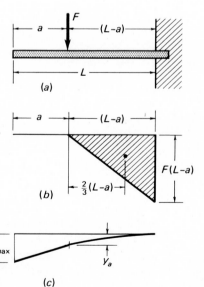

FIGURE 12-26  (a) Beam diagram of a cantilever beam with load at any point. (b) Moment diagram. (c) Elastic curve.

Deflection under the load: the equation for the deflection $y_a$, under load $F$ in Fig. 12-26c, can be determined by considering a cantilever beam of length $L - a$ with a concentrated load at its free end. This deflection is given by an equation of the form of Eq. (12-16) with a span of $L - a$.

$$y_a = \frac{F(L - a)^3}{3EI} \qquad (12\text{-}18)$$

# 12-13 DEFLECTION OF A CANTILEVER BEAM (UNIFORM LOAD)

Maximum deflection: by Eq. (12-15),

$$y_{max} = \frac{\bar{x}A}{EI}$$

From Fig. 12-27b, $A = \frac{1}{3}$ of the enclosing rectangle.

$$A = \frac{1}{3}L\frac{WL}{2} = \frac{WL^2}{6}$$

$$\bar{x} = \frac{3L}{4}$$

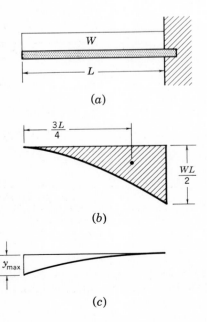

(a)

(b)

(c)

FIGURE 12-27 (a) Beam diagram of a cantilever beam with uniform load. (b) Moment diagram. (c) Elastic curve.

Substituting,

$$y_{max} = \frac{\dfrac{3L}{4}\left(\dfrac{WL^2}{6}\right)}{EI}$$

$$y_{max} = \frac{WL^3}{8EI} \qquad (12\text{-}19)$$

## 12-14 DEFLECTION OF BEAMS WITH COMBINED LOADS

When a beam carries more than one load (or type of load) the deflection at a given point is determined by calculating the deflection at that point due to each load and adding these results. Table 12-2 gives maximum deflections for various beam loadings.

   To demonstrate this principle, consider the beam in Fig. 12-28. The load $F$ causes the beam to deflect, as shown in Fig. 12-29$a$. The effect of the uniform load $W$ is shown in Fig. 12-29$b$. The actual beam deflection due to both loads is obtained by adding the individual effects, as in Fig. 12-29$c$.

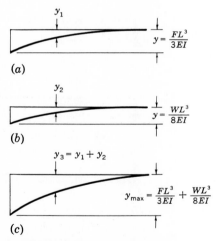

(a)

$$y = \frac{FL^3}{3EI}$$

(b)

$$y = \frac{WL^3}{8EI}$$

(c)

$$y_{max} = \frac{FL^3}{3EI} + \frac{WL^3}{8EI}$$

$y_3 = y_1 + y_2$

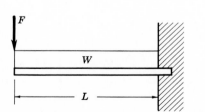

FIGURE 12-28   Cantilever beam with load at free end and uniform load.

FIGURE 12-29   (a) Elastic curve—deflection due to $F$. (b) Elastic curve—deflection due to $W$. (c) Elastic curve—combined effect.

**Sample Problem 7**   An S12 × 31.8 simply supported beam, 12 ft long, carries a total uniform load of 8000 lb and a concentrated load of 14 000 lb at midspan (see Fig. 12-30).

**TABLE 12-2**   SELECTED MAXIMUM DEFLECTION FORMULAS

| Beam and loading | Maximum deflection | Location |
|---|---|---|
| | $\dfrac{FL^3}{48EI}$ | Midspan |
| | $\dfrac{5WL^3}{384EI}$ | Midspan |
| | $\dfrac{23FL^3}{648EI}$ | Midspan |
| | $\dfrac{F}{24EI}(3L^2a - 4a^3)$ | Midspan |
| | $\dfrac{19FL^3}{384EI}$ | Midspan |
| | $\dfrac{FL^3}{3EI}$ | Free end |
| | $\dfrac{F}{6EI}(2L^3 - 3aL^2 + a^3)$ | Free end |
| | $\dfrac{WL^3}{8EI}$ | Free end |

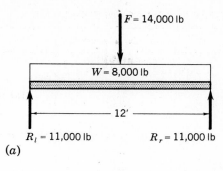

(a)

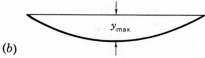

(b)

FIGURE 12-30  (a) Beam diagram for Sample Problem 7. (b) Elastic curve.

(a)  Find the maximum deflection of the beam.
(b)  If permissible deflection is limited to $\frac{1}{360}$ of the span (AISC code), is this beam acceptable based on deflection?

**Solution a:**  The deflection at the center of the beam due to load $F$ is given by Eq. (12-11) or Table 12-2.

$$y_F = \frac{FL^3}{48EI}$$

In this case,

$$F = 14\,000 \text{ lb}$$
$$L = 12 \text{ ft} = 144 \text{ in}$$

and for an S12 × 31.8,

$$I_x = 218 \text{ in}^4 \text{ (App. B, Table 6)}$$
$$E = (30)(10^6) \text{ psi}$$

Substituting into Eq. (12-11),

$$y_F = \frac{14\,000(144)^3}{48(30 \times 10^6)(218)} = 0.133 \text{ in}$$

The deflection at midspan due to $W$ is given by Eq. (12-13) or Table 12-2.

$$y_W = \frac{5WL^3}{384EI}$$

Where $W = 8000$ lb,

$$y_W = \frac{5(8000)144^3}{384(30 \times 10^6)(218)} = 0.048 \text{ in}$$

The maximum deflection due to both loads is

$$y_{max} = y_F + y_W = 0.133 + 0.048 = 0.181 \text{ in}$$

**Solution b:**  The maximum permissible deflection is

$$y = \frac{L}{360} = \frac{12(12)}{360} = \frac{144}{360} = 0.40 \text{ in}$$

Since actual $y_{max} = 0.181$ in, the beam is acceptable, based on deflection.

## 12-15  DESIGN OF A BEAM

The fundamental factors upon which beam analysis depends are:

1.  Loading—arrangement and magnitude
2.  Span
3.  Type of beam support (simple, cantilever, etc.)
4.  Allowable stresses (bending, shear)
5.  Permissible deflection
6.  Shape of beam cross section
7.  Size of beam

The most common design procedure involves determining the size of beam. A suitable beam must have the required strength in bending and in shear. A further limitation may be imposed with respect to maximum deflection. For example, the AISC code limits maximum deflection for beams carrying plastered ceilings to $\frac{1}{360}$ of the span length. The following sample problem, which includes a limitation on deflection, will help to clarify this procedure.

**Sample Problem 8**  Design a W-shape, A36 structural steel cantilever beam with a span of 12 ft carrying a uniform load of 670 lb/ft and a concentrated load of 16 500 lb at its midpoint. The allowable stresses, as given by the AISC code, are 24 000 psi in bending and 14 500 psi in web shear. The permissible deflection is $\frac{1}{360}$ of span .

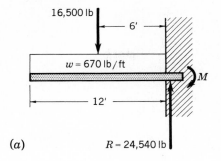

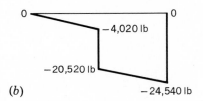

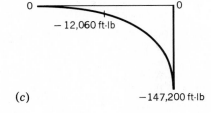

FIGURE 12-31 (a) Beam diagram for Sample Problem 8. (b) Shear force diagram. (c) Moment diagram.

**Solution:**

**Bending** (*See Fig.* 12-31a):

$$\text{Allowable } s = 24\,000 \text{ psi}$$
$$M_{\text{max}} = 670(12)(6) + 16\,500(6) = 48\,200 + 99\,000$$
$$= 147\,200 \text{ ft} \cdot \text{lb} \quad \text{(at wall)} \quad \text{(see Fig. 12-31c)}$$
$$s = \frac{M}{S}$$
$$S = \frac{M}{s} = \frac{147\,200(12)}{24\,000} = 73.6 \text{ in}^3 \text{ (minimum)}$$

Selecting on the basis of economy from the Section Modulus Table in App. B (Table 9), we find that a W16 × 50 (S = 80.8 in³) beam meets the bending requirement.

***Deflection:***

$$\text{Allowable deflection} = \frac{12(12)}{360} = 0.40 \text{ in}$$

$$y_{max} = \frac{WL^3}{8EI} + \frac{F}{6EI}(2L^3 - 3aL^2 + a^3)$$

$$W = 670(12) = 8040 \text{ lb} \quad a = 6(12) = 72 \text{ in} \quad L = 12(12) = 144 \text{ in}$$

$$0.40 = \frac{8040(144)^3}{8(30 \times 10^6)I} + \frac{16\,500}{6(30 \times 10^6)I}[2(144)^3 - 3(72)(144)^2 + (72)^3]$$

$$= \frac{100}{I} + \frac{170.8}{I} = \frac{270.8}{I}$$

$$I = \frac{270.8}{0.40} = 677 \text{ in}^4 \text{ (minimum)}$$

The moment of inertia $I$ for the W16 × 50 is 657 in$^4$, which is not sufficient for the deflection requirement. However, a W18 × 50 beam will satisfy both bending and deflection requirements ($S = 89.1$ in$^3$ and $I = 802$ in$^4$). It should be noted that other beams would satisfy the bending and deflection requirements, such as W16 × 58, W14 × 68, W18 × 55, and W12 × 85, but these are more expensive beams since their weights exceed the 50 lb/ft of the W18 × 50.

***Web Shear:*** Check the W18 × 50 in web shear.

$$\text{Depth} = d = 18 \text{ in}$$
$$\text{Web thickness} = t = 0.358 \text{ in}$$
$$V_{max} = 24\,540 \text{ lb} \quad \text{(see Fig. 12-31}b)$$
$$s_v = \frac{V}{td} = \frac{24\,540}{0.358(18)} = 3810 \text{ psi}$$
$$\text{Allowable } s_v = 14\,500 \text{ psi} \quad \text{therefore OK}$$

Use W18 × 50.

# 12-16 LATERAL BUCKLING

Beams with long spans tend to buckle laterally, wrinkle locally, or twist under heavy loading. Intermediate lateral (side) supports are used to reduce such failures. If lateral supports are not used, the AISC recommends* reducing the allowable bending stress ($s$) to a value given by (use only for W, S, HP, and Channel shapes)

---

* For additional information, see *Manual of Steel Construction*, 7th ed., AISC, Specifications, Section 1.5.1.4.6a.

$$s = \frac{12\ 000\ 000\ C_b}{ld/A_f} \tag{12-20}$$

where $l$ = unsupported length of simple beam or twice the span
of cantilever beam, in; m

$d$ = depth of beam, in; m

$A_f$ = area of flange in compression, in²; m²

$C_b$ = 1, for simple and cantilever beams

The AISC code further specifies that the reduced allowable bending stress shall not exceed $0.60s_y$. Equation (12-20) is used only when $ld/A_f$ exceeds

555 for A36 structural steel
475 for A529 structural steel
475 for A572, Grade 42 structural steel
445 for A572, Grade 45 structural steel
400 for A572, Grade 50 structural steel
365 for A572, Grade 55 structural steel
335 for A572, Grade 60 structural steel
310 for A572, Grade 65 structural steel
400 for A242, A441, A588 structural steel

## PROBLEMS

*12-1.  A simple beam 150 by 200 mm rough sawn in cross section and 3 m long carries a concentrated load of 36 kN at the center. Find the largest bending stresses developed at sections 0.9, 1.5, and 2.4 m from the left end.

12-2.  A wooden beam 12 ft long is supported at the ends. If the rough-sawn cross section is 6 by 8 in and the allowable bending stress 1100 psi, what total uniform load can the beam support?

12-3.  In Prob. 12-2, if the allowable bending stress is 1200 psi and the rough-sawn cross section of the beam is 6 by 9 in, what is the maximum length of the span if the total uniform load is 6000 lb?

*12-4.  What should be the diameter of the hollow axle for the railway car shown in Fig. Prob. 12-4? The safe bending stress is 70 MPa. The outside diameter of the axle is twice the inside diameter.

12-5.  What is the factor of safety based on the ultimate strength of the Class 60 cast-iron beam shown in Fig. Prob. 12-5? Consider the weight of the beam as a uniform load.

12-6.  A total uniform load of 16 000 lb is carried by a beam of Douglas fir. If the span is 14 ft, select a dressed size whose nominal width is 8 in.

12-7.  How far apart may the supports for an A36 structural steel W8

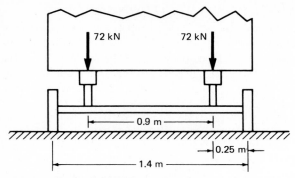

FIGURE PROBLEM 12-4

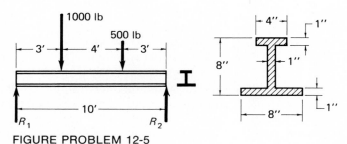

FIGURE PROBLEM 12-5

× 35 beam be placed if it supports a load of 8000 lb at the middle of the span? Use the AISC code.

**12-8.** What is the maximum bending stress in an A36 structural steel S 15 × 50 beam of 24-ft span if it supports a uniform load of 800 lb in addition to its own weight, and loads of 6000 and 9000 lb, respectively, at 6 and 16 ft from the right support?

**12-9.** What should be the size of A36 structural steel S-shape beams spaced 4 ft on centers to carry a balcony load of 225 psf? The balcony is constructed as a cantilever and extends 8 ft beyond the wall. Use the AISC code. (*Hint:* Find the load on one beam.)

**12-10.** Select an A36 structural steel W-shape beam 24 ft long, over-hanging each support 4 ft, to carry a load of 16 000 lb at the left end, 12 000 lb at the center, and 8000 lb at the right end. Use the AISC code.

**12-11.** A floor is supported by A529 structural steel S12 × 35 beams spaced 8 ft center to center. What uniformly distributed floor load (psf) will they support with a span of 18 ft? Use the AISC code.

**\*12-12.** What should be the depth (rough sawn) of a western hemlock beam, simply supported, full width of 100 mm and 3.6 m long, to carry equal concentrated loads of 4.5 kN at the third points?

**12-13.** What must be the diameter of a steel shaft 10 ft long between the bearings if the allowable bending stress is 10 000 psi? A pulley located 3 ft from one end if subjected to a total downward pull of 1600 lb (steel weighs 490 lb/ft³).

**12-14.** What will be the maximum stress in the box girder of Fig. Prob. 12-14? The girder is made up by connecting four 2- by 12-in (rough-sawn) planks as shown. The span is 12 ft, and the uniformly distributed load is 1250 lb per linear foot.

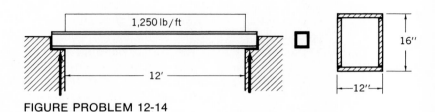

**FIGURE PROBLEM 12-14**

**12-15.** An AISI 1020 steel pin in a pin-connected truss is to be considered as a simple beam of 8-in span. It resists a variable force of 8000 lb at its center. What should be the diameter of this pin?

**\*12-16.** The maximum force between a connecting rod and the crank-pin is 540 kN. This force may be assumed to be uniformly distributed against the pin which is 200 mm long. What must be the diameter of the AISI 1045 steel pin if shock loading is assumed?

**\*12-17.** In a scaffolding used in construction, there is a plank 50 by 300 mm rough sawn used to support the workmen. The plank is held up by means of brackets 2.4 m apart. If a person having a mass of 91 kg stands at the middle and a person having a mass of 68 kg stands 0.3 m from the left end, could Eastern white pine be used for the plank? $\rho = 705$ kg/m³ for Eastern white pine.

**\*12-18.** What will be the value of equal loads placed at the quarter points of a rough-sawn Douglas fir beam 150 mm wide, 250 mm deep, and 3.6 m long to develop the maximum allowable bending stress?

**12-19.** Select a dressed-size Ponderosa pine beam 16 ft long to carry a uniform load of 200 lb/ft, a concentrated load of 1200 lb 6 ft from the right end, and 800 lb 4 ft from the right end.

**12-20.** A wooden beam, dressed size 6 by 8 in and 12 ft long, supported at the ends, carries a load of 2000 lb at the middle of the span.

*a.* What is the unit shearing stress at the neutral axis?

*b.* What is the unit shearing stress at a point 3 ft from the end and 2 in from the neutral axis?

*c.* What is the maximum bending stress in the beam?

**12-21.** A simple beam, dressed size 6 by 6 in and 6 ft long, carries a concentrated load at the middle of the span. What is the load if it is determined by the allowable shearing stress of 120 psi?

**12-22.** A rectangular Sitka spruce beam, dressed size 8 by 12 in with a 12-ft simple span, carries a uniform load of 1600 lb/ft. What is the unit horizontal shearing stress at the following points?

*a.* At the neutral axis and at the end of the beam

*b.* Four inches from the neutral axis and 1 ft from the support

*c.* Two inches from the neutral axis and at a quarter point

**12-23.** *a.* What will be the total uniform load that can be placed on a Mountain hemlock beam, dressed size 10 by 14 in and 8 ft long, supported at the ends, if the allowable shear stress alone is considered?

*b.* What will be the stress due to bending as a result of this load?

**12-24.** What should be the proper length of a dressed-size 10- by 12-in oak beam carrying a uniform load of 1800 lb per linear foot? Solve for both a simple span and a cantilever. Consider both horizontal shear and bending. Use an allowable shear stress of 120 psi and an allowable bending stress of 1450 psi.

**12-25.** What should be the spacing of $\frac{1}{2}$-in bolts in a beam consisting of a 4- by 6-in (rough-sawn) fir timber bolted to a 6- by 6-in (rough-sawn) timber? The pieces will be placed so that the cross section of the beam will be 6 by 10 in. The span and loading are shown in Fig. Prob. 12-25. The allowable shear stress in the bolts is 10 000 psi.

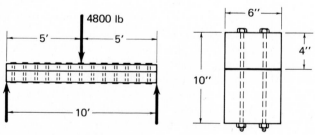

FIGURE PROBLEM 12-25

**12-26.** A shape is built up by bolting together three rough-sawn 50 mm by 200-mm planks 3.6 m long, as shown in Fig. Prob. 12-26. What should be the spacing of 10-mm bolts to resist the shearing stresses? The bolts are made of AISI 1020 steel, and a safety factor of 5 is specified.

**12-27.** What is the maximum shearing stress developed in an A36 structural steel S10 × 35 beam that carries a uniform load over a simple span of 12 ft? Use the AISC code. Determine load from allowable bending stress.

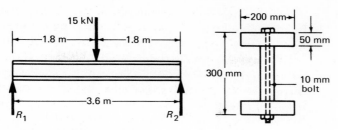

FIGURE PROBLEM 12-26

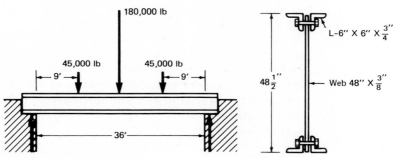

FIGURE PROBLEM 12-28

**12-28.** A plate girder used as a simple beam 36 ft long must support a concentrated load of 180 000 lb at the middle of the span and 45 000 lb at each quarter point (Fig. Prob. 12-28). The girder is composed of a 48- by $\frac{3}{8}$-in web plate and four 6- by 6- by $\frac{3}{4}$-in flange angles, all made of A36 structural steel. What should be the bolt spacing if $\frac{7}{8}$-in A307 bolts are used? Use the AISC code. Assume bearing-type connections.

**12-29.** A W12 × 72 A36 structural steel beam is used for a simple span of 12 ft and carries a uniformly distributed load. What is the maximum shearing stress at the base of the upper flange? Consider the flange as a rectangle. Use the AISC code. Determine load based on allowable bending stress.

**12-30.** A Douglas fir beam 10 by 12 in dressed size and 14 ft long is supported at the ends. The beam carries a total uniform load of 9000 lb and a concentrated load at the middle of the span of 3600 lb.
    *a.* What will be the maximum bending stress, shearing stress, and deflection?
    *b.* Are the values within safe limits?

**12-31.** What safe uniformly distributed load will a 6- by 10-in yellow pine beam carry if the span is 10 ft? Design for both dressed and full sizes. Use an allowable bending stress of 1750 psi and an allowable shear stress of 120 psi.

**12-32.** For what length of beam will the total uniform load be determined by shear only and for what length by bending? Assume a Ponderosa pine beam and 8- by 10-in dressed-size cross section.

**12-33.** A W12 × 79 A36 structural steel beam carries a uniformly distributed load of 48 000 lb on a simple span of 24 ft. Compute the maximum bending stress, shearing stress, and deflection. Use the AISC code.

**12-34.** A Mountain hemlock beam, 4 by 8 in dressed size and 7 ft long, is used as a cantilever to support an end load of 900 lb. What are the maximum bending stress, shearing stress, and deflection?

**12-35.** A floor carrying a load of 140 psf is supported by A441 structural steel S12 × 35 beams spaced 6 ft from center to center. The span of the beams is 21 ft. Use the AISC code.
  *a.* Does the deflection exceed the allowable?
  *b.* What are the maximum bending and shearing stresses?

**\*12-36.** A steel shaft 250 mm in diameter resting in bearings 1.05 m apart carries a flywheel having a mass of 18.14(10³) kg midway between the bearings. Treating the shaft as a simple beam, find the maximum values of bending stress, shearing stress, and deflection.

**12-37.** A 4-in steel shaft 10 ft long and supported at the ends carries a pulley, weighing 200 lb, midway between the supports. There is a belt pull in a downward direction of 500 lb acting on the pulley.
  *a.* What will be the resultant deflection?
  *b.* Determine the horizontal shear stress and the bending stress developed.

**12-38.** *a.* Select the most economical A36 structural steel S-shape beam for the loading shown in Fig. Prob. 12-38. Use the AISC code.
  *b.* Is this beam safe in vertical web shear?
  *c.* What is the maximum deflection?

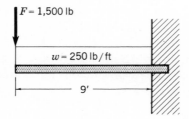

$F = 1,500$ lb

$w = 250$ lb/ft

9'

FIGURE PROBLEM 12-38

**12-39.** Select the most economical A36 structural steel S-shape beam for the loading in Fig. Prob. 12-39. Use the AISC code.

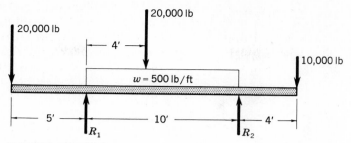

20,000 lb

20,000 lb

4'

10,000 lb

$w = 500$ lb/ft

5'

10'

4'

$R_1$

$R_2$

FIGURE PROBLEM 12-39

**\*12-40.**   A Class 20, hollow, circular, cast-iron post with an outside diameter of 75 mm and a wall thickness of 12.5 mm is proposed for the loading shown in Fig. Prob. 12-40. Does this post satisfy a required safety factor of 7? Neglect the weight of the post.

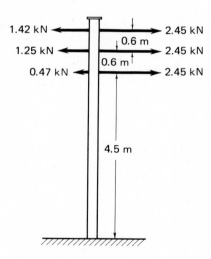

1.42 kN ← → 2.45 kN

0.6 m

1.25 kN ← → 2.45 kN

0.6 m

0.47 kN ← → 2.45 kN

4.5 m

FIGURE PROBLEM 12-40

# CHAPTER 13

## Torsion, Shafts, Shaft Couplings, and Keys

### 13-1 TORSION

In Fig. 13-1, $F_1$ and $F_2$ are the tensions on a belt, with $F_2 > F_1$. The difference in tensions tends to rotate the shaft clockwise. A cable is wound around the shaft, as shown, with an attached weight $W$. This weight tends to rotate the shaft in a counterclockwise direction. When a shaft is acted upon by two equal and opposite twisting moments in parallel planes, it is said to be in *torsion*. When such a condition exists, the shaft may be either stationary or rotating uniformly. The twisting moment, or torque, is generally expressed in inch-pounds or newton-meters.

If, in Fig. 13-1, $F_2 = 250$ lb, $F_1 = 100$ lb, and the pulley radius $r = 6$ in, the torque $T = (F_2 - F_1)r = (250 - 100)(6) = 900$ in·lb.

If the radius of the shaft is 1 in and the moments are equal, $W$ must equal 900 lb. This assumes negligible bearing friction.

In the previous illustration, the twisting moment, or torque, is constant for the length of shaft between the pulley and the cable. However, if a shaft has several pulleys attached to it, the torque will be different in different sections of the shaft.

In Fig. 13-2, pulley $B$ drives the shaft in clockwise rotation and provides power for pulleys $A$ and $C$. Pulley $B$ transmits a torque $T_b = (F_3 - F_4)r_b = (1500 - 500)(4) = 4000$ in·lb, which is available to pulleys $A$ and $C$. Thus, the torque at pulley $A$, transmitted along the shaft between $A$ and $B$, is

$$T_a = (F_5 - F_6)r_a = (300 - 100)(6) = 1200 \text{ in·lb}$$

Similarly, the shaft from $B$ to $C$ transmits the torque

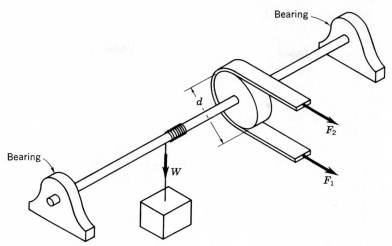

FIGURE 13-1   Torsional system.

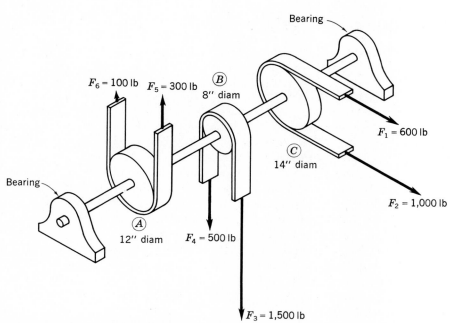

FIGURE 13-2   Torque transmission.

$$T_c = (F_2 - F_1)r_c = (1000 - 600)(7) = 2800 \text{ in} \cdot \text{lb}$$

Since bearing friction has been assumed to be negligible, no torque exists in the shaft between the right bearing and pulley $C$ and also between the left bearing and pulley $A$.

## 13-2 **TORSIONAL SHEARING STRESS**

When a shaft is in torsion, any cross section of the shaft tends to slip by or shear across the adjacent face. The resistance of the fibers per unit area of the shaft is called the *torsional shearing stress*. In Fig. 13-3a, when a couple is applied to the lever, an element $AB$, originally straight, is twisted into the position $AC$. Radius $OB$ of the end section is rotated into the position $OC$. Any portion of material originally at $B'$ is rotated to $C'$. Now, the shear stress resisting that change is proportional to the strain since $G = s_s/\epsilon_s$. But the elongation $B'C'$ is proportional to the distance $OB'$ from the center of the shaft. Then, within the elastic limit of the material, the torsional shearing stress at any point of the cross section of a shaft is proportional to the distance of that point from the center of the section.

Since the center of the section has no shear stress, the axis of the shaft is a neutral axis. If the radius $OB$ is 3 in, and the shear stress at $B$ is 6000 psi, and if $B'$ is 1.5 in from the center, the stress at $B'$ is 3000 psi (see Fig. 13-3b).

An enlarged cross section of the shaft is shown in Fig. 13-4. Consider a thin ring of material (cross-hatched) whose area is $a$ at an average radius $r$. The shearing force acting on area $a$ is $F_r = as_{sr}$, where $s_{sr}$ represents the shear stress at radius $r$. The torque associated with this shearing force is

$$T_r = F_r r = s_{sr} a r$$

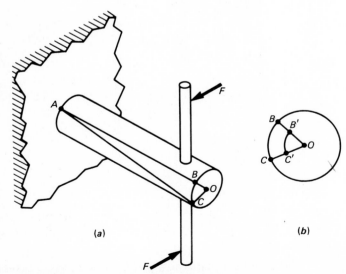

FIGURE 13-3   Torsional strain in a shaft.

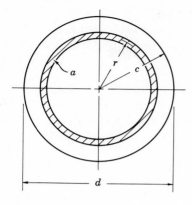

FIGURE 13-4   Shaft cross section.

Now, let $s_s$ be the maximum shearing stress that occurs on the extreme outer fiber at radius $c$. Since shear stress is proportional to radial distance,

$$\frac{s_{sr}}{r} = \frac{s_s}{c} \qquad \text{or} \qquad s_{sr} = \frac{s_s r}{c}$$

Then, by substitution,

$$T_r = \frac{s_s(ar^2)}{c}$$

The total shearing torque on the entire cross section is the sum of the torques on all rings such as $a$. Thus,

$$T = \frac{s_s}{c}[\Sigma ar^2]$$

The term in brackets is the moment of inertia of the area of the circle about a centroidal axis *perpendicular to the plane of the circle*. This is called the *polar moment of inertia* and will be represented by $J$. Then

$$T = \frac{s_s J}{c}$$

or

$$s_s = \frac{Tc}{J} \qquad\qquad (13\text{-}1)$$

where  $s_s$ = maximum shearing stress due to torsion at the outer fiber of a uniform, solid, or hollow circular member,* psi; Pa (usually expressed as $10^6$ Pa or MPa)

* Equations for other shapes may be found in R. J. Roark and W. C. Young, *Formulas for Stress and Strain*, 5th ed., McGraw-Hill Book Company, New York, 1975.

$T$ = torque, in·lb; N·m

$c$ = distance from neutral axis to the extreme fiber, in; m

$J = I_x + I_y$ = polar moment of inertia about a centroidal axis perpendicular to cross section, in⁴; m⁴

The student should note the similarity between the torsion formula Eq. (13-1) and the flexure formula Eq. (12-1). In the same manner as for bending, it is convenient to group the moment of inertia and the $c$-distance terms to form $S' = J/c$, where $S'$ is the polar section modulus. Thus, Eq. (13-1) may be written as

$$s_s = \frac{T}{S'} \tag{13-2}$$

For a solid shaft of *circular cross section*,

$$J = I_x + I_y = \frac{\pi d^4}{64} + \frac{\pi d^4}{64} = \frac{\pi d^4}{32}$$

$$c = \frac{d}{2}$$

$$S' = \frac{J}{c} = \frac{\pi d^4/32}{d/2} = \frac{\pi d^3}{16}$$

Then

$$s_s = \frac{T}{S'} = \frac{T}{\pi d^3/16} = \frac{16T}{\pi d^3} \tag{13-3}$$

For a *hollow circular shaft* of inside diameter $d_i$ and outside diameter $d_o$,

$$J = \frac{\pi d_o^4}{32} - \frac{\pi d_i^4}{32} = \frac{\pi(d_o^4 - d_i^4)}{32}$$

$$c = \frac{d_o}{2}$$

$$S' = \frac{J}{c} = \frac{\pi(d_o^4 - d_i^4)/32}{d_o/2} = \frac{\pi}{16}\left(\frac{d_o^4 - d_i^4}{d_o}\right)$$

Then

$$s_s = \frac{T}{S'} = \frac{T}{\dfrac{\pi}{16}\left(\dfrac{d_o^4 - d_i^4}{d_o}\right)} = \frac{16Td_o}{\pi(d_o^4 - d_i^4)} \tag{13-4}$$

The previous equations apply to pure torsional systems *within the elastic limit* of the material. The torsional strength of a shaft depends on the maximum safe torque it can transmit. The strength of two shafts of the same material may be compared by finding the ratio of their maximum safe torques.

*Sample Problem 1  Find the diameter of a solid AISI 1045 steel shaft which is to transmit a torque of 18 kN·m if the factor of safety based on the ultimate shearing strength is 10.

Solution:   From App. B, Table 1, the ultimate $s_s = 480(10^6)$ Pa for AISI 1045. Therefore, the allowable $s_s = 480(10^6)/10 = 48(10^6)$ Pa. From Eq. (13-3) for a circular cross section,

$$s_s = \frac{16T}{\pi d^3} \quad \text{or} \quad d^3 = \frac{16T}{\pi s_s}$$

$$d^3 = \frac{16(18)(10^3)}{\pi(48)(10^6)} = 1.91(10^{-3})$$

$$d = 1.24(10^{-1}) \text{ m} = 124 \text{ mm}$$

Sample Problem 2  A solid shaft with a diameter of 3 in has the same cross-sectional area as a hollow shaft of $d_o = 5$ in and $d_i = 4$ in. If the shafts are made of the same material, which shaft is stronger (torsionally) and by what factor?

Solution:   It is unnecessary to know the allowable shear stress since only the ratio of the torques is required, not the actual values of the torques. For the solid shaft [Eq. (13-3)],

$$s_s = \frac{16T_1}{\pi d^3} \quad \text{or} \quad T_1 = \frac{\pi d^3 s_s}{16}$$

$$T_1 = \frac{\pi(3^3)s_s}{16} = 5.3 s_s$$

For the hollow shaft [Eq. (13-4)],

$$s_s = \frac{16T_2 d_o}{\pi(d_o^4 - d_i^4)} \quad \text{or} \quad T_2 = \frac{\pi(d_o^4 - d_i^4)s_s}{16 d_o}$$

$$T_2 = \frac{\pi(5^4 - 4^4)s_s}{16(5)} = \frac{\pi(625 - 256)s_s}{16(5)} = 14.5 s_s$$

$$\frac{T_2}{T_1} = \frac{14.5 s_s}{5.3 s_s} = 2.73$$

The hollow shaft is 2.73 times as strong in torsion as the solid shaft.

Sample Problem 3  What maximum safe torque, considering torsional shear only, may be applied in tightening a 1″-8 UNC bolt whose allowable shear stress is 8800 psi?

Solution:   Root diameter of threads = 0.8466 in (App. B, Table 3). From Eq. (13-3),

$$s_s = \frac{16T}{\pi d^3} \quad \text{or} \quad T = \frac{\pi d^3 s_s}{16}$$

$$T = \frac{\pi (0.8466^3)(8800)}{16} = 1050 \text{ in} \cdot \text{lb (max safe torque)}$$

## 13-3  ANGLE OF TWIST

In some types of machinery, it is desirable to know the amount of angle of twist in a shaft due to the torque. Let Fig. 13-5 represent a part of a

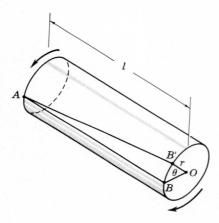

FIGURE 13-5  Angle of twist of a shaft subjected to uniform torque.

shaft subjected to a uniform torque throughout. $AB$ represents an element of the untwisted shaft, and $AB'$ is the same element after torque has been applied. Owing to the torque, the radius $OB$ assumes a position $OB'$. The angle $BOB'$ is the angle of twist $\theta$. The displacement from $B$ to $B'$ is a shearing displacement. For the fiber represented by $AB$, the deformation in the length $l$ is $BB'$. The unit deformation (strain) is

$$\epsilon_s = \frac{BB'}{l} \quad \text{and} \quad BB' = r\theta \quad \text{Then,} \ \epsilon_s = \frac{r\theta}{l} \quad \text{(refer to Fig. 8-3c)}$$

where $\theta$ is the angle $BOB'$ expressed in radians. Also, from the relations from Hooke's law,

$$\epsilon_s = \frac{s_s}{G}$$

where $G$ is the modulus of rigidity (modulus of elasticity in shear). Then

$$\frac{r\theta}{l} = \frac{s_s}{G}$$

Note that $r = c$. Therefore,

$$\theta = \frac{s_s l}{Gc} \tag{13-5}$$

The above expression gives the angle of twist in terms of the fiber stress. However, the design of shafts usually involves the angle of twist expressed in terms of torque; thus, from Eq. (13-1),

$$s_s = \frac{Tc}{J}$$

Substituting for $s_s$ in Eq. (13-5),

$$\theta = \frac{Tcl}{JGc} = \frac{Tl}{JG} \tag{13-6}*$$

where $\theta$ = angle of twist, radian
$\quad T$ = torque, in · lb; N · m
$\quad l$ = length of shaft subjected to torque, in; m
$\quad J$ = polar moment of inertia, in⁴; m⁴
$\quad G$ = modulus of rigidity, psi; Pa (usually expressed as
$\qquad 10^9$ Pa or GPa)

Attention is called to the fact that the angle of twist increases with the length of the shaft and decreases as the radius increases.

## 13-4 POWER TRANSMISSION

When a force $F$ that is constant in magnitude and direction acts on a body through a distance $L$ in the direction in which the force acts, the work done is the product of the force times the distance, or

$$\text{Work} = F(L)$$

When $F$ is in pounds and $L$ in feet, work is in foot-pounds. For example, if $F = 20$ lb and $L = 10$ ft,

$$\text{Work} = 20(10) = 200 \text{ ft · lb} = 2400 \text{ in · lb}$$

* Equation (13-6) has the same limitation as Eq. (13-1).

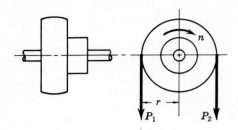

**FIGURE 13-6** Forces on rotating pulley.

In Fig. 13-6, the driving force is $P_2 - P_1 = F$ on the pulley of radius $r$. Suppose the shaft is making $n$ revolutions per minute (rpm). In one revolution, the force $F$ acts through the distance $2\pi r$ (the circumference of the pulley). The work per revolution is $2\pi Fr$. But $Fr$ is the moment, or torque, $T$. Then, in $n$ revolutions, the total work done is $2\pi Tn$.

The total work done can also be found from the relationship that the work done is the product of torque and the angle, in radians, through which the torque acts, or

$$\text{Work} = T\theta$$

For $n$ revolutions,

$$\text{Work} = T \times \left( n\,\text{rev.} \times \frac{2\pi\,\text{rad}}{1\,\text{rev.}} \right) = 2\pi Tn$$

$$\text{since Power} = \frac{\text{work}}{\text{time}}$$

$$\text{Power} = \frac{2\pi Tn}{\text{time}}$$

For U.S. customary units, the usual power unit is horsepower (hp). One horsepower is defined as 33 000 ft·lb/min, or 33 000 × 12 in·lb/min. Thus,

$$\text{hp} = \frac{2\pi Tn}{33\,000(12)} = \frac{Tn}{63\,000} \tag{13-7a}$$

where $T$ = torque, in·lb
$n$ = rpm

For SI metric units, the power unit is the watt (W). One watt is equal to 1 N·m/s. Thus,

$$\text{W} = \frac{2\pi Tn}{60} = \frac{Tn}{9.55} \tag{13-7b}$$

where W = watt (usually expressed in terms of kilowatts)

$T$ = torque, N·m

$n$ = rpm

*Sample Problem 4** A 50-mm-diameter shaft of AISI 2340 steel is to be driven at 1800 rpm. A safety factor of 10 is specified based on the ultimate torsional shearing strength.

(a)  What maximum power may be safely transmitted?

(b)  What will be the angle of twist in a 1-m (20-diameters) length of this shaft when the maximum safe power is transmitted?

**Solution:**  From App. B, Table 1, for AISI 2340 steel,

$$\text{Ultimate } s_s = 690(10^6) \text{ Pa}$$

$$\text{Allowable } s_s = \frac{690(10^6)}{10} = 69(10^6) \text{ Pa}$$

$$d = 50 \text{ mm} = 50(10^{-3}) \text{ m}$$

$$n = 1800 \text{ rpm}$$

(a)  From Eq. (13-3),

$$s_s = \frac{16T}{\pi d^3} \qquad \text{or} \qquad T = \pi \frac{d^3 s_s}{16}$$

$$T = \frac{\pi(50 \times 10^{-3})^3(69 \times 10^6)}{16} = 1694 \text{ N·m} = 1.694(10^3) \text{ N·m}$$

From Eq. (13-7b),

$$W = \frac{Tn}{9.55} = \frac{(1.694 \times 10^3)(1800)}{9.55} = 319(10^3)W$$
$$= 319 \text{ kW} \qquad \text{(maximum safe power)}$$

(b)  $l = 1$ m, $G = 80(10^9)$ Pa (App. B, Table 1), $c = 25$ mm $= 25(10^{-3})$ m.

Using Eq. (13-5),

$$\theta = \frac{s_s l}{Gc} = \frac{(69 \times 10^6)(1)}{(80 \times 10^9)(25 \times 10^{-3})} = 0.0345 \text{ rad}$$

or, since $\pi$ rad = 180°,

$$\theta = \frac{180}{\pi}(0.0345) = 57.3(0.0345) = 1.98° \qquad \text{say, } \theta = 2°$$

Note that a limitation of 1° (0.0175 rad) twist in 20 diameters of length is often specified. If this specification were enforced in this problem, the shaft would be stressed to only $35(10^6)$ Pa and would transmit 162 kW at a torque of 860 N·m.

**Sample Problem 5** A solid shaft 8 in in diameter has the same cross-sectional area as a hollow shaft of the same material, with inside diameter of 6 in.

    (a) Compare the horsepower transmission of these shafts at the same revolutions per minute.

    (b) Compare the angle of twist in equal lengths of these shafts when stressed to the same intensity.

**Solution:** Find the outside diameter $d_o$ of the hollow shaft. Since the cross-sectional areas are equal,

$$\frac{\pi d^2}{4} = \frac{\pi}{4}(d_o{}^2 - d_i{}^2)$$

$$\frac{\pi(8^2)}{4} = \frac{\pi}{4}(d_o{}^2 - 6^2)$$

$$8^2 = d_o{}^2 - 6^2$$

$$d_o{}^2 = 64 + 36 = 100$$

$$d_o = 10 \text{ in}$$

(a) From Eq. (13-3) for the solid shaft,

$$s_s = \frac{16T}{\pi d^3} \qquad \text{or} \qquad T = \frac{\pi d^3 s_s}{16}$$

$$T = \frac{\pi(8^3)s_s}{16}$$

From Eq. (13-7a) for the solid shaft,

$$\text{hp}_{\text{solid}} = \frac{Tn}{63\,000} = \frac{\pi(8^3)s_s n}{63\,000(16)} = \frac{\pi s_s n}{(63\,000)(16)} (512)$$

From Eq. (13-4) for the hollow shaft,

$$s_s = \frac{16T d_o}{\pi(d_o{}^4 - d_i{}^4)} \qquad \text{or} \qquad T = \frac{\pi(d_o{}^4 - d_i{}^4)s_s}{16 d_o}$$

$$T = \frac{\pi(10^4 - 6^4)s_s}{16(10)}$$

From Eq. (13-7a) for the hollow shaft,

$$\mathrm{hp}_{\mathrm{hollow}} = \frac{Tn}{63\,000} = \frac{\pi(10^4 - 6^4)s_s n}{63\,000(16)(10)} = \frac{\pi s_s n}{63\,000(16)}\left(\frac{10\,000 - 1300}{10}\right)$$

$$= \frac{\pi s_s n}{63\,000(16)}(870)$$

$$\frac{\mathrm{hp}_{\mathrm{hollow}}}{\mathrm{hp}_{\mathrm{solid}}} = \frac{870}{512} = 1.7$$

The hollow shaft can transmit 1.7 times the horsepower that the solid shaft can transmit.

(b)  For the solid shaft, $c = d/2 = \frac{8}{2} = 4$ in. From Eq. (13-5),

$$\theta_{\mathrm{solid}} = \frac{s_s l}{Gc} = \frac{s_s l}{G(4)}$$

For the hollow shaft, $c = d_o/2 = \frac{10}{2} = 5$ in. Using Eq. (13-5),

$$\theta_{\mathrm{hollow}} = \frac{s_s l}{Gc} = \frac{s_s l}{G(5)}$$

$$\frac{\theta_{\mathrm{hollow}}}{\theta_{\mathrm{solid}}} = \frac{1/5}{1/4} = \frac{4}{5} = 0.80$$

The twist in the hollow shaft is only 80 percent of the twist in the solid shaft.

## 13-5  SHAFT COUPLINGS

A number of devices are used to connect sections of coaxial shafting. The simplest and most widely used of these devices is the bolted flanged coupling. This type of coupling provides a clear application of torsional shearing stress induced by power transmission. Figure 13-7 represents such a flange coupling. The torque is transmitted from one shaft to the other as follows.

1. From shaft $A$ to key $B$
2. From key $B$ to the left side of the coupling $C$
3. From the left side of the coupling $C$ to the six bolts $D$
4. From the bolts $D$ to the right side of the coupling $E$
5. From the right side of the coupling $E$ to the key $F$
6. From key $F$ to the other shaft $G$

The transmitted torque will develop a resisting torque in the bolts equal and opposite to it. The resisting torque in each bolt will be

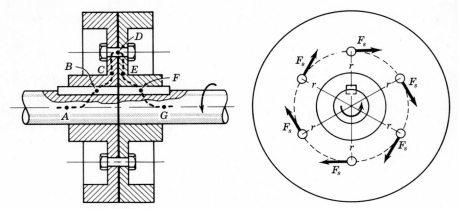

FIGURE 13-7   Torque transmission in flange coupling.

$$T_r = \frac{T}{n_b}$$   (assuming that each bolt will take an equal share)

where $T_r$ = resisting torque in each bolt
  $T$ = total transmitted torque
  $n_b$ = number of bolts

But torque equals force times radius. Therefore,

$$T_r = Fr$$

where $F$ = resisting force in each bolt
  $r$ = bolt-circle radius

**Shear, Bolts:**   The shear stress developed will be

$$s_s = \frac{F}{A_s}$$

where $s_s$ = shear stress, psi; Pa (usually expressed as $10^6$ Pa or MPa)
  $F$ = load causing shear, lb; N
  $A_s$ = cross-sectional area of bolt, in²; m²

**Bearing, Bolts:**   In addition to causing shear in the bolts, it should be noted that the force $F$ will also cause bearing between the bolts and the web of the flange coupling. The compressive stress in the bolt will be

$$s_c = \frac{F}{A_c}$$

where $s_c$ = compressive stress, psi; Pa (usually expressed as $10^6$ Pa or MPa)
  $F$ = load causing bearing, lb; N
  $A_c$ = projected area of bolt in the web, in²; m²

***Shear, Hub:***   Further analysis of the coupling should include the consideration of possible failure due to shearing the hub from the web. Such failure might occur at the section where the web joins the hub, shown in Fig. 13-8$a$. The area resisting shear is the cylindrical area (striped area) in Fig. 13-8$b$ which is $t$ in [m] wide and $d_{hub}$ in [m] in diameter. The shear force on this area is $F = T/r_{hub}$.
Therefore, the shear stress is

$$s_s = \frac{F}{A_s}$$

where $s_s$ = shear stress, psi; Pa (usually expressed as $10^6$ Pa or MPa)
  $F = T/r_{hub}$ = shearing force on this area, lb; N
  $A_s = \pi d_{hub}t$, in²; m²      (see Fig. 13-8$b$)

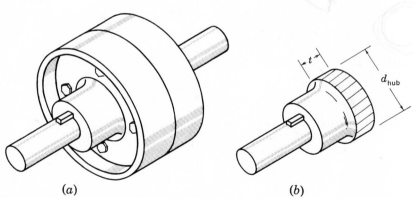

(a)                                (b)

FIGURE 13-8   Hub shear in flange coupling.

# 13-6 KEYS

When torque is transmitted by means of couplings, gears, or pulleys, some method must be used to fasten these devices to their shafts. One commonly used device is the key. Figure 13-9 shows a rectangular key fastening a spur gear to a shaft. A clockwise torque is transmitted from the shaft through the key to the gear.

***Shear, Key:***   There is a tendency for the key to shear on the cross-hatched rectangular area (Fig. 13-10$a$) at the surface of the shaft. The

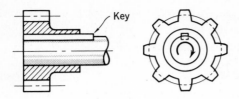

FIGURE 13-9   Spur gear keyed to a shaft.

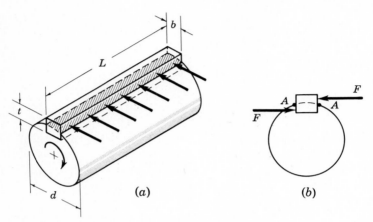

(a)                          (b)

FIGURE 13-10   Shear analysis of a key.

force $F$ which produces shear at section $AA$ (Fig. 13-10$b$) is equal to $T/r_{shaft}$. The shear stress developed in the key will then be equal to

$$s_s = \frac{F}{A_s}$$

where $s_s$ = shear stress, psi; Pa (usually expressed as $10^6$ Pa or MPa)

$F = \dfrac{T}{r_{shaft}}$ = force causing shear in key, lb; N

$A_s = bL$ = shear area, in²; m²

***Bearing, Key:***   The analysis of the key may be carried further to determine the bearing stress. The force $F$ will tend to crush the key at the cross-hatched area (Fig. 13-11$a$). This force may be taken as the same force in the shear analysis of the key. Therefore, the bearing stress developed in the key is

$$s_c = \frac{F}{A_c}$$

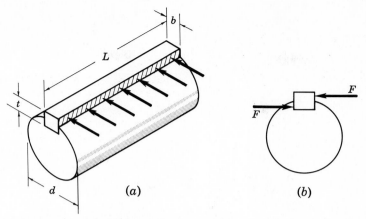

FIGURE 13-11   Bearing analysis of a key.

where $s_c$ = bearing stress, psi; Pa (usually expressed as $10^6$ Pa or MPa)

$$F = \frac{T}{r_{\text{shaft}}} = \text{force causing bearing, lb; N}$$

$$A_c = \frac{tL}{2} = \text{bearing area, in}^2; \text{m}^2$$

**\*Sample Problem 6**  A flange coupling with four bolts transmits 19 kW at 200 rpm. The bolt-circle diameter is 150 mm. The thickness of the flange web is 20 mm, and the flange-hub diameter is 110 mm. The shafts joined by the coupling are 60 mm in diameter. The bolts and coupling are made of AISI 1020 steel. A safety factor of 4, based on ultimate, is specified.

(a)   Determine the required diameter of the bolts.
(b)   Is the hub safe in shear where it joins the web?

**Solution:**

$$T = \frac{9.55 \text{ W}}{n} = \frac{9.55(19 \times 10^3)}{200} = 907 \text{ N} \cdot \text{m}$$

$$F_{\text{bolt}} = \frac{T}{n_b r_{bc}} = \frac{907}{4(75 \times 10^{-3})} = 3.02(10^3)\text{N}$$

From App. B, Table 1,

$$\text{Ultimate } s_s = 345(10^6) \text{ Pa}$$
$$\text{Ultimate } s_c = 450(10^6) \text{ Pa}$$

then,  $$\text{Allowable } s_s = \frac{345(10^6)}{4} = 86.2(10^6) \text{ Pa}$$

$$\text{Allowable } s_c = \frac{450(10^6)}{4} = 112(10^6) \text{ Pa}$$

### (a) Shear:

$$A_s = \frac{F_{\text{bolt}}}{s_s} = \frac{3.02(10^3)}{86.2(10^6)} = 0.035(10^{-3}) = 35(10^{-6}) \text{ m}^2$$

$$A_s = 0.785d^2$$

$$d^2 = \frac{35(10^{-6})}{0.785} = 44.6(10^{-6})$$

$$d = 6.7(10^{-3}) \text{ m} = 6.7 \text{ mm (minimum)}$$

### (b) Bearing:

$$A_c = \frac{F_{\text{bolt}}}{s_c} = \frac{3.02(10^3)}{112(10^6)} = 27(10^{-6}) \text{ m}^2$$

$$A_c = dt$$

$$d = \frac{27(10^{-6})}{20(10^{-3})} = 1.35(10^{-3}) \text{ m} = 1.35 \text{ mm (minimum)}$$

From these results, the minimum safe bolt diameter is 6.7 mm. Use the next largest available standard size, $d = 8$ mm.*

$$F_{\text{hub}} = \frac{T}{r_{\text{hub}}} = \frac{907}{55(10^{-3})} = 16.5(10^3)\text{N}$$

Find the shear stress in the hub where it joins the web, and compare that with the allowable stress.

$$A_s = \pi d_{\text{hub}} t = \pi(110 \times 10^{-3})(20 \times 10^{-3}) = 6.91(10^{-3}) \text{ m}^2$$

$$s_s = \frac{F_{\text{hub}}}{A_s} = \frac{16.5(10^3)}{6.91(10^{-3})} = 2.39(10^6) \text{ Pa} = 2.39 \text{ MPa}$$

Since the actual $s_s$ is below the allowable $s_s$ of 86.2 MPa, the hub is safe in shear.

*Sample Problem 7* In the above problem, a 16- by 10-mm flat key of AISI 1020 steel is used to fasten each side of the coupling to its shaft. Use a safety factor of 4 based on ultimate. What length of key is required?

### Solution:

$$F_{\text{key}} = \frac{T}{r_{\text{shaft}}} = \frac{907}{30(10^{-3})} = 30.2(10^3)\text{N}$$

*Standard size taken from Table 4, "Preferred ISO Metric Bolt Lengths and Diameters," E. Oberg and F. Jones, *Machinery's Handbook*, 19th Ed., New York, Industrial Press, Inc., 1971, p. 1168.

*Shear:*

$$A_s = \frac{F_{key}}{s_s} = \frac{30.2(10^3)}{86.2(10^6)} = 350(10^{-6}) \text{ m}^2$$

$$bL = 350(10^{-6})$$

$$b = 16 \text{ mm} = 16(10^{-3}) \text{ m}$$

$$L = \frac{350(10^{-6})}{16(10^{-3})} = 21.9(10^{-3})\text{m} = 21.9 \text{ mm (minimum)}$$

*Bearing:*

$$A_c = \frac{F_{key}}{s_c} = \frac{30.2(10^3)}{112(10^6)} = 0.27(10^{-3}) = 270(10^{-6})\text{m}^2$$

$$A_c = \frac{t}{2}L$$

$$t = 10 \text{ mm} = 10(10^{-3}) \text{ m}$$

$$L = \frac{2(270)(10^{-6})}{10(10^{-3})} = 54(10^{-3})\text{m} = 54 \text{ mm (minimum)}$$

The minimum safe length of key is 54 mm.

**Sample Problem 8**  Two hollow shafts are joined by a flange coupling according to the following specifications.

|  |  |
|---|---|
| *Shaft:* | 302 stainless steel |
|  | 5 in OD |
|  | $2\frac{1}{2}$ in ID |
| *Key:* | AISI 1045 steel |
|  | 1 by 1 by 8 in |
| *Coupling:* | Class 60 cast iron |
|  | hub diameter = 8 in |
|  | web thickness = $1\frac{1}{2}$ in |
|  | bolt-circle diameter = 11 in |
|  | bolt diameter = $\frac{3}{4}$ in |
|  | number of bolts = 6 |
|  | material of bolts—AISI 1045 Steel |
|  | factor of safety on all parts = 6 (based on ultimate) |

What maximum horsepower can this arrangement transmit at 350 rpm?

**Solution:**  In order to determine the maximum horsepower for this power-transmission arrangement, it is necessary to calculate the maximum safe torque which the combination can transmit. The maximum safe torque will be the torque which the weakest component can transmit. To determine this torque, the following calculations will be made.

1.  Shaft in torsional shear

2. Key in shear
3. Key in bearing
4. Hub in shear at the web
5. Bolts in shear
6. Bolts in bearing with web

## 1. *302 Stainless Shaft—Torsion:*

$$\text{Ultimate } s_s = 110\,000 \text{ psi}$$

$$\text{Allowable } s_s = \frac{110\,000}{6} = 18\,330 \text{ psi}$$

$$T = s_s S'$$

$$S' = \frac{J}{c} = \frac{\pi}{16}\left[\frac{d_0{}^4 - d_i{}^4}{d_0}\right] \qquad \text{(see Sec. 13-2)}$$

$$= \frac{\pi}{16}\left(\frac{5^4 - 2.5^4}{5}\right) = \frac{\pi}{16}\left(\frac{625 - 39.1}{5}\right)$$

$$= \frac{\pi}{16}\left(\frac{585.9}{5}\right)$$

$$= 23 \text{ in}$$

$$T_{\text{shaft}} = 18\,330(23) = 421\,000 \text{ in·lb}$$

## 2. *AISI 1045 Key in Shear:*

$$\text{Ultimate } s_s = 70\,000 \text{ psi}$$

$$\text{Allowable } s_c = \frac{70\,000}{6} = 11\,670 \text{ psi}$$

$$T = F_{\text{key}}(r_{\text{shaft}})$$

$$F_{\text{key}} = A_s s_s$$

$$A_s = bL = 1(8) = 8 \text{ in}^2$$

$$F_{\text{key}} = 8(11\,670) = 93\,300 \text{ lb}$$

$$T = 93\,300\,(2.5) = 234\,000 \text{ in·lb}$$

## 3. *AISI 1045 Key in Bearing:*

$$\text{Ultimate } s_c = 95\,000 \text{ psi}$$

$$\text{Allowable } s_c = \frac{95\,000}{6} = 15\,830 \text{ psi}$$

$$T = F_{\text{key}}(r_{\text{shaft}})$$

$$F_{\text{key}} = A_c s_c$$

$$A_c = \frac{tL}{2} = \frac{1}{2}(8) = 4 \text{ in}^2$$

$$F_{\text{key}} = 4(15\,830) = 63\,300 \text{ lb}$$

$$T = 63\,300(2.5) = 158\,000 \text{ in·lb}$$

From calculations 2 and 3, $T_{\text{key}} = 158\,000$ in·lb

4. **Class 60 Cast-Iron Hub in Shear:**

$$\text{Ultimate } s_s = 65\,000 \text{ psi}$$

$$\text{Allowable } s_s = \frac{65\,000}{6} = 10\,830 \text{ psi}$$

$$T = F_{\text{hub}}(r_{\text{hub}})$$

$$F_{\text{hub}} = A_s s_s$$

$$A_s = \pi d_{\text{hub}} t = \pi(8)1.5 = 37.7 \text{ in}^2$$

$$F_{\text{hub}} = 37.7(10\,830) = 408\,000 \text{ lb}$$

$$T_{\text{hub}} = 408\,000(4) = 1\,632\,000 \text{ in·lb}$$

5. **AISI 1045 Bolts in Shear:**

$$\text{Allowable } s_s = 11\,670 \text{ psi} \qquad \text{(from calculation 2)}$$

$$T = n_b F_{\text{bolt}} r_{bc}$$

$$F_{\text{bolt}} = A_s s_s$$

$$A_s = \frac{\pi(d_{\text{bolt}})^2}{4} = \frac{\pi}{4}(0.75)^2 = 0.442 \text{ in}^2$$

$$F_{\text{bolt}} = 0.442(11\,670) = 5160 \text{ lb}$$

$$n_b = 6 \qquad r_{bc} = 5\tfrac{1}{2} \text{ in}$$

$$T = 6(5160)(5.5) = 170\,000 \text{ in·lb}$$

6. **AISI 1045 Bolts in Bearing against Class 60 Cast-iron Web:** When bearing occurs between materials with different compressive strengths, the permissible stress is that of the weaker material.

$$\text{AISI 1045 bolts} \qquad \text{Ultimate } s_c = 95\,000 \text{ psi}$$
$$\text{Class 60 CI web} \qquad \text{Ultimate } s_c = 170\,000 \text{ psi}$$

Therefore, use allowable stress of bolts.

$$\text{Allowable } s_c = 15\,830 \text{ psi} \qquad \text{(see calculation 3)}$$

$$T = n_b F_{\text{bolt}} r_{bc}$$

$$F_{\text{bolt}} = A_c s_c$$

$$A_c = d_{\text{bolt}} t = 0.75(1.5) = 1.125 \text{ in}^2$$

$$F_{\text{bolt}} = 1.125(15\,830) = 17\,800 \text{ lb}$$

$$n_b = 6 \quad r_{bc} = 5\tfrac{1}{2} \text{ in}$$

$$T = 6(17\,800)5.5 = 587\,000 \text{ in·lb}$$

From calculations 5 and 6, $T_{\text{bolt}} = 170\,000$ in·lb.

SUMMARY OF TORQUE CALCULATIONS

| | Allowable torque, in·lb |
|---|---|
| Shaft | 421 000 |
| Key | 158 000 |
| Hub | 1 632 000 |
| Bolts | 170 000 |

The maximum safe torque for this power-transmission arrangement is 158 000 in·lb. The maximum allowable horsepower is

$$\text{hp} = \frac{Tn}{63\ 000} = \frac{158\ 000(350)}{63\ 000} = 880$$

## PROBLEMS

*13-1.  What torque can a 75-mm-diameter solid Monel shaft transmit safely under steady load conditions?

13-2.  A $1\frac{3}{4}$-in-diameter solid shaft of 302 stainless steel transmits a torque of 31 000 in·lb. Assuming varying load conditions, is this shaft satisfactory?

*13-3.  An AISI 1045 steel shaft transmits a torque of 1.1 kN·m under shock loading.
   a.   What diameter solid shaft is required?
   b.   What size hollow shaft is required if the outside diameter is twice the inside diameter?

13-4.  What is the length of a $\frac{1}{2}$-in AISI 1095 steel rod that can be twisted through one-half revolution without exceeding a shearing stress of 22 000 psi?

*13-5.  An AISI 1045 steel rod 20 mm in diameter and 300 mm long is subjected to a torque of 180 N·m. What will be the angle of twist?

13-6.  a.   Find the horsepower which a 3-in shaft can transmit at 200 rpm if the torsional shearing stress is not to exceed 7000 psi.
   b.   What material would you recommend for the shaft (assume the stress specified in part a includes shock loading conditions)?

*13-7.  Find the diameter of a solid AISI 1020 steel shaft to transmit 110 kW at 2200 rpm under shock loading.

*13-8.  Design a hollow shaft with $d_0 = 1.8d_i$ for the conditions in Prob. 13-7.

13-9.  At what revolutions per minute would you operate a 12-in-diameter AISI 1020 steel shaft to transmit 6000 hp at a maximum shear stress of 5000 psi?

13-10.  A 302 stainless-steel shaft has the following specifications. Outside diameter = 0.500 in

Inside diameter = 0.375 in
Length = 3.25 in
Loading = varying
rpm = 10 000
Find *a*. Maximum safe horsepower. *b*. Angle of twist (degrees).

**\*13-11.** A gear transmitting 18 kW at 130 rpm is fastened to a 75-mm-diameter AISI 1045 steel shaft by means of an AISI 1045 steel flat key 20 mm wide by 12 mm thick. If the permissible shear and bearing stresses are $55(10^6)$ and $110(10^6)$ Pa, respectively, what length of key is required?

**13-12.** A torque of 90 000 in·lb is transmitted from a 6-in AISI 1095 steel shaft to a pulley by means of two AISI 1095 steel square keys, each 6 in long. If $s_s$ = 10 000 psi and $s_c$ = 18 000 psi, respectively, what should be the dimensions of the keys?

**13-13.** A $\frac{3}{4}$-in-diameter AISI 1045 steel shaft transmits 10 hp at 500 rpm to a pulley which is keyed to the shaft by an AISI 1045 steel flat key $\frac{3}{16}$ in wide by $\frac{1}{8}$ in thick and 4 in long. What are the safety factors of the key based on the allowable stresses of 8000 psi in shear and 16 000 psi in bearing?

**13-14.** What torque can be transmitted by the eight steel bolts shown in the coupling of Fig. Prob. 13-14? The bolts, $\frac{7}{8}$ in in diameter, are in bearing against the $\frac{1}{2}$-in web of the coupling. The bolt-circle diameter is 11 in. Allowable stresses are as in Prob. 13-12.

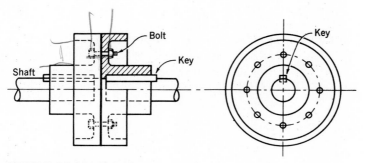

FIGURE PROBLEM 13-14

**\*13-15.** An AISI 1045 steel shaft transmits 373 kW at 200 rpm. The flange couplings use 10 AISI 1020 steel bolts 20 mm in diameter arranged in a circle with 300 mm diameter. The web of the coupling is 25 mm thick. What are the stresses in the bolts?

**13-16.** An AISI 1095 steel shaft 1 in in diameter is twisted through an angle of 2° in a length of 10 ft. What will be the shearing stress?

**\*13-17.** An AISI 1045 steel shaft 3 m long is subjected to a torque of 200 N·m. If the shearing stress is not to exceed 83 MPa, and if the angle of twist may not exceed 0.035 rad, what diameter is needed?

**13-18.** A hollow AISI 1045 steel shaft 5 in OD and 4 in ID transmits 180 hp at a speed of 240 rpm. The length of shaft subjected to the torque is 10 ft.

*a.* What is the angle of twist?

*b.* What is the shearing stress?

# CHAPTER
# 14
# Combined Stresses

## 14-1 PRINCIPLE OF SUPERPOSITION

It has been shown that either direct axial loading (Chap. 7) or bending (Chap. 12) can induce tensile (or compressive) stresses. Furthermore, we have seen that a shearing stress may result from either direct shearing forces (Chap. 7) or from torsion (Chap. 13).

Stresses of the *same* kind which act simultaneously on a given area (or point) may be added to give their combined effect or resultant. If such stresses are collinear (act along the same line) they may be added algebraically. If the stresses are not collinear, they must be added vectorially. These procedures resemble the methods used earlier in this text (Chap. 2) to resolve concurrent force systems into resultants.

This principle of superposition will be applied in this chapter to the following situations.

1. Combined tension or compression
    (a) Combined axial and bending stresses in beams
    (b) Eccentrically loaded short columns
    (c) Eccentrically loaded machine members
2. Combined direct and torsional shear
    (a) Eccentrically loaded bolted joints

Members which are simultaneously loaded in tension (or compression) and shear are designed by considering the combined effect of these *different* kinds of stress. Of the several methods available for such combinations, two are demonstrated in this chapter: the maximum-principal-stress method and the maximum-shear-stress method. These methods will be applied to:

3. Combined tension and shear
    (a) Combined bending and torsional shearing stresses which occur in shafts

339

## 14-2 COMBINED AXIAL AND BENDING STRESSES

Consider a simply supported beam with a uniformly distributed load which is subjected to an axial tensile force $F$, as in Fig. 14-1. The problem

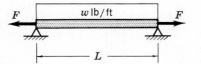

□ FIGURE 14-1 Beam subjected to combined axial and bending stresses.

of determining the combined stress at any point in the beam may be subdivided by examining the effect of the axial load and the effect of the uniform load separately and then superimposing these effects to give the combined result.

The axial force induces a tensile stress on all cross sections of the beam given by

$$s_1 = \frac{F}{A}$$

where $A$ is the cross-sectional area of the beam. This stress is assumed to act equally at all points of a given cross section, as shown in Fig. 14-2.

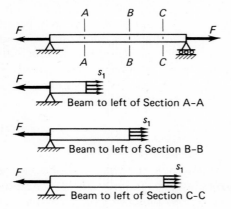

FIGURE 14-2 Uniform stress due to axial load.

The uniform load $w$ causes the beam to bend and induces a stress at each cross section which varies from maximum tension at the bottom fibers through zero at the neutral axis to maximum compression at the top fibers. The bending stresses can be determined from

$$s_2 = \frac{Mc}{I} \quad \text{(see Chap. 12)}$$

Since a design is usually concerned with critical conditions, the extreme fiber values of $s_2$ are most often required. Furthermore, the largest of these extreme fiber values tends to occur at the cross section where $M$ (bending moment) is maximum. For the case of a uniformly distributed load on a simple span, $M_{max}$ occurs at the middle cross section of the beam. The bending-stress distribution at this cross section is shown in Fig. 14-3.

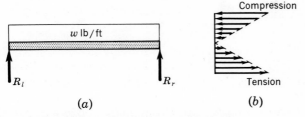

(a)          (b)

FIGURE 14-3   Bending stress distribution.

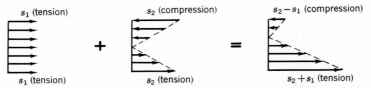

FIGURE 14-4   Combined stress distribution by addition of axial and bending stresses.

The resultant combined stresses are determined by algebraically adding the individual axial and bending stresses,* as in Fig. 14-4. Then

$$s = s_1 \pm s_2$$

or
$$s = \frac{F}{A} \pm \frac{Mc}{I} \tag{14-1}$$

In the case of axial tension and bending of a simply supported beam, when the bottom fibers are considered, both $F/A$ and $Mc/I$ are tensile stresses; thus, the plus sign is proper. When the axial stress is tension, while $Mc/I$ is compression in the top fibers, the values must be subtracted. Figure 14-5 shows a plot of combined stresses for top and bottom fibers at all sections. The axial stress $F/A$ could be either tension or compression, depending upon the nature of the load $F$; and we have seen that $Mc/I$ can be either tension or compression, depending upon which fiber of the beam is considered. In general, when $F/A$ and $Mc/I$ are both tension (or both compression), they should be added, but when one is tension and the other is compression, they should be subtracted. The resulting

* Valid for small deflections.

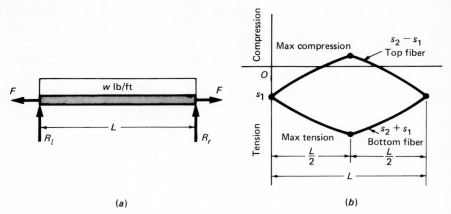

FIGURE 14-5   (a) Simply supported beam with uniform load and axial load. (b) Distribution of stresses in top and bottom fibers of the beam.

combined value should always be identified by stating whether it is tension or compression. These procedures will be clarified in the sample problems which follow.

**Sample Problem 1** A simply supported, W 12 × 65 beam, 16 ft long, carries concentrated loads of 6000 lb at each quarter point and is subjected to an axial tensile force of 25 000 lb applied at the end sections.

(a)   Find maximum combined tensile stress and maximum combined compressive stress.

(b)   If it were necessary to make a 1½-in hole in the web of this beam at the center cross section so that a water pipe could be accommodated, where on this cross section would you recommend that the hole center be located?

**Solution a:**   The weight of the beam is 65(16) = 1040 lb. The total vertical load on the beam is 6000 + 6000 + 6000 = 18 000 lb. Since the weight of the beam is only (1040/18 000)(100) = 5.8 percent of the total vertical load, it may be neglected without excessive error (approximately 4 percent error occurs here).

Figure 14-6 shows the beam with its shear-force and bending-moment diagrams. Note that the 25 000 lb axial force does not affect these diagrams. For the W 12 × 65, $A = 19.1$ in², $S = 88.0$ in³, and $d = 12.12$ in (App. B, Table 4).

***Direct Stress:***

$$s_1 = \frac{F}{A} = \frac{25\ 000}{19.1} = 1310 \text{ psi}\quad \text{(tension)}$$

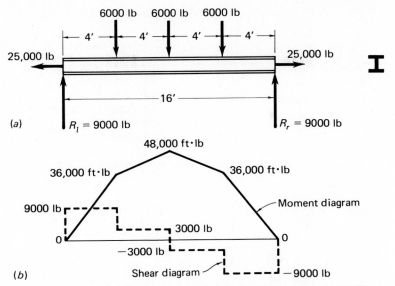

FIGURE 14-6   (a) Beam diagram for Sample Problem 1. (b) Shear-force and bending moment diagrams.

### Bending Stress:

$$s_2 = \frac{Mc}{I} = \frac{M}{S}$$

The shear diagram indicates that the center cross section is the location of the maximum moment. A free-body diagram of the left half of the beam facilitates calculating $M_{\text{max}}$ (Fig. 14-7).

$\Sigma M = 0$ about an axis through the center cross section gives

$$M_{\text{max}} = 9000(8) - 6000(4) = 72\,000 - 24\,000 = 48\,000 \text{ ft·lb}$$

$$s_2 = \frac{M}{S} = \frac{48\,000(12)}{88.0} = 6550 \text{ psi}$$

$$s_2 = 6550 \text{ psi} \begin{cases} \text{tension at bottom fiber} \\ \text{compression at top fiber} \end{cases}$$

Therefore, from Eq. (14-1), at center cross section,

Top fiber $s = s_2 - s_1 = 6550 - 1310$
Top fiber $s = 5240 \text{ psi}$   (maximum compression)
Bottom fiber $s = s_2 + s_1 = 6550 + 1310$
Bottom fiber $s = 7860 \text{ psi}$   (maximum tension)

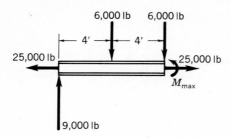

FIGURE 14-7 Free-body diagram of left half of beam for Sample Problem 1.

**Solution b:** The distribution of combined stress at the center cross section is shown in Fig. 14-8.

The location of zero combined stress would be the most preferable position for a hole in the web. The level of zero stress can be found from Fig. 14-8 by similar triangles.

$$h = \left(\frac{7860}{7860 + 5240}\right)(12.12) = \left(\frac{7860}{13\ 100}\right)(12.12) = 7.27 \text{ in}$$

The center of the $1\frac{1}{2}$-in hole should be located 7.27 in above the bottom of the lower flange.

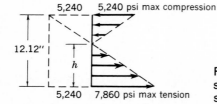

FIGURE 14-8 Location of point of zero stress from distribution of combined stress at center cross section.

**\*Sample Problem 2** A 0.45-m cantilever member is subjected to a load $F = 27$ kN, as shown in Fig. 14-9a. Find the maximum tensile and compressive stresses. The centroid of the area is 50 mm from the top.

**Solution:** Force $F$ can be resolved into vertical and horizontal components $F_y$ and $F_x$.

$$F_y = F \sin 30° = 27(0.5) = 13.5 \text{ kN} = 13.5(10^3) \text{ N}$$
$$F_x = F \cos 30° = 27(0.866) = 23.4 \text{ kN} = 23.4(10^3) \text{ N}$$

The beam shown in Fig. 14-9b may now be treated as a combined-stress problem.

*Direct Stress:*

$F_x$ causes uniform compression at all sections.

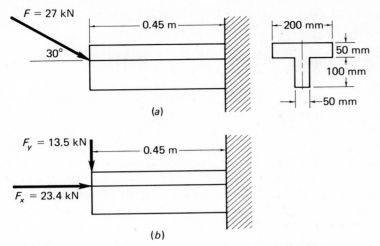

FIGURE 14-9 (a) Beam diagram for Sample Problem 2.
(b) Resolution of force F into horizontal and vertical components.

$$A = 200(50) + 50(100) = 15\,000 \text{ mm}^2 = 15(10^3) \text{ mm}^2 = 15(10^{-3}) \text{ m}^2$$

$$s_1 = \frac{Fx}{A} = \frac{23.4(10^3)}{15(10^{-3})} = 1.56(10^6) \text{ Pa}$$

**Bending Stress:**

$$s_2 = \frac{Mc}{I} \qquad M_{\max} = F_y(L) = (13.5 \times 10^3)(0.45)$$

$$= 6.075(10^3) \text{ N} \cdot \text{m (at wall)}$$

$$I_x = \frac{200(50)^3}{12} + 10\,000(25)^2 + \frac{50(100)^3}{12} + 5000(50)^2$$

$$I_x = 2.08(10^6) + 6.25(10^6) + 4.17(10^6) + 12.5(10^6)$$
$$= 25(10^6) \text{ mm}^4 = 25(10^{-6}) \text{ m}^4$$

Top fiber $s_2 = \dfrac{(6.075 \times 10^3)(50 \times 10^{-3})}{25(10^{-6})} = 12.1(10^6) \text{ Pa (tension)}$

Bottom fiber $s_2 = \dfrac{(6.075 \times 10^3)(100 \times 10^{-3})}{25(10^{-6})} = 24.3(10^6) \text{ Pa (compression)}$

**Combined Stresses:**

Top fiber $s = s_2 - s_1 = 12.1(10^6) - 1.56(10^6) = 10.54(10^6) \text{ Pa (tension)}$
$$\text{Say, } s = 10.5 \text{ MPa (tension)}$$

Bottom fiber $s = s_2 + s_1 = 24.3(10^6)$
$$+ 1.56(10^6) = 25.86(10^6) \text{ Pa (compression)}$$
$$\text{Say, } s = 25.9 \text{ MPa (compression)}$$

## 14-3 ECCENTRICALLY LOADED SHORT COMPRESSION MEMBERS

The reader will recall from Sec. 2-8 that a member under compression which does not tend to buckle or bend is called a short compression member.

Let us examine a short compression member that has a load $F$ applied eccentric to the centroid of its cross section, Fig. 14-10$a$. Now imagine two equal and opposite forces $F_1$ equal to the original load $F$, applied at the centroid, as indicated in Fig. 14-10$b$. The loading has not changed, since the sum of the additional loads equals zero. We have effectively replaced the eccentric force system of Fig. 14-10$a$ by a downward force $F_1$ at the centroid, acting to compress the member, and, the upward force $F_1$ and the force $F$, which together form a couple $Fe$ which subjects the member to a clockwise moment, Fig. 14-10$c$. The distance $e$, called the

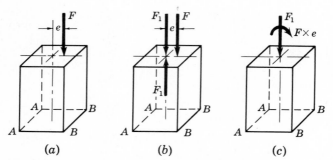

$(a)$                   $(b)$                   $(c)$

FIGURE 14-10   (a) Eccentrically loaded short compression member. (b) Introduction of equal and opposite forces at centroid. (c) Equivalent system for eccentrically loaded short compression member.

*eccentricity*, is the perpendicular distance from the centroid to the line of action of the force $F$. This problem can now be solved by applying the principles of combined axial and bending stresses, as in Sec. 14-2. The compressive stress due to the compressive load $F_1 = F$ is

$$s_1 = \frac{F}{A}$$

The flexural stress due to the moment $Fe$ is

$$s_2 = \frac{Mc}{I} = \frac{Fec}{I}$$

Then the combined stress at the base is

$$s = s_1 \pm s_2$$

or
$$s = \frac{F}{A} \pm \frac{Fec}{I}$$                           (14-2)

At edge $AA$ the negative sign will be used since the direct stress due to force $F$ is compression and the bending stress due to the moment is tension. For edge $BB$ the stresses will be added, since both the direct and bending stresses are compressive.

**Sample Problem 3** A short compression member 6 by 8 in in cross section and 10 in high has a load of 25 000 lb applied $1\frac{1}{2}$ in from the centroid of the cross section, as indicated in Fig. 14-11. Determine the stresses at sides $AB$ and $CD$.

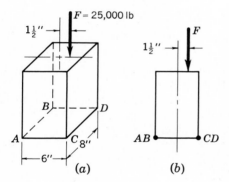

FIGURE 14-11   Diagram for Sample Problems 3 and 4.

**Solution:** The direct compressive effect of the 25 000-lb force will cause a compressive stress on the area $ABCD$.

$$s_1 = \frac{F}{A}$$

$$s_1 = \frac{25\ 000}{6(8)} = \frac{25\ 000}{48} = 521 \text{ psi (compression)}$$

As a result of the eccentricity, there will be a moment equal to $Fe$ tending to bend the short column. The bending stress due to this moment will be

$$s_2 = \frac{Fec}{I}$$

where $F = 25\,000$ lb

$\quad e = 1\frac{1}{2}$ in

$\quad c = 3$ in

$\quad I = \dfrac{bh^3}{12} = \dfrac{8(6)^3}{12} = 144$ in$^4$    (Note that the eccentric load tends to rotate the base about a centroidal axis perpendicular to the 6-in side)

$\quad s_2 = \dfrac{25\,000(1\frac{1}{2})(3)}{144} = 782$ psi

At side $AB$ there will be compression due to the direct action of the load $F$ and tension due to the bending effect. Since the stresses are not the same type, we use the negative sign in Eq. (14-2).

$$s_{ab} = \frac{F}{A} - \frac{Fec}{I}$$

$$s_{ab} = 521 - 782 = -261 \text{ psi (tension)}$$
$$\text{Say, } s_{ab} = 260 \text{ psi (tension)}$$

The negative sign indicates that the bending tensile stress will dominate and the resultant stress is tension.

At side $CD$ there will be compression due to the direct action of the load $F$ and compression due to the bending effect. Since both stresses are of the same type, we use the positive sign in Eq. (14-2).

$$s_{cd} = \frac{F}{A} + \frac{Fec}{I}$$

$$s_{cd} = 521 + 782 = 1303 \text{ psi (compression)}$$
$$\text{Say, } s_{cd} = 1300 \text{ psi (compression)}$$

**Sample Problem 4**  If, for the conditions given in Sample Problem 3, the maximum compressive stress is specified as 1000 psi, what maximum load $F$ can be applied?

**Solution:**  We have seen, in Sample Problem 3, that the stress at side $AB$ will be tension. Therefore, we shall consider only side $CD$ where a compressive stress occurs.

From Eq. (14-2),

$$s = s_1 + s_2 = \frac{F}{A} + \frac{Fec}{I}$$

$$1000 = \frac{F}{48} + \frac{F(3)(1\frac{1}{2})}{144} = 0.0208F + 0.0313F$$

$$0.0521F = 1000$$

$$F = 19\,200 \text{ lb (maximum)}$$

## 14-4 ECCENTRIC LOADING OF MACHINE MEMBERS

The previous analysis for an eccentrically loaded short compression member can be extended to eccentric loading of machine members, such as the press frame of Fig. 14-12a.

Let us determine the stresses developed on plane $AA$ at edges $a$ and $b$ as a result of the force $F$ applied to the jaws of the C frame.

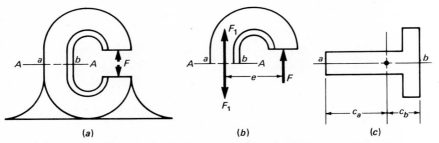

FIGURE 14-12 (a) Eccentrically loaded C frame. (b) Introduction of equal and opposite forces at centroid of section $AA$. (c) Section $AA$.

Figure 14-12b shows the portion of the press frame above the plane $AA$. At the centroid of the cross section, imagine that two equal and opposite forces $F_1 = F$ are applied. Then the downward force $F_1$ and the punching force $F$ form a counterclockwise couple, or counterclockwise bending moment, equal to $Fe$. This moment will create a compressive stress at edge $a$ and a tensile stress at edge $b$. The upward force $F_1$ creates tension evenly distributed across plane $AA$. As a result, tensile stresses are developed at edges $a$ and $b$. The resultant stresses at $a$ and $b$ will be the sum of stresses due to tension and bending.

At edge $a$ due to direct tension,

$$s_1 = \frac{F}{A}$$

At edge $a$ due to bending (compression),

$$s_2 = \frac{Mc_a}{I} = \frac{Fec_a}{I}$$

where $c_a$ = distance from neutral axis (centroid) to edge $a$, in
$I$ = moment of inertia of the cross section, in$^4$

Therefore, the combined stress at edge $a$ will be

$$s = s_1 - s_2$$

or

$$s = \frac{F}{A} - \frac{Fec_a}{I}$$

Whether the resultant stress is tension or compression will be governed by the values of $s_1$ and $s_2$.
At edge $b$ due to bending (tension),

$$s_2 = \frac{Mc_b}{I} = \frac{Fec_b}{I}$$

where $c_b$ = distance from neutral axis (centroid) to edge $b$, in
$I$ = moment of inertia of the cross section, in$^4$

At edge $b$ due to direct tension,

$$s_1 = \frac{F}{A}$$

Therefore, the combined stress at edge $b$ will be

$$s = s_1 + s_2$$

or

$$s = \frac{F}{A} + \frac{Fec_b}{I}$$

This resultant stress will be tension since both $s_1$ and $s_2$ are tensile stresses.

**Sample Problem 5** A punch-press frame, as in Fig. 14-13$a$, applies a maximum load of $F = 15\ 000$ lb. The distance $l$ from the line of action of the force to the inside surface of the frame is 12 in. The dimensions of the cross section $AA$ are as given in Fig. 14-13$b$. Determine the stresses on the inside and outside surfaces of this section.

**Solution:** Since the force $F$ is eccentric to the centroid of the cross section $AA$, this problem must be analyzed for combined stress. On section $AA$ there will be a direct tensile force $F$ pulling the section apart and a bending moment equal to $F$ times the perpendicular distance from the line of action of the force to the centroid of the section $AA$. The direct

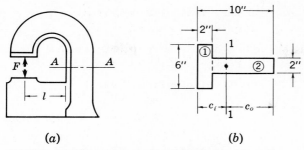

(a)                                   (b)

FIGURE 14-13   Diagram for Sample Problem 5.

tensile force will create a tensile stress on both the inside and outside surfaces. The bending moment will cause a tensile stress on the inside surface and a compressive stress on the outside surface. Therefore, the stress on the inside surface will be

$$s_i = \frac{F}{A} + \frac{Fec_i}{I}$$

and the stress on the outside surface will be

$$s_0 = \frac{F}{A} - \frac{Fec_0}{I}$$

$$F = 15\,000 \text{ lb}$$
$$A = A_1 + A_2 = (6 \times 2) + (8 \times 2) = 12 + 16 = 28 \text{ in}^2$$

In order to determine $e$, the location of the centroid from the inside surface must be found. Taking the moment of areas about this side

$$c_i = \frac{(12 \times 1) + (16 \times 6)}{28} = \frac{12 + 96}{28} = \frac{108}{28} = 3.86 \text{ in}$$

Therefore, $e = 12 + 3.86 = 15.86$ in.

To find $I$ we must find the moment of inertia of the entire figure about axis 1-1.

$$I_1 \text{ about axis 1-1} = \frac{6(2)^3}{12} + 12(2.86)^2 = 4 + 98.2 = 102.2$$

$$I_2 \text{ about axis 1-1} = \frac{2(8)^3}{12} + 16(2.14)^2 = 85.3 + 73.3 = 158.6$$

$$I = 260.8 \text{ in}^4 \qquad \text{Say, } I = 261 \text{ in}^4$$

$$s_i = \frac{15\ 000}{28} + \frac{15\ 000(15.86)(3.86)}{261}$$

$$= 536 + 3520 = 4056\ \text{psi} \qquad \text{(tension)}$$

$$\text{Say, } s_i = 4060\ \text{psi} \qquad \text{(tension)}$$

$$s_0 = \frac{15\ 000}{28} - \frac{15\ 000(15.86)(6.14)}{261}$$

$$= 536 - 5600 = -5064\ \text{psi} \qquad \text{(compression)}$$

$$\text{Say, } s_0 = 5060\ \text{psi} \qquad \text{(compression)}$$

## 14-5 ECCENTRICALLY LOADED BOLTED JOINTS

Whenever possible, a bolted joint should be constructed so that the line of action of the resultant force on the joint passes through the centroid of the area of the group of bolts.

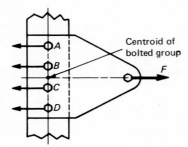

FIGURE 14-14   Load on bolted joint without eccentricity.

Load $F$, Fig. 14-14, has its line of action passing through the centroid of the area of the group of bolts $A, B, C$, and $D$. Each of the bolts will then resist a load of $F/4$, which could cause shear failure of the bolts. Thus, the resisting shear stress developed will be

$$s_1 = \frac{F}{4A}$$

where $A$ = cross-sectional area of one bolt, in²

In Fig. 14-15a, the line of action of force $F$ does not pass through the centroid of the bolt group. The line of action is eccentric to the centroid by the distance $e$. This results in a direct pull on the bolts, plus a counterclockwise moment about the centroid due to the eccentricity of the applied force. The direct shear stress developed is

$$s_1 = \frac{F}{4A}$$

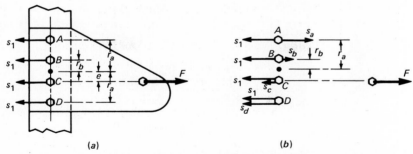

FIGURE 14-15   (a) Eccentrically loaded bolted joint. (b) Direct and rotational shear stresses developed in the bolts.

The moment of the force $F$ is

$$M = Fe$$

In order to maintain equilibrium, the sum of the resisting moments developed in the bolts must be equal and opposite to $Fe$. Hence, the resisting moments act in a clockwise direction.

$$Fe = M = M_a + M_b + M_c + M_d$$

where $M_a$ = resisting moment of bolt $A$, in·lb
$M_b$ = resisting moment of bolt $B$, in·lb
$M_c$ = resisting moment of bolt $C$, in·lb
$M_d$ = resisting moment of bolt $D$, in·lb

The resisting moment $M_a$ on bolt $A$ is

$$M_a = F_a r_a$$

where $F_a$ = rotational resisting force of bolt $A$, lb
$r_a$ = moment arm from line of action of $F_a$ to the centroid, in

But the rotational resisting force will be a function of the resisting shear stress developed and the area of the bolt.

$$F_a = A_a s_a$$

where $A_a$ = area of the bolt, in²
$s_a$ = resisting rotational shear stress in bolt $A$, psi

Therefore,

$$M_a = F_a r_a = A_a s_a r_a$$

Since the bolts are the same size and the distance of bolt $D$ from the centroid is the same as that of bolt $A$, the resisting moment of bolt $D$ will be equal to that of bolt $A$.

$$M_a = M_d$$

The rotational shear stress developed in bolt $D$ will be equal and opposite to that in bolt $A$.

$$s_a = -s_d$$

Similarly, the resisting moment developed by bolt $B$ equals

$$M_b = F_b r_b = A_b s_b r_b$$

where $M_b$ = resisting moment of bolt $B$, in·lb
$\quad F_b$ = rotational resisting force of bolt $B$, lb
$\quad r_b$ = moment arm from line of action of $F_b$ to centroid, in
$\quad A_b$ = area of bolt $B$, in²
$\quad s_b$ = resisting rotational shear stress developed in bolt $B$, psi

The resisting moment of bolt $C$ will be equal to that of bolt $B$ since the area of the bolts and the distance from the centroid are the same. The stress in bolt $C$ will be equal and opposite to that in bolt $B$.

$$s_b = -s_c$$

Therefore,

$$Fe = A_a s_a r_a + A_b s_b r_b + A_c s_c r_b + A_d s_d r_a$$
$$= 2(A_a s_a r_a) + 2(A_b s_b r_b)$$

Figure 14-15$b$ indicates the directions in which the stresses will act. The direct shear stress and rotational shear stress on each bolt can be added directly since their lines of action are parallel. The resultant shear stresses in the bolts will be

Bolt $A$: $s_s = s_1 - s_a$
Bolt $B$: $s_s = s_1 - s_b$
Bolt $C$: $s_s = s_1 + s_c$
Bolt $D$: $s_s = s_1 + s_d$

**Sample Problem 6** What is the greatest shearing stress in the bolted connection shown in Fig. 14-16? Bolts are $\frac{7}{8}$ in in diameter.

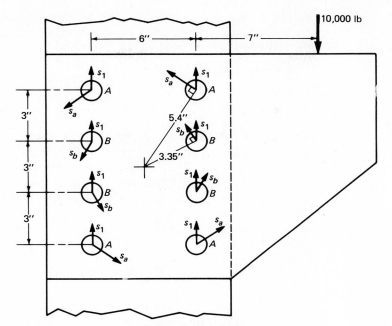

FIGURE 14-16    Diagram for Sample Problem 6.

**Solution:**    The load produces a direct shearing stress which is assumed to be the same for all bolts. The area for a $\frac{7}{8}$-in bolt $= 0.6$ in². Then, the direct shear stress on each bolt is

$$s_1 = \frac{10\ 000}{(0.6)8} = 2080 \text{ psi}$$

The load also tends to rotate the plate in a clockwise direction. The moment due to the load is

$$M = Fe = 10\ 000(10) = 100\ 000 \text{ in·lb}$$

For equilibrium, the resisting moment of the bolts must be equal to 100 000 in·lb. The shearing stresses in the bolts will be proportional to their respective distances from the centroid of the group. These distances are 5.4 in for bolts $A$ and 3.35 in for bolts $B$ (distances calculated from other dimensions in the figure).

Let $s_a$ represent the rotational shearing stress due to the moment in the outermost bolts $A$; then the stress in bolts $B$ will be $s_b = (3.35/5.4)s_a$. Taking moments about the centroid of the group, we have

$$Fe = 4(A_a s_a r_a) + 4(A_b s_b r_b)$$

$$100\,000 = 4[(0.6)s_a(5.4)] + 4\left[(0.6)\left(\frac{3.35}{5.4}\right)s_a(3.35)\right]$$

$$= 12.96 s_a + 5.0 s_a$$

$$17.96 s_a = 100\,000$$

$$s_a = 5570 \text{ psi}$$

The actual stress will be the resultant of 2080 and 5570 psi. Before this resultant can be determined, the directions of the stresses must be known.

The 2080 psi is an upward stress for each bolt. The 5570 psi for each bolt $A$ is perpendicular to the radius arm of the bolt. The directions vary with the position of the bolt, as can be seen from the figure. By inspection, it is evident that at the upper right-hand bolt $A$ and the lower right-hand bolt $A$ the resultant is greatest.

The stress of 5570 psi makes an angle $\theta$ with the horizontal that is equal to the angle that the radius makes with the vertical (see Fig 14-17a).

$$\tan \theta = \frac{3}{4.5} = 0.667$$

$$\theta = 33.7°$$

Resolving $s_a = 5570$ psi into its vertical and horizontal components (Fig. 14-17b), we have

$$s_{ay} = 5570 \sin 33.7° = 3080 \text{ psi}$$

$$s_{ax} = 5570 \cos 33.7° = 4630 \text{ psi}$$

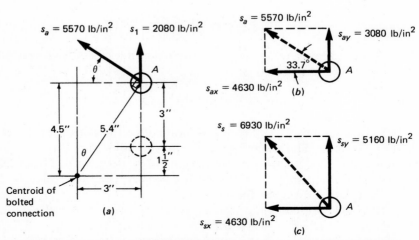

FIGURE 14-17   (a) Stresses developed in bolt A. (b) Components of rotational shear stress in bolt A. (c) Resultant shear stress in bolt A.

The vertical and horizontal components of the resultant are (Fig. 14-17c)

$$s_{sy} = 2080 + 3080 = 5160 \text{ psi}$$
$$s_{sx} = 4630 \text{ psi}$$

The resultant is

$$s_s = \sqrt{5160^2 + 4630^2} = 6930 \text{ psi} \quad \text{(maximum)}$$

## 14-6 SHEAR STRESS DUE TO TENSION OR COMPRESSION

If an axial force is applied to a short compression member, the stress, which might be either tension or compression, is uniformly distributed over the normal cross section $AA$. There is no component of force parallel to the section, and consequently, there is no shearing stress produced on the section.

If a section $BB$ making some angle $\alpha$ with the normal section is chosen, as shown in Fig. 14-18, then the axial force may be resolved into two components, one at right angles to section $BB$, called $F_n$, and the other parallel to section $BB$, called $F_s$. Then

$$F_n = F \cos \alpha$$
$$F_s = F \sin \alpha$$

The area resisting these forces will be the area of the inclined section and equal to $A/\cos \alpha$, where $A$ is the area of the normal section. Then

$$s_n = \frac{F_n}{A_n} = \frac{F \cos \alpha}{A/\cos \alpha} = \frac{F \cos^2 \alpha}{A}$$

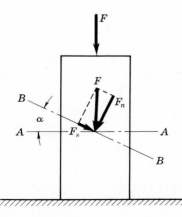

FIGURE 14-18   Member in axial compression.

This value is a maximum when the angle $\alpha = 0°$. That is, the maximum tension or compression will take place on the normal cross section. Also,

$$s_s = \frac{F_s}{A_s} = \frac{F \sin \alpha}{A/\cos \alpha} = \frac{F}{A}(\sin \alpha \cos \alpha) = \frac{F}{A}\frac{\sin 2\alpha}{2}$$

The value of $\sin 2\alpha$ is largest when $\alpha = 45°$; therefore, the value of $s_s$ is a maximum on a plane making an angle of $45°$ with the normal cross section.

## 14-7 TENSION OR COMPRESSION DUE TO SHEAR

When a member such as the shaft in Fig. 14-19$a$ is subjected to pure torsion, direct shear stresses $s_1$ are developed on a very small square surface area. From Sec. 12-2 we know that, to maintain equilibrium, other shear stresses, $s_2$ equal to $s_1$, are induced on the section at right angles to $s_1$, Fig. 14-19$b$. Analysis of the $45°$ plane $AA$ (Fig. 14-20$a$) indicates that

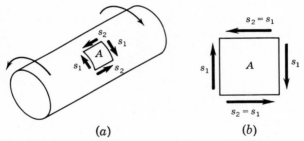

(a)   (b)

FIGURE 14-19   (a) Shaft in pure torsion. (b) Shear stresses on small surface section A.

tensile stresses are acting to pull the section apart (Fig. 14-20$b$). On $45°$ plane $BB$, compressive stresses are acting to compress the section (Fig. 14-20$c$). No shear exists on these $45°$ planes. If some other planes, such

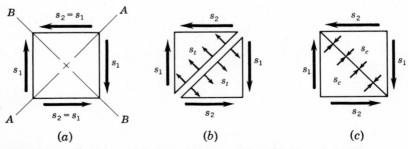

(a)   (b)   (c)

FIGURE 14-20   (a) Shear stresses on small surface section. (b) Tension at plane $AA$ due to shear. (c) Compression at plane $BB$ due to shear.

as 15° or 20° planes, were analyzed, it would be found that combinations of shear and tension or compression exist.

## 14-8 COMBINED BENDING AND TORSION

Shafts which transmit or receive power by means of gear, belt, or chain drives develop not only a torsional shear stress but a bending stress due to the forces acting on the shaft. Such a shaft is shown in Fig. 14-21.

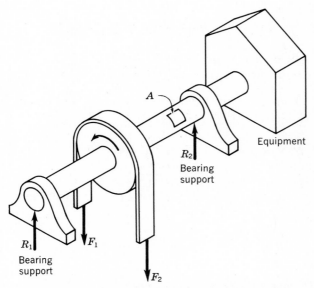

FIGURE 14-21   Shaft subjected to combined bending and torsion.

Let us examine the stresses acting on an extremely small section $A$ of the surface of the shaft. When this outer fiber section is at the bottom, as a result of the rotation of the shaft, there exists both a torsional and tensile effect on the section. Torsion will create a shear stress $s_s$ and the bending will cause a tensile stress $s_t$ on this section (Fig. 14-22).

In Secs. 14-6 and 14-7 we have seen how shear can result in tensile and compressive stresses and how tensile and compressive loads can

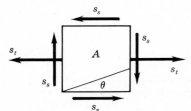

FIGURE 14-22   Tensile and shear stresses on small surface section of shaft.

create shear stresses. In a similar manner, the combined effect of the shear and tensile stresses on section $A$ (Fig. 14-22) will result in the following.

1.   A maximum resultant tensile (or compressive) stress, greater than the tensile (or compressive) stress due to bending alone, called the *principal stress*.
   This maximum tensile (or compressive) stress is given by

$$(s_t)_{max} = \frac{s_t}{2} + \sqrt{s_s^2 + \left(\frac{s_t}{2}\right)^2} \qquad (14\text{-}3)$$

   The angle $\theta$ of the plane on which the maximum resultant tensile (or compressive) stress will occur can be found from

$$\tan 2\theta = \frac{2s_s}{s_t} \qquad (14\text{-}4)$$

For design purposes, Eq. (14-3) is recommended when *brittle* materials are involved.

2.   A maximum value of shear stress greater than the shear stress due to torsion alone. This *maximum shear stress* is given by

$$(s_s)_{max} = \sqrt{s_s^2 + \left(\frac{s_t}{2}\right)^2} \qquad (14\text{-}5)$$

For design purposes, Eq. (14-5) is recommended when *ductile* materials are used. Also, the *ASME Boiler and Pressure Vessel Code* specifies that members subjected to combined stresses are to be designed based on maximum shear stress.
   For a solid or hollow shaft simultaneously subjected to torsion and bending, the shear stress due to the transmitted torque is

$$s_s = \frac{Tc}{J} = \frac{T}{S'}$$

while the tensile stress due to bending is

$$s_t = \frac{Mc}{I} = \frac{M}{S}$$

where $S' = 2S$ for solid or hollow circular cross sections. Substituting this information into Eq. (14-3),

$$(s_t)_{max} = \frac{M}{2S} + \sqrt{\frac{T^2}{(S')^2} + \frac{M^2}{4S^2}}$$

but $$S' = 2S \qquad \text{and} \qquad (S')^2 = 4S^2$$

$$(s_t)_{max} = \frac{M}{2S} + \frac{1}{2S}\sqrt{T^2 + M^2}$$

$$M_e = (s_t)_{max} S = \frac{M}{2} + \frac{1}{2}\sqrt{T^2 + M^2} \qquad (14\text{-}6)$$

Equation (14-6) is used to calculate $M_e$, the equivalent bending moment. Similarly, Eq. (14-5) can be transformed into the equation for equivalent torque.

$$T_e = (s_s)_{max}S' = \sqrt{T^2 + M^2} \qquad (14\text{-}7)$$

where $M_e$ = equivalent bending moment, in·lb; N·m
$\quad T_e$ = equivalent torque, in·lb; N·m
$\quad M$ = actual bending moment, in·lb; N·m
$\quad T$ = actual torque, in·lb; N·m
$\quad S$ = section modulus of a solid shaft $\pi d^3/32$

$\qquad$ or a hollow shaft $\dfrac{\pi}{32}\left(\dfrac{d_0{}^4 - d_i{}^4}{d_0}\right)$ , in³; m³

$\quad S'$ = polar section modulus of a solid shaft $\pi d^3/16$

$\qquad$ or a hollow shaft $\dfrac{\pi}{16}\left(\dfrac{d_0{}^4 - d_i{}^4}{d_0}\right)$ , in³; m³

The equivalent bending moment $M_e$ will always be greater than $M$, the actual bending moment, owing to the combined effect of bending with torsion. Similarly, $T_e$, the equivalent torque, is always greater than $T$, the actual torque.

*Sample Problem 7** What diameter shaft is required to transmit a torque of 2.7 kN·m with a maximum bending moment of 2.0 kN·m:

(a)  If the allowable shear stress is 55 MPa?
(b)  If the allowable tensile stress is specified as 75 MPa?

**Solution a:**  The maximum shear-stress equation or equivalent torque expression is called for. From Eq. (14-7),

$$T_e = (s_s)_{max}S' = \sqrt{T^2 + M^2}$$

$$(55 \times 10^6)\frac{\pi d^3}{16} = \sqrt{(2.7 \times 10^3)^2 + (2.0 \times 10^3)^2}$$

$$= \sqrt{(7.29 \times 10^6) + (4.0 \times 10^6)} = \sqrt{11.29 \times 10^6}$$
$$= 3.36 \times 10^3 \, \text{N} \cdot \text{m}$$
$$d^3 = \frac{(3.36 \times 10^3)(16)}{(55 \times 10^6)\pi} = 0.311 \times 10^{-3}$$
$$d = 0.678 \times 10^{-1} \, \text{m} = 67.8 \, \text{mm}$$

Use 70-mm-diameter shaft if material is ductile.*

**Solution b:**   The maximum resultant tensile stress equation or equivalent bending-moment expression is called for. From Eq. (14-6),

$$M_e = (s_t)_{max}S = \frac{M}{2} + \frac{1}{2}\sqrt{T^2 + M^2}$$
$$(75 \times 10^6)\frac{\pi d^3}{32} = \frac{2.0 \times 10^3}{2} + \frac{1}{2}(3.36 \times 10^3) = 2.68 \times 10^3$$
$$d^3 = \frac{2.68(10^3)(32)}{75(10^6)\pi} = 0.364 \times 10^{-3}$$
$$d = 0.714 \times 10^{-1} \, \text{m} = 71.4 \, \text{mm}$$

Use 75-mm-diameter shaft if material is brittle.*

**Sample Problem 8**   A $1\frac{1}{2}$-in Monel shaft (Fig. 14-21) transmits 30 hp at 250 rpm to a piece of equipment. Power is transmitted to the shaft by means of a belt drive. The shaft is supported on bearings 18 in apart and the belt pulley is centrally located. The total belt pull on the shaft is 600 lb. Is the shaft satisfactory if a safety factor of 4 based on the ultimate is required?

**Solution:**   From App. B, Table 1, ultimate $s_s = 56\,000$ psi for Monel. Since Monel is a ductile material, the maximum shear-stress expression (Eq. 14-5) will be used

$$(s_s)_{max} = \sqrt{s_s^2 + \left(\frac{s_t}{2}\right)^2}$$
$$s_s = \frac{T}{S'} \qquad \text{hp} = \frac{Tn}{63\,000} \qquad T = \frac{63\,000 \, \text{hp}}{n}$$
$$T = \frac{63\,000(30)}{250} = 7560 \, \text{in} \cdot \text{lb}$$
$$S' = \frac{\pi d^3}{16} = \frac{\pi(1.5)^3}{16} = 0.663 \, \text{in}^3$$

---

*Standard size taken from *Standard Metric Dimensions, Steel Mill Products,* United States Steel Corp. publication, 1977.

$$s_s = \frac{7560}{0.663} = 11\ 400 \text{ psi}$$

$$s_t = \frac{M}{S} \qquad M = \frac{FL}{4} = \frac{600(18)}{4} = 2700 \text{ in} \cdot \text{lb}$$

$$S = \frac{\pi d^3}{32} = 0.332 \text{ in}^3$$

$$s_t = \frac{2700}{0.332} = 8140 \text{ psi}$$

$$(s_s)_{\text{max}} = \sqrt{11\ 400^2 + \left(\frac{8140}{2}\right)^2} = \sqrt{11\ 400^2 + 4070^2}$$

$$(s_s)_{\text{max}} = \sqrt{(130 + 16.6)(10^6)}$$
$$= \sqrt{(146.6)(10^6)} = 12\ 100 \text{ psi}$$

$$\text{Allowable stress} = \frac{56\ 000}{4} = 14\ 000 \text{ psi}$$

Therefore, the shaft is satisfactory.

## PROBLEMS

**\*14-1.** A beam has an axial tensile load of 45 kN and concentrated loads of 6.75 kN at the three quarter points of the beam. What would the maximum and minimum stresses be on a 100- by 150-mm rough-sawn beam simply supported on a 2.1-m span?

**14-2.** A simply supported rough-sawn timber beam 10 in wide and 12 ft long carries a uniform load of 700 lb/ft and an axial tensile load of 24 000 lb. What is the proper depth for the beam if the allowable stress is 1100 psi?

**\*14-3.** Determine the maximum stress at the center cross section of the link shown in Fig. Prob. 14-3.

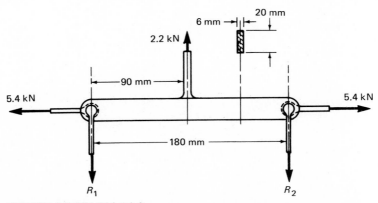

FIGURE PROBLEM 14-3

**14-4.** An S18 × 54.7 simply supported beam carries a concentrated load of 25 000 lb at the center of a 15-ft span, and a total uniform load of 4500 lb, including the weight of the beam. What maximum axial tensile force may be applied to this beam if the allowable stress is 20 000 psi?

**14-5.** Find the maximum and minimum stresses in the wide-flanged member used as an eccentrically loaded short column, as shown in Fig. Prob. 14-5.

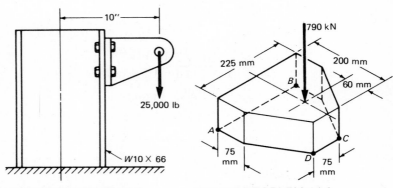

FIGURE PROBLEM 14-5          FIGURE PROBLEM 14-6

**\*14-6.** The short column shown in Fig. Prob. 14-6 is subjected to an eccentric load of 790 kN. Determine the stresses at edges *AB* and *CD*.

**14-7.** A rectangular concrete footing, 3 by 6 ft, supports two columns whose loads act as shown in Fig. Prob. 14-7. Calculate the base pressure at each corner of the footing. Include the weight of the footing.

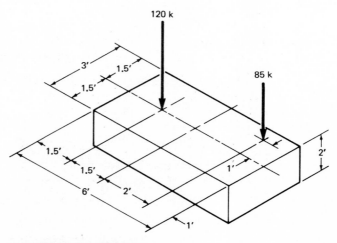

FIGURE PROBLEM 14-7

**14-8.** For the Class 20 cast-iron frame shown in Fig. Prob. 14-8, determine the maximum tensile and compressive stresses that would develop at section *AA* due to an applied load of $F = 12\ 000$ lb.

**14-9.** Find the maximum safe load $F$ that can be applied to the Class 20 cast-iron machine frame of Fig. Prob. 14-9. Assume shock loading.

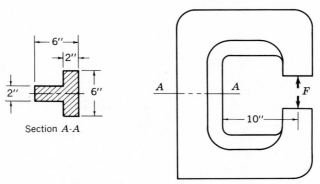

FIGURE PROBLEM 14-8 and 14-9

**\*14-10.** What should be the diameter of the clamp at section *AA* for the arrangement shown in Fig. Prob. 14-10, if the allowable stress is 105 MPa?

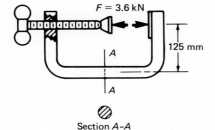

FIGURE PROBLEM 14-10

**\*14-11.** Determine the stresses acting at $A$ and $B$ for the machine link shown in Fig. Prob. 14-11.

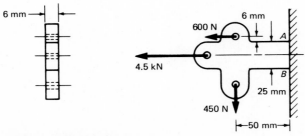

FIGURE PROBLEM 14-11

**14-12.**   Determine the maximum force $F$ that the structural steel channel, loaded as shown in Fig. Prob. 14-12, can take if the allowable tensile and compressive stresses are 20 000 psi.

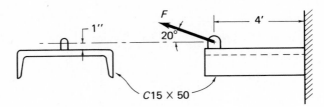

FIGURE PROBLEM 14-12

**\*14-13.**   The davits in Fig. Prob. 14-13 are to be designed to support the lifeboat, which has a mass of 816 kg and has a capacity of 40 people. Assume an average mass of 77 kg per person. The davits are aluminum alloy 6061-T6 with a hollow circular cross section of 150 mm OD and 100 mm ID. Determine the factor of safety $N_u$ for the davits.

FIGURE PROBLEM 14-13

**14-14.**   The allowable tensile and compressive stresses at section $AA$ of Fig. Prob. 14-14 are 11 000 psi. How large may the eccentricity $e$ be?

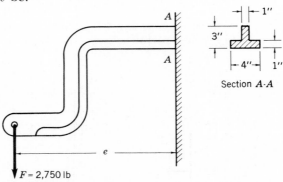

FIGURE PROBLEM 14-14

**14-15.**    What shear stress will be developed in the bolts in Fig. Prob. 14-15? The bolts are ¾-in in diameter and are spaced 2½ in from center to center.

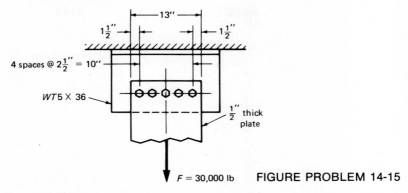

$F = 30,000$ lb      **FIGURE PROBLEM 14-15**

**14-16.**    The bracket in Fig. Prob. 14-16 is connected by ⅞-in-diameter bolts to a vertical member.

    *a.*    Determine the shearing stress in each bolt.

    *b.*    Omitting the center bolt, repeat part *a*.

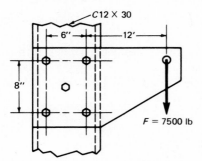

**FIGURE PROBLEM 14-16**

**14-17.**    In order to support a platform, it was found necessary to provide a bracket bolted to a column, as shown in Fig. Prob. 14-17. If $s_s$ is limited to 10 000 psi, what is a proper bolt size?

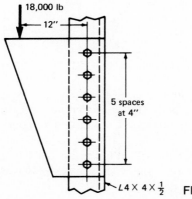

$L4 \times 4 \times \frac{1}{2}$    **FIGURE PROBLEM 14-17**

**14-18.** For Fig. Prob. 14-18, if the bolts are 1 in in diameter and $\theta = 0°$, find the maximum shear stress and identify the bolt or bolts in which it occurs. $F = 10\ 000$ lb.

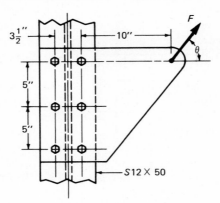

FIGURE PROBLEM 14-18
through 14-20

**14-19.** Same as Fig. Prob. 14-18, except $\theta = 30°$.

**14-20.** For Fig. Prob. 14-20, what maximum force $F$ at $\theta = 90°$ may be applied to the 1-in-diameter bolts, if the allowable shear stress is 15 000 psi?

**14-21.** The $\frac{1}{2}$-in bolts in Fig. Prob. 14-21 are spaced $2\frac{1}{2}$ in center to center.

  *a.* What maximum force $F$ may be applied if the allowable shear stress is 15 000 psi?

  *b.* What maximum force $F$ may be applied if bolt $D$ is removed?

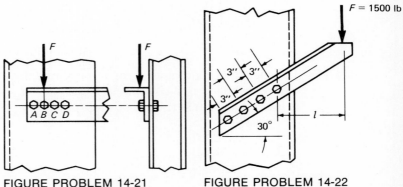

FIGURE PROBLEM 14-21          FIGURE PROBLEM 14-22

**14-22.** The allowable shear stress is 9000 psi for the $\frac{5}{8}$-in bolts in Fig. Prob. 14-22. What is the largest permissible distance $l$?

**14-23.** A steel shaft 3 in in diameter, making 150 rpm, is supported in bearings 6 ft apart. A load of 750 lb acts downward at the middle of the shaft due to belt pull. What horsepower can be transmitted by the shaft if the maximum shearing stress is 6000 psi?

**\*14-24.** Find the diameter of a shaft subjected to a bending moment of 680 N·m and a twisting moment of 900 N·m, if the bending stress does not exceed 70 MPa and the shear stress does not exceed 55 MPa.

**14-25.** Find the diameter of a steel shaft in bearings 6 ft apart to transmit 150 hp at 300 rpm. There is a load of 900 lb at the middle, and the maximum allowable stress is 10 000 psi in bending and 7000 psi in shear.

**\*14-26.** A 100-mm shaft is transmitting 22 kW at 180 rpm and is subjected to a bending moment of 950 N·m. What will be the maximum tensile and shearing stresses developed?

**14-27.** If the bending moment is equal to the twisting moment, what should be the diameter, expressed in terms of the allowable stress, of a solid shaft that is to transmit 120 hp at 125 rpm?
 *a.*   The shaft is made of a ductile material.
 *b.*   The shaft is made of a brittle material.

**\*14-28.** Determine the size of a hollow steel shaft whose inside diameter is equal to one-half the outside diameter, if the bending moment developed is 1.4 kN·m and the torque transmitted is 1.7 kN·m. The allowable shear stress is 55 MPa.

**\*14-29.** Same as Prob. 14-28 for a Class 40 cast-iron shaft with a factor of safety of 10 based on the ultimate stresses. Use SI units.

# CHAPTER
# 15
## Columns

## 15-1 INTRODUCTION

Short compression members which are subjected to axial loads can be treated by $s = F/A$, as was done in Chap. 7. When a short compression member is eccentrically loaded, it can be dealt with by $s = F/A \pm Fec/I$, as in Chap. 14. In both cases, the equations given produce reasonably reliable solutions for actual members. Minor discrepancies may appear due to the following.

1. Nonhomogeneous material
2. Unforeseen or accidental misalignment of loading
3. Slight variations in the straightness of the member
4. Presence of unknown initial stress in the member

These effects are usually negligible in short compression members as well as in tension members, torsion members, and beams. For relatively long compression members (columns) with axial loads, the above effects are of prime importance in determining and limiting the loads which such members may carry. The student may demonstrate this by a simple experiment with a piece of ordinary gray cardboard (such as the backing of a pad of paper), say $8\frac{1}{2}$ by 11 in, and several textbooks. When the full length of cardboard is used to support one end of a book (as in Fig. 15-1a), it will probably bend or buckle if a second book is added. After determining the book load for the 11-in length, cut the cardboard in half to produce two pieces $8\frac{1}{2}$ by $5\frac{1}{2}$ in. If the test is repeated using one $8\frac{1}{2}$- by $5\frac{1}{2}$-in piece with the $5\frac{1}{2}$-in length supporting the books, approximately four books will be safely carried (see Fig. 15-1b). Thus, merely halving the length of the compression member will permit a compressive load four times as large to be safely supported.

Column failures are characterized by sudden bending or local buckling. This *column action*, combined with the uncertain effects which

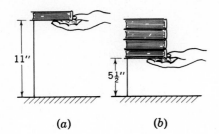

FIGURE 15-1 Demonstration of column action.

(a)        (b)

contribute to failure, makes it difficult to calculate the actual stresses developed in the column material. Because of these special problems, column-design formulas have been developed to find the maximum safe axial load and to relate this load to the shape and size of the column, and the properties of the material.

## 15-2 SLENDERNESS RATIO

The experiment discussed in the previous section demonstrates the importance of column length as a determinant of safe load. Length alone is not the only important factor. This can be shown by taking the unused $8\frac{1}{2}$- by $5\frac{1}{2}$-in piece of cardboard and rolling it into a cylindrical shape $5\frac{1}{2}$ in long and about 2 in in diameter (use tape to keep cylinder from unrolling). If the experiment is repeated with this new shape, your supply of books might be exhausted before failure occurs. Apparently, the cross-sectional shape and size of the column play an important role in determining the safe load.

If you think back to the first two experiments, you may recall that the cardboard seemed to prefer to fail in a particular direction. That is, in both cases failure occurred as shown in Fig. 15-2. A horizontal cross section through the cardboard would look something like Fig. 15-3. In each of the first two experiments, the cardboard tended to buckle about axis $YY$. In neither case did buckling occur about axis $XX$. The reason for this is that the moment of inertia about axis $YY$ is *less* than the moment of inertia about axis $XX$. To emphasize this point, let us calculate these moments of inertia and compare the results.

$$I_x = \frac{bh^3}{12} = \frac{0.03(8.5)^3}{12} = 1.54 \text{ in}^4$$

$$I_y = \frac{bh^3}{12} = \frac{8.5(0.03)^3}{12} = 0.0000191 \text{ in}^4$$

From these figures it is apparent that there is much less resistance to buckling about the $YY$ axis.

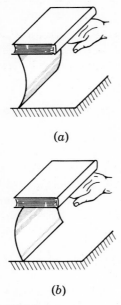

(a)

(b)

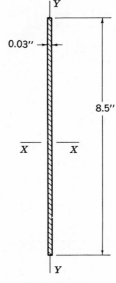

FIGURE 15-2   Direction of buckling in demonstration.

FIGURE 15-3   Cross section of cardboard.

Thus far, we have found that the important determinants of column strength for a given material are as follows.

1.   Length of column
2.   Cross-sectional shape
3.   Moment of inertia

However, from a known cross-sectional shape, the moment of inertia can be determined. This means that both of these may be grouped together as one factor. The single factor that is used in column design which is related to both cross-sectional area and moment of inertia is called *radius of gyration*, the symbol for which is $r$. (The symbol $k$ is sometimes used in place of $r$ for radius of gyration.) Our listing of column-strength determinants for a given material reduces to the following.

1.   Length of column $l$
2.   Radius of gyration $r$

If both of these terms are measured in the same units, their ratio is called *slenderness ratio*. Then

$$\text{Slenderness ratio} = \frac{l}{r} \qquad (15\text{-}1)$$

## 15-3 RADIUS OF GYRATION

To determine the radius of gyration of a plane figure such as a rectangle, let us consider the following problem. For the rectangle in Fig. 15-4a, calculate the moment of inertia about a centroidal axis and calculate the radius of gyration for that axis.

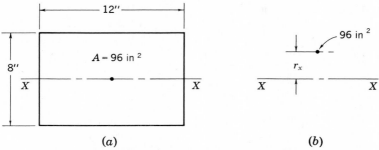

FIGURE 15-4 (a) Rectangular area with centroidal axis XX.
(b) Representation of radius of gyration about axis XX.

*Moment of Inertia:*

$$I_x = \frac{bh^3}{12} = \frac{12(8)^3}{12} = 512 \text{ in}^4$$

The radius of gyration is the distance from an axis to a point where all the area of a plane figure may be imagined to be concentrated so that the moment of inertia is left unchanged. Moment of inertia can be thought of as the "second moment of an area" and was expressed in Chap. 10 as

$$I = \Sigma a \bar{y}^2$$

where $\bar{y}$ was the distance from the centroid of each segment of area to the axis. If all the area of the rectangle of Fig. 15-4a is concentrated at a point, as in Fig. 15-4b, then from the above definition the distance from that point to the axis is $r$, the radius of gyration. Therefore,

$$I = Ar^2$$

or
$$r = \sqrt{\frac{I}{A}} \qquad (15\text{-}2)$$

where  $I$ = moment of inertia, in$^4$
  $A$ = total area of figure, in$^2$
  $r$ = radius of gyration, in

*Radius of Gyration:*

$$r_x = \sqrt{\frac{I_x}{A}} = \sqrt{\frac{512}{96}} = \sqrt{5.33} = 2.31 \text{ in}$$

Let us now calculate $I$ and $r$ for the same figure about the $YY$ centroidal axis (Fig. 15-5).

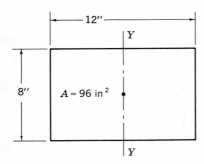

FIGURE 15-5   Rectangular area with centroidal axis *YY*

*Moment of Inertia:*

$$I_y = \frac{bh^3}{12} = \frac{8(12)^3}{12} = 1152 \text{ in}^4$$

*Radius of Gyration:*

$$r_y = \sqrt{\frac{I_y}{A}} = \sqrt{\frac{1152}{96}} = \sqrt{12} = 3.46 \text{ in}$$

From these calculations we can see that the axis which gives the smallest moment of inertia also gives the least radius of gyration. The same conclusion results from examination of Eq. (15-2).

In the experiments with cardboard columns, buckling occurred about the axis with the smallest moment of inertia. Thus, buckling of columns tends to occur about the axis with the least radius of gyration.

In determining the slenderness ratio of a column, the least radius of gyration is used.

$$\text{Slenderness ratio} = \frac{l}{r} \qquad (15\text{-}1)$$

where $l =$ unsupported length of column, in
  $r = $ *least* radius of gyration, in

Values of $r$ for common cross sections are given in App. B, Table 12.

## 15-4 CATEGORIES OF COLUMNS

Compression members are often subdivided into three categories according to slenderness ratio.

1. Short compression members
2. Intermediate columns
3. Long slender columns

These subdivisions can be visualized from an experimental curve, plotting $F/A$ (at failure) vs. $l/r$. This experiment can be done by compression testing various lengths of a member of standard cross section. Figure 15-6 shows a curve of $F/A$ (at failure) vs. $l/r$ which might result from such an experiment. The broken lines indicate the probable variation which might be expected in experimental results. The experimental curve seems to have three distinct portions which correspond to the three categories mentioned earlier.

For low values of slenderness ratio, the experimental curve shows a horizontal straight-line portion. This means that $F/A$ is constant in this range regardless of the slenderness ratio of the specimen. Specimens in this range are called *short compression members*, and their values of $F/A$ upon failure are determined from the ultimate stress of the material.

$$\frac{F}{A} = s_u \tag{15-3}$$

Since for short compression members the $l/r$ ratio has no effect on

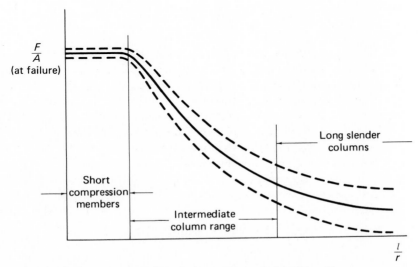

FIGURE 15-6  Expected ranges for $F/A$ (at failure) vs. $l/r$.

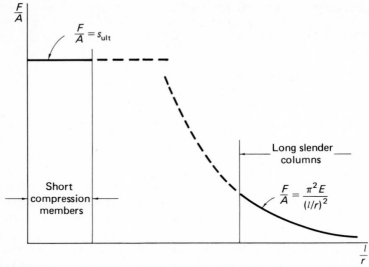

FIGURE 15-7 Plot of $F/A$ (at failure) vs. $l/r$ for short compression members and long, slender columns.

strength, it is clear that *column action* does not become significant until a certain $l/r$ value is exceeded. Briefly, then, short compression members require no special treatment as columns, but instead are simply handled by $s = F/A$.

For members with large values of $l/r$, the experimental curve of Fig. 15-6 shows that very small $F/A$ values may cause failure. Such members are categorized as *slender columns*. Experimental curves for slender columns closely fit an equation proposed by Euler,

$$\frac{F}{A} = \frac{\pi^2 E}{(l/r)^2} \qquad (15\text{-}4)$$

This equation takes account of the primary source of failure of slender columns: column action, or buckling.

In Fig. 15-7, Eqs. (15-3) and (15-4) are plotted as solid lines in the regions where they apply to columns and are extended as broken lines where they do not describe experimental results.

*Intermediate columns* are members whose $l/r$ values lie in the range between short compression members and slender columns. The intermediate column range is characterized by a sharp decrease in the $F/A$ value, causing failure as the $l/r$ value is increased. The actual shape of the curve describing intermediate columns is difficult to determine, since minor differences in specimens cause large changes in $F/A$ values. Because of this sensitivity to uncertainties, no one equation can be identified as the "correct" formula. Instead, several equally "correct" formulas are used for intermediate columns. The most important feature of such formulas is that they can be successfully used for design and selection of actual intermediate columns. In other words, they are justified because they work. The most common forms of intermediate-column equations are the following.

1. The straight-line form:

$$\frac{F}{A} = s - C\left(\frac{l}{r}\right) \tag{15-5}$$

2. The parabolic form:

$$\frac{F}{A} = s - C\left(\frac{l}{r}\right)^2 \tag{15-6}$$

Figure 15-8 shows the two forms of intermediate-column equations plotted as solid lines in their range of application and as broken lines outside this range. Notice that both of these equations can be adapted to represent the experimental intermediate-column range. The constant term $C$ in each equation can be given an appropriate value to ensure a good fit to experimental results.

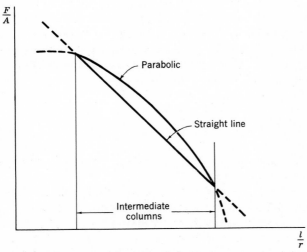

FIGURE 15-8 Plot of $F/A$ (at failure) vs. $l/r$ for two types of intermediate column formulas.

## 15-5 END CONDITIONS

The strength of a column is dependent upon the way in which the ends of the column are held, in addition to the slenderness ratio. The various end conditions may be classified into the following four groups.

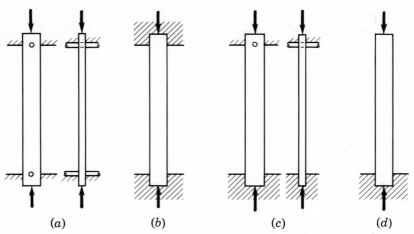

**FIGURE 15-9** End conditions: (*a*) both ends pinned; (*b*) both ends fixed; (*c*) one end fixed, one end pinned; (*d*) one end fixed, one end free.

1. Both ends pinned or hinged (Fig. 15-9*a*)
2. Both ends fixed (Fig. 15-9*b*)
3. One end fixed, one end pinned (Fig. 15-9*c*)
4. One end fixed, one end free (Fig. 15-9*d*)

Other things being equal, column-type 2 is strongest, type 3 is next in strength, type 1 next, and type 4 is the least strong. Each of these conditions may be approximated in actual practice, although certain end fixtures which appear to be securely fixed often permit some lateral or rotational movement. To account for the effect on the strength of a column due to the end conditions, the factor $K$ is introduced. The factor is used in conjunction with the slenderness ratio $l/r$ to produce an *effective* slenderness ratio. Thus,

$$\text{Effective slenderness ratio} = \frac{Kl}{r} \qquad (15\text{-}7)$$

Values of $K$ recommended by the Column Research Council for use when end conditions are approximated in actual design are given in Table 15-1.

## 15-6 COLUMN FORMULAS (METALS)

A great variety of column-design formulas of the straight-line, parabolic, Euler, and other types have been proposed and used. A few of the more popular equations are presented in this section. The student may wonder which is the best formula to use. The answer to this puzzle is that an *appropriate* formula must be used. In actual practice, a column design is

**TABLE 15-1**   RECOMMENDED VALUES OF $K$ FOR VARIOUS COLUMN
END CONDITIONS

| End Conditions | Recommended $K$ Value* |
|---|---|
| 1. Both ends pinned or hinged (Fig. 15-9*a*) | 1.0 |
| 2. Both ends fixed (Fig. 15-9*b*) | 0.65 |
| 3. One end fixed, one end pinned (Fig. 15-9*c*) | 0.8 |
| 4. One end fixed, one end free (Fig. 15-9*d*) | 2.1 |

* *Note:* Theoretical values of $K$ for these conditions are 1.0, 0.5, 0.7, and 2.0, respectively.

often subjected to codes, laws, or company practices which dictate the design equations to be used, so that the designer has little or no choice. In situations which are not so restricted, the designer will select an equation, based on experience, which best fits the conditions of the design. The student will usually not be asked to make such a choice, but rather the problems in this chapter will specify the formulas to be applied.

Several sets of column-design equations for metal members are given in Table 15-2. All the equations are based on the *safe* load $F$. It is interesting to note that many of the sets of equations are not categorized specifically for short, intermediate, and slender columns. If you refer to Figs. 15-7 and 15-8, you will find that the parabolic curve (typical of the AISC equation for $Kl/r \leq C_c$) flattens out and becomes nearly horizontal in the low $l/r$ range. Thus, use of this equation for short compression members becomes valid. In a like manner, we can establish the reasoning behind the use of the various column equations that do not adhere strictly to the previously established categories.

As discussed in Sec. 15-5, $K$ represents an end-condition factor.

With reference to the AISC equations, $C_c$ represents the maximum slenderness ratio at which a column may be designed by the parabolic equation. Above $C_c$, Euler's form is used. Take note that Table 15-2 indicates the value of $C_c$ for various $s_y$.

The use of the AISC parabolic equation ($Kl/r \leq C_c$) involves a factor of safety $N$. This is to be calculated from the given equation and has a range of 1.67 when $l/r = 0$, to 1.92 when $l/r = C_c$. Because A36 structural steel is most commonly used for columns and since pinned ends is the most frequent design condition, Fig. 15-10 has been included so that $N$ may be rapidly determined for various values of $l/r$.

**Sample Problem 1**   Determine the allowable axial compressive load $F$ which the AZ61A-F magnesium-alloy T section shown in Fig. 15-11 can carry if its ends are pinned and its length is 6 ft.

**Solution:**   For this T section, the moments of inertia have been calculated in Sec. 10-9.

$$I_x = 57.9 \text{ in}^4 \qquad I_y = 38.7 \text{ in}^4 \qquad A = 20 \text{ in}^2$$

**TABLE 15-2** DESIGN EQUATIONS FOR AXIALLY LOADED METAL COLUMNS

| Designation | Short Compression Members | Intermediate Columns | Long, Slender Columns |
|---|---|---|---|
| Machine design for any steel | $\max \dfrac{F}{A} = \dfrac{s_y}{N_y}$ $\dfrac{l}{r} < 40$ | $\dfrac{F}{A} = \dfrac{s_y}{N_y}\left[1 - \dfrac{s_y\left(\dfrac{Kl}{r}\right)^2}{4\pi^2 E}\right]$ $40 \leqslant \dfrac{l}{r} < \pi\sqrt{\dfrac{2E}{s_y(K)^2}}$ | $\dfrac{F}{A} = \dfrac{\pi^2 E}{N_y\left(\dfrac{Kl}{r}\right)^2}$ $\dfrac{l}{r} \geqslant \pi\sqrt{\dfrac{2E}{s_y(K)^2}}$ |
| AISC for structural steels | $\dfrac{F}{A} = \dfrac{s_y}{N}\left[1 - \dfrac{\left(\dfrac{Kl}{r}\right)^2}{2C_c{}^2}\right]$ | | $\dfrac{F}{A} = \dfrac{149\,000\,000}{\left(\dfrac{Kl}{r}\right)^2}$ |

| $s_y$(ksi)* | $C_c$ |
|---|---|
| 36 | 126.1 |
| 42 | 116.7 |
| 46 | 111.6 |
| 50 | 107.0 |

| Designation | Short Compression Members | | Long, Slender Columns |
|---|---|---|---|
| | $** N = \dfrac{5}{3} + \dfrac{3\left(\dfrac{Kl}{r}\right)}{8C_c} - \dfrac{\left(\dfrac{Kl}{r}\right)^3}{8C_c{}^3}$ $C_c = \sqrt{\dfrac{2\pi^2 E}{s_y}}$ [see $C_c$ values at left] $\dfrac{Kl}{r} \leqslant C_c$ | | $\dfrac{Kl}{r} > C_c$     [max $\dfrac{l}{r} = 200$] |
| Magnesium alloy AZ61A-F [for pinned ends] | | $\dfrac{F}{A} = \dfrac{42\,800}{1 + \dfrac{42\,800\left(\dfrac{l}{r}\right)^2}{64.4 \times 10^6}}$ | |
| Cast iron [for pinned ends] | | $\dfrac{F}{A} = 9000 - 40\left(\dfrac{l}{r}\right)$ $\dfrac{l}{r} \leqslant 70$ | |

* See Table 9-1 for recommended steels and their $s_y$.
** *Note:* Values of $N$ for A36 structural steel and $K = 1$ can be determined from the graph in Fig. 15-10.

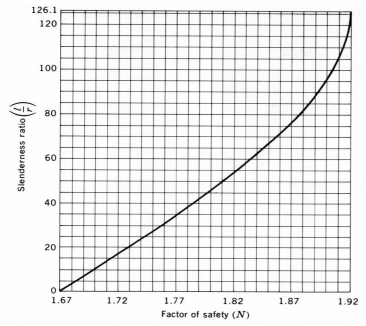

FIGURE 15-10   Factor of safety (N) vs. slenderness ratio (l/r) for A36 structural steel columns, K = 1.

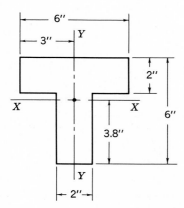

FIGURE 15-11   Diagram for Sample Problem 1.

The least radius of gyration will be referred to the axis with the smallest moment of inertia. Therefore, from Eq. (15-2),

$$r_y = \sqrt{\frac{I_y}{A}} = \sqrt{\frac{38.7}{20}} = \sqrt{1.935} = 1.39 \text{ in}$$

$$l = 6(12) = 72 \text{ in}$$

$$\frac{l}{r} = \frac{72}{1.39} = 51.8$$

From Table 15-2,

$$\frac{F}{A} = \frac{42\ 800}{1 + \dfrac{42\ 800\left(\dfrac{l}{r}\right)^2}{64.4 \times 10^6}} = \frac{42\ 800}{1 + \dfrac{42\ 800(51.8)^2}{64.4 \times 10^6}}$$

$$= \frac{42\ 800}{1 + 1.78} = 15\ 400$$

$$F = 15\ 400(20) = 308\ 000\ \text{lb (maximum)}$$

**Sample Problem 2** Figure 15-12 shows the cross section of a column (pinned ends) made by bolting two C15 × 40 sections to two 16- by $\frac{13}{16}$-in plates. Find the safe load it will carry when 20 ft long. Use the AISC formula. The material is A36 structural steel.

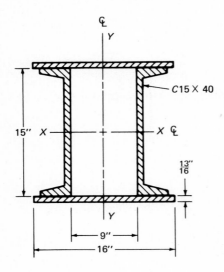

FIGURE 15-12 Diagram for Sample Problem 2.

**Solution:** For the column,

$$I_x = 2(I_x)_{\text{channel}} + 2(I_x + Ad^2)_{\text{plate}}$$

But $I_x$ for plate will be very small and can be neglected.

$$I_x = 2(349) + 2(16)\left(\frac{13}{16}\right)\left(7.5 + \frac{13}{32}\right)^2 = 698 + 1625$$

$$= 2323\ \text{in}^4$$

$$I_y = 2\left[\frac{13(16)^3}{16(12)}\right] + 2[9.23 + 11.8(4.5 + 0.778)^2]$$

$$= 554 + 676 = 1230\ \text{in}^4$$

Least radius of gyration,

$$r_y = \sqrt{\frac{I_y}{A}}$$

$$A = 2(11.8) + 2(13) = 49.6 \text{ in}^2$$

$$r_y = \sqrt{\frac{1230}{49.6}} = \sqrt{24.8} = 4.98 \text{ in}$$

From Table 15-2 for A36 structural steel, $s_y = 36\ 000$ psi (Table 9-1) and $C_c = 126.1$. Since $K = 1.0$ for pinned ends and $r_y = 4.98$ in,

$$\frac{Kl}{r} = \frac{1.0(20)(12)}{4.98} = 48.2 \text{ (which is less than } C_c)$$

Using

$$\frac{F}{A} = \frac{s_y}{N} \left[ 1 - \frac{\left(\dfrac{Kl}{r}\right)^2}{2C_c{}^2} \right]$$

$$N = \frac{5}{3} + \frac{3(48.2)}{8(126.1)} - \frac{(48.2)^3}{8(126.1)^3} = 1.667 + 0.143 - 0.007 = 1.80$$

Or  $N = 1.80$ (from graph, Fig. 15-10)

Then

$$\frac{F}{A} = \frac{36\ 000}{1.80} \left[ 1 - \frac{(48.2)^2}{2(126.1)^2} \right] = 20\ 000(1 - 0.073) = 18\ 500 \text{ psi}$$

$$F = 18\ 500(49.6) = 918\ 000 \text{ lb (maximum)}$$

**Sample Problem 3**  Select the most economical A36 structural steel W-shape column (pinned ends), 20 ft long, to carry an axial load of 300 000 lb. Use the AISC specifications.

**Solution:**  In this type of problem, there are two unknowns, $A$ and $r$, in the equation. Therefore, a trial-and-error solution is called for.

Assume that the column to be selected will have a $Kl/r < C_c$. From Table 15-2

$$\frac{F}{A} = \frac{s_y}{N} \left[ 1 - \frac{\left(\dfrac{Kl}{r}\right)^2}{2C_c{}^2} \right]$$

This formula is limited to a range of $Kl/r$ from 0 to 126.1 ($C_c$ = 126.1 for $s_y$ = 36 ksi). Assume $Kl/r$ will fall in the middle of the range, say 63. Then

$$N = \frac{5}{3} + \frac{3\left(\dfrac{Kl}{r}\right)}{8C_c} - \frac{\left(\dfrac{Kl}{r}\right)^3}{8C_c{}^3} = 1.67 + 0.19 - 0.02 = 1.84$$

Or $\quad N = 1.84$ (from graph, Fig. 15-10)

$$\frac{F}{A} = \frac{36\,000}{1.84}\left[1 - \frac{(63)^2}{2(126.1)^2}\right] = 17\,200 \text{ psi}$$

$$A = \frac{F}{17\,200} = \frac{300\,000}{17\,200} = 17.45 \text{ in}^2$$

Now select a W-shape section whose area is approximately equal to the above value. There may be several sections with approximately this area in the tables. Since we shall eventually be interested in the most economical column, check one of lighter weight.

Try W12 × 65, $A$ = 19.1 in², $r_y$ = 3.02 in.

$$\frac{Kl}{r} = \frac{1(20)(12)}{3.02} = 79.5$$

$$N = 1.67 + \frac{3(79.5)}{8(126.1)} - \frac{(79.5)^3}{8(126.1)^3} = 1.67 + 0.24 - 0.03 = 1.88$$

Or $\quad N = 1.88$ (from graph, Fig. 15–10)

$$\frac{F}{A} = \frac{36\,000}{1.88}\left[1 - \frac{(79.5)^2}{2(126.1)^2}\right] = \frac{36\,000}{1.88}(0.801) = 15\,340$$

$$F = 15\,340(19.1) = 293\,000 \text{ lb}$$

This column does not meet the 300 000-lb specification, although it comes quite close.

For our second trial, we want $F$ to increase slightly without increasing $A$, if possible. Therefore, a section should be chosen with a larger $r_y$, for approximately the same $A$. Thus, the next most economical section would be W12 × 72, $A$ = 21.2 in², $r_y$ = 3.04 in.

$$\frac{Kl}{r} = \frac{1(20)(12)}{3.04} = 79$$

$$N = 1.67 + \frac{3(79)}{8(126.1)} - \frac{(79)^3}{8(126.1)^3} = 1.88$$

Or $\quad N = 1.88$ (from graph, Fig. 15-10)

$$\frac{F}{A} = \frac{36\,000}{1.88}\left[1 - \frac{(79)^2}{2(126.1)^2}\right] = \frac{36\,000}{1.88}(0.804) = 15\,400$$

$$F = 15\,400(21.2) = 326\,000 \text{ lb}$$

The W12 $\times$ 72 is the most economical W-shape column section for the conditions specified. The Manual of Steel Construction (AISC) contains tables which permit rapid selection of economical column sections.

**Sample Problem 4** Design an 18-in-long linkage rod of circular cross section, assuming a varying axial compressive load of 4500 lb if the material is AISI 1045 steel. Use machine-design formulas.

**Solution:** From App. B, Table 1, for AISI 1045 steel,

$$s_y = 60\,000 \text{ psi} \qquad E = 30 \times 10^6 \text{ psi}$$

Assume pinned ends ($K = 1.0$). From App. B, Table 2,

$$N_y = 3$$

First determine the $l/r$ ratio which separates intermediate and slender columns.

$$\frac{l}{r} = \pi \sqrt{\frac{2E}{s_y(K)^2}} = \pi \sqrt{\frac{2(30 \times 10^6)}{60\,000(1)^2}} = \pi(31.6) = 99.3$$

At this stage, the actual value of $l/r$ for the rod is not known. Let us assume that it will turn out to be an intermediate column ($40 \leq l/r < 99.3$). From Table 15-2,

$$\frac{F}{A} = \frac{s_y}{N_y}\left[1 - \frac{s_y\left(\dfrac{Kl}{r}\right)^2}{4\pi^2 E}\right] \qquad \begin{array}{l} A = 0.785d^2 \\ r = 0.25d \end{array}$$

$$\frac{4500}{0.785d^2} = \frac{60\,000}{3}\left[1 - \frac{60\,000\left(\dfrac{1 \times 18}{0.25d}\right)^2}{4(3.14)^2(30 \times 10^6)}\right]$$

$$= 20\,000\left(1 - \frac{0.263}{d^2}\right) = 20\,000 - \frac{5260}{d^2}$$

$$\frac{4500}{0.785} = 20\,000d^2 - 5260$$

$$20\,000d^2 = 5730 + 5260 = 10\,990$$

$$d^2 = 0.550$$

$$d = 0.742 \qquad \text{use } d = \tfrac{3}{4}\text{ in}$$

Check $l/r$:

$$r = 0.25d = 0.25(0.75) = 0.1875 \text{ in}$$

$$\frac{l}{r} = \frac{18}{0.1875} = 96$$

This confirms our use of the intermediate-column formula. If $l/r$ had been larger than 99.3, the calculation would have had to be redone using the slender-column equation.

## 15-7 COLUMN FORMULAS (TIMBER)

The National Forest Products Association (NFPA), formerly called the National Lumber Manufacturers Association, recommends Eq. (15-8) for the design of simple solid columns made of stress-grade lumber. The formula may be used to design square, rectangular, and circular cross sections for pin-ended or square-end conditions.

$$\frac{F}{A} = \frac{0.30E}{\left(\frac{l}{b}\right)^2} \; ; \; \frac{F}{A} \leq s_{\text{allowable}} \qquad \frac{l}{b} \leq 50 \qquad (15\text{-}8)$$

where $F$ = axial compressive load on column, lb

$A$ = cross-sectional area of column, in²

$s_{\text{allowable}}$ = allowable compressive stress parallel to grain, psi

$l$ = unsupported length of column, in

$b \begin{cases} = \text{length of shortest side for rectangular cross section, in} \\ = 0.886 \text{ times the diameter for circular cross section, in} \end{cases}$

$E$ = modulus of elasticity, psi

**Sample Problem 5** Find the safe load that a nominal 6- by 6-in Douglas fir column is permitted to carry, using the NFPA formula. The column is 8 ft long.

*Solution:*

$$A = 30.3 \text{ in}^2$$
$$b = 5.5 \text{ in}, \qquad l = 8(12) = 96 \text{ in}$$
$$E = 1.2 \times 10^6 \text{ psi}$$

$$\frac{F}{A} = \frac{0.30E}{\left(\frac{l}{b}\right)^2} \qquad\qquad F = \frac{A(0.30)E}{\left(\frac{l}{b}\right)^2}$$

$$F = \frac{30.3(0.30)(1.2 \times 10^6)}{\left(\frac{96}{5.5}\right)^2} = \frac{10.9 \times 10^6}{305}$$

$$F = 35\,700 \text{ lb (maximum safe load)}$$

**Sample Problem 6** Find the diameter of a Sitka spruce compression member 28 in long to carry a load of 1000 lb. Use the NFPA formula.

**Solution:**    For spruce, $s = 875$ psi, $E = 1.3 \times 10^6$ psi. For a circular cross section, $b = 0.886d$.

$$\frac{F}{A} = \frac{0.3E}{(l/b)^2}$$

$$A = 0.785d^2$$

$$\frac{1000}{0.785d^2} = \frac{0.3(1.3)(10^6)}{(28/0.886d)^2} = \frac{(0.39)(10^6)d^2}{1000}$$

$$d^4 = \frac{1000(1000)}{0.785(0.39)(10^6)} = 3.27$$

$$d = 1.34 \text{ in (minimum)}$$

Check the stress with maximum allowable stress:

$$\frac{F}{A} = \frac{1000}{0.785(1.34)^2} = 710 \text{ psi}$$

This does not exceed 875 psi; therefore, the calculated $d$ is acceptable.

## PROBLEMS

**15-1.**    An S15 × 42.9 section is used as a 12-ft (pinned ends) column. Find the safe load it will carry, using AISC formulas. The material is A36 structural steel.

**15-2.**    Find the safe load that a C15 × 40 section will carry when used as a 6-ft column (fixed ends) using AISC formulas. The material is A36 structural steel.

**15-3.**    Find the safe load that a L6 × 6 × 1 section will carry when used as a column (pinned ends) 15 ft long. Use AISC formulas. The material is A36 structural steel.

**\*15-4.**    A 12-mm-diameter, 302 stainless-steel pin, 200 mm long with pinned ends, is to support a varying load of 9.0 kN in compression. Is the member satisfactory?

**15-5.**    Determine the maximum safe length of a hollow cast-iron column whose outside diameter is 8 in and inside diameter is 6 in, with an axial load of 139 000 lb.

**15-6.**    A hollow round cast-iron column 20 ft long is to carry a load of 250 tons. If the external diameter is 15 in, what is the inside diameter?

**15-7.**    Find the safe load that a column (pinned ends) 24 ft long, consisting of a 12- by $\frac{1}{2}$-in web plate, four L6 × 4 × $\frac{3}{4}$ sections, and two 13- by $\frac{5}{8}$-in cover plates, will support. Short legs of the angles are to be riveted to the web plate. Use the AISC code. All material is A36 structural steel.

**15-8.**    Find the dimensions of a square AZ61A-F magnesium-alloy strut, pin ended, 10 ft long to carry a compressive load of 20 000 lb.

**15-9.** A hollow rectangular strut, pin ended, 7 ft long, is made of AZ61-A-F magnesium alloy. The outside dimensions of the strut are 3 by 4 in and the wall thickness is $\frac{1}{2}$ in. Determine the safe axial compressive load it can support.

**15-10.** A piece of seamless AISI 1045 steel tubing is used as a brace which requires it to support an axial compressive load of 12 000 lb. The tube outside diameter is 2.875 in and the wall thickness is 0.203 in. Find the maximum permissible length.

**15-11.** The upper chord compression member of a truss must safely carry an axial load of 22 kips. The member will be 11 ft long. Select two equal-leg A36 structural steel angles which when welded back to back will meet AISC specifications. Assume pinned-end conditions.

**15-12.** What should be the diameter of a solid AISI 1095 steel rod, pin ended, 8 ft long, which is required to support a compressive load of 20 000 lb?

**15-13.** Select the most economical S section to carry an axial load of 50 000 lb for a column length of 12 ft with pinned ends. Use AISC formulas. The material is A36 structural steel.

**15-14.** Select the most economical W section, according to AISC, to carry a central compressive load of 225 000 lb if the member is 15 ft long with pinned ends. The material is A36 structural steel.

**15-15.** A link with pinned ends, 15.75 in long, is subject to a compressive load of 4700 lb. The link is made of AISI 2340 steel, the cross section is rectangular, and the width is twice the depth. Find the dimensions of the cross section, assuming varying load.

**15-16.** Choose the proper W section for a 15-ft column (pinned ends) to carry a central load of 230 000 lb under AISC specifications. The material is A36 structural steel.

**15-17.** A Sitka spruce column, 10 ft long and 6 by 6 in dressed in cross section, supports a load of 9000 lb. Is this arrangement safe according to the NFPA specifications?

**15-18.** What safe load will an Eastern white pine post, 12 ft long and 4 by 6 in dressed in cross section, support according to the NFPA formulas?

**15-19.** Find the dressed size of a square column of Douglas fir, 20 ft long, to carry a load of 100 000 lb. Use NFPA formula.

**15-20.** Find the dressed size of a square Ponderosa pine timber to be used as a column 25 ft long to carry a load of 75 000 lb. Use the NFPA formula.

**15-21.** A floor beam which carries a uniform load of 1000 lb/ft on a 12-ft simple span is supported by an 18-ft-long column at the left end and by an 11-ft-long column at the right end. What size Douglas fir columns should be used for the supports according to the NFPA formula?

# CHAPTER

# 16

# Indeterminate Beams

## 16-1 TYPES OF STATICALLY INDETERMINATE BEAMS

The beams discussed in Chap. 11 were statically determinate; that is, the unknown reactions and moments were found by using the conditions for static equilibrium: $\Sigma F_x = 0$, $\Sigma F_y = 0$, and $\Sigma M = 0$. Certain types of beams require more equations for their solution than are available from the static equilibrium formulas because there are too many unknowns. Such beams are called *statically indeterminate*. Other methods are used to find the unknown reactions and moments on these beams. Some typical indeterminate beams are shown in Fig. 16-1.

## 16-2 BEAM WITH ONE END FIXED, ONE END SUPPORTED

Figure 16-2a shows a beam with the right end fixed in a wall and the left end on a support; the beam carries a uniform load. The elastic curve for this beam is shown in Fig. 16-2b. If the support were removed, the beam would be a cantilever with a maximum downward deflection at the free end. In the actual beam (Fig. 16-2), there is no deflection at the left end. Therefore, the supporting force $R_l$ exactly counteracts the deflection which would occur without the support. This fact can be used to help solve this type of indeterminate beam.

Imagine that the left support is removed and the beam becomes a cantilever, as in Fig. 16-3a. The elastic curve of the cantilever (Fig. 16-3b) shows the deflection $y_L$ at the free end. From Chap. 12, Eq. (12-19),

$$y_L = \frac{WL^3}{8EI} \qquad \text{(for the free end of a cantilever with uniform load)}$$

Now we may ask what force at the free end is required to eliminate $y_L$

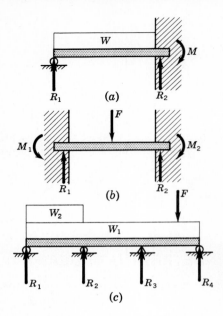

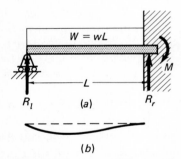

FIGURE 16-1    Some typical statically indeterminate beams: (a) One end fixed, one end supported; $R_1$, $R_2$, and $M$ are unknown. (b) Both ends fixed; $R_1$, $R_2$, $M_1$, and $M_2$ are unknown. (c) Continuous beam; $R_1$, $R_2$, $R_3$, and $R_4$ are unknown.

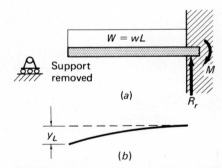

FIGURE 16-2    (a) Beam with one end fixed, one end supported; $R_l$, $R_r$, and $M$ are unknown. (b) Elastic curve for the beam.

FIGURE 16-3    (a) Beam of Fig. 16-2 with left support removed. (b) Elastic curve for this condition.

(because in the actual beam, Fig. 16-2, $y_L = 0$). Imagine the same beam without the uniform load (as in Fig. 16-4a), but with an upward force $R_l$ at the free end which produces a deflection $y_L{}'$ (Fig. 16-4b). From Chap. 12, Eq. (12-16),

$$y_L{}' = \frac{R_l L^3}{3EI} \qquad \text{(for the free end of a cantilever with load at end)}$$

Since $y_L$ is to be eliminated by the effect of $R_l$, then

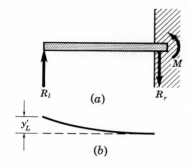

FIGURE 16-4 (a) Beam of Fig. 16-2 with only one support force acting. (b) Elastic curve for this condition.

$$y_L = y_L'$$

or
$$\frac{WL^3}{8EI} = \frac{R_lL^3}{3EI}$$

from which
$$R_l = \frac{3}{8}W$$

With this new piece of information, the solution of the actual beam may proceed. Figure 16-5a shows the beam with only two unknowns, $R_r$ and $M$. The remainder of the solution follows the usual steps of finding $R_r$ from $\Sigma F_y = 0$ and finding $M$ from $\Sigma M = 0$. The shear diagram (Fig. 16-5c) indicates the location of maximum bending moments as before, while the moment diagram (Fig. 16-5d) shows how the moments vary along this beam.

The principle of compensated deflection at the end support may be used for this type of beam with any loading. In one of the following sample problems, a beam of this kind with a concentrated load is solved. This method is easily extended to beams in which the support is caused to settle by the loading, as in the following illustrative problem.

**Sample Problem 1** A machine base rests on two W8 × 31 beams, each of which is mounted with one end embedded in concrete and the other end resting on a vibration damping pad (see Fig. 16-6). After the machine is put into place, the supported end settles $\frac{1}{2}$ in. Find the reaction on one of the beams due to the vibration pad, and find the maximum stress in the beam.

**Solution:** The uniform load on one beam is

$$W_1 = \frac{10(2000)}{2} = 10\,000\ \text{lb} \qquad \text{(load is shared by two beams)}$$

$$W_2 = 20(31) = 620\ \text{lb} \qquad \text{(weight of beam)}$$

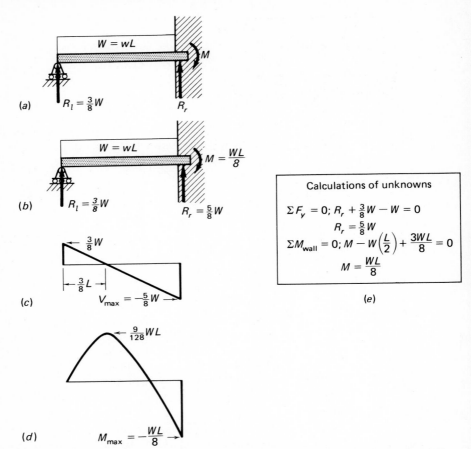

FIGURE 16-5   (a) Beam of Fig. 16-2. (b) Beam with computed values of $R_l$, $R_r$, and $M$. (c) Shear-force diagram. (d) Bending moment diagram. (e) Calculations for $R_r$ and $M$.

Therefore,

$$W = 10\,000 + 620 = 10\,620\,\text{lb}$$

The deflection which would occur with the support removed is given by

$$y_L = \frac{WL^3}{8EI} = \frac{10\,620(240)^3}{8(30)(10^6)(110)} = 5.56\,\text{in}$$

The upward deflection of an unloaded beam due to a concentrated reaction force $R_l$ at the left end is

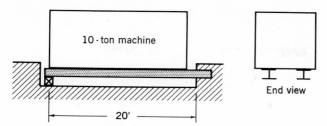

FIGURE 16-6    Diagram for Sample Problem 1.

$$y_L' = \frac{R_l L^3}{3EI} = \frac{R_l}{3}\left[\frac{(240)^3}{(30)(10^6)(110)}\right] = 0.0014R_l$$

The left end of the actual beam, when fully loaded, settles $\frac{1}{2}$ in; therefore,

$$y_L = y_L' + \frac{1}{2}$$

from which $5.56 = 0.0014R_l + 0.5$

$$R_l = 3620 \text{ lb} \qquad \text{(reaction of vibration pad)}$$

The problem may now be solved in the usual way.

$\Sigma F_y = 0$

$$R_l + R_r - 10\,620 = 0$$
$$R_r = 7000 \text{ lb}$$

$\Sigma M_{\text{wall}} = 0$

$$M - 10\,620(10) + 3620(20) = 0$$
$$M = 33\,800 \text{ ft} \cdot \text{lb} \qquad \text{(at wall)}$$

The shear diagram (Fig. 16-7b) indicates a maximum moment at $x$ feet from the left support. By similar triangles,

$$\frac{x}{3620} = \frac{20 - x}{7000}$$
$$7000x = 72\,400 - 3620x$$
$$x = \frac{72\,400}{10\,620} = 6.8 \text{ ft}$$

The bending moment at this section in the beam is

$$M_{6.8} = 3620(6.8) - 10\,620\left(\frac{6.8}{20}\right)(3.4)$$
$$= 24\,620 - 12\,280 = 12\,340 \text{ ft} \cdot \text{lb}$$

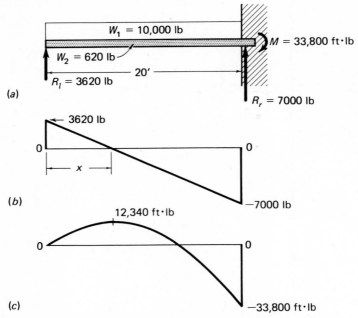

FIGURE 16-7   (a) Beam diagram for Sample Problem 1.
(b) Shear force diagram. (c) Bending moment diagram.

The maximum bending moment occurs at the wall; $M_{max} = 33\ 800$ ft·lb (see Fig. 16-7c).

For a W8 × 31, the section modulus $S = 27.4$ in³. Thus,

$$s = \frac{M}{S} = \frac{33\ 800(12)}{27.4} = 14\ 800 \text{ psi} \qquad \text{(maximum stress)}$$

**Sample Problem 2**   A beam has one end fixed and the other end resting on a support. The beam carries a concentrated load of 10 000 lb at the center of its 12-ft span. Select the most economical standard A36 structural steel S beam (Fig. 16-8).

**Solution:**   If the support is removed, the deflection at the free end is given by Eq. (12-17).

$$y_L = \frac{F}{6EI} (2L^3 - 3aL^2 + a^3)$$

$$= \frac{10\ 000}{6EI} [2(144)^3 - 3(72)(144)^2 + (72)^3]$$

$$= \frac{31.1(10^8)}{EI}$$

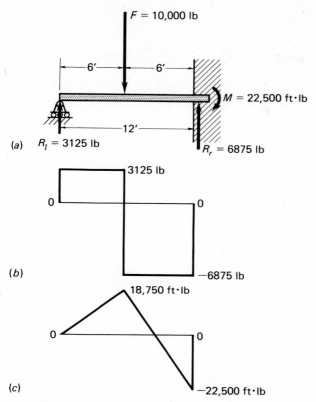

FIGURE 16-8    (a) Beam diagram for Sample Problem 2.
(b) Shear-force diagram. (c) Bending moment diagram.

The upward force $R_l$ at the support would produce a deflection on an unloaded beam equal to

$$y_L' = \frac{R_l(144)^3}{3EI} = \frac{995\,000R_l}{EI}$$

Equating these deflections,

$$\frac{31.1(10^8)}{EI} = \frac{995\,000R_l}{EI}$$

$$R_l = 3125 \text{ lb}$$

$$\Sigma F_y = 0$$

$$R_l + R_r - 10\,000 = 0$$

$$R_r = 6875 \text{ lb}$$

$$\Sigma M_{wall} = 0$$

$$M - 10\,000(6) + 3125(12) = 0$$

$$M = 60\,000 - 37\,500$$

$$= 22\,500 \text{ ft} \cdot \text{lb} \quad \text{(at wall)}$$

$$M_6 = 3125(6) = 18\,750 \text{ ft} \cdot \text{lb} \quad \text{(at 6-ft section)}$$

$$M_{max} = 22\,500 \text{ ft} \cdot \text{lb} \quad \text{(at wall)}$$

$$s = \frac{M}{S}$$

$$S = \frac{M}{s} = \frac{22\,500(12)}{24\,000} = 11.25 \text{ in}^3$$

Select the S7 × 20 ($S = 12.1 \text{ in}^3$).

## 16-3 BEAM WITH BOTH ENDS FIXED

Beams of this type may be treated by imagining them to be separated into three portions forming a simply supported beam in the central portion and cantilever beams at both ends. The point of imagined separation is the point where the bending-moment diagram crosses zero. This point, called the *point of inflection of the elastic curve*, locates a section in the beam where the moment is zero. Since the end of a simply supported beam and the free end of a cantilever beam have zero bending moments, the imagined separation is valid if it is done at a zero-moment section.

***Beam with Concentrated Load at Center of Span:*** Consider the beam in Fig. 16-9. There are two unknown forces $R_l$ and $R_r$, and two unknown

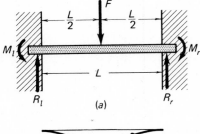

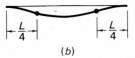

FIGURE 16-9   (a) Beam with both ends fixed and concentrated load at center of span. (b) Elastic curve showing location of inflection points.

moments $M_l$ and $M_r$. The points of inflection are located on the elastic curve. These locations are derived in more advanced books. The problem is to express the reaction forces and moments in terms of $F$ and $L$.

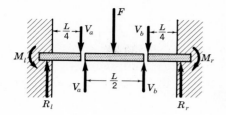

FIGURE 16-10 Beam of Fig. 16-9
separated at the inflection points.

Imagine this beam to be separated at the points of inflection into three portions, as shown in Fig. 16-10. Although the beam appears to be cut into three portions, the shear forces acting at the separation sections are not zero and must be considered in order to maintain equivalence to the original beam. $V_a$ represents the shear force in the beam at a distance $L/4$ from the left wall. When we separate the beam at this section, $V_a$ is treated as an external effect of the adjacent portion. $V_b$ is treated similarly. We have converted a statically indeterminate beam into three statically determinate beams.

Let us examine the central portion (Fig. 16-11). The portion is a

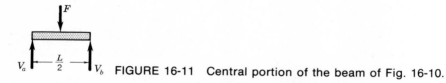

FIGURE 16-11 Central portion of the beam of Fig. 16-10.

simply supported beam $L/2$ long, with a concentrated load $F$ at the center of the span. From the symmetry, we note that

$$V_a = \frac{F}{2}$$

and

$$V_b = \frac{F}{2}$$

Furthermore, it was shown in Chap. 11 that the maximum bending moment for this type of beam occurs at midspan and equals the load times one-fourth of the span. Since the span in Fig. 16-11 is $L/2$, the maximum moment in this portion is

$$M = \frac{F(L/2)}{4} = \frac{FL}{8}$$

Now consider one of the cantilever portions. Either end portion may be

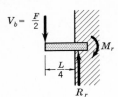

FIGURE 16-12   End portion of the beam of Fig. 16-10.

analyzed and the results applied to both ends because of the symmetrical loading on the original beam. The right end is shown in Fig. 16-12. Let us apply the conditions for static equilibrium to this portion of the original beam, noting that $V_b = F/2$ from the previous step.

$$\Sigma F_y = 0$$

$$R_r - \frac{F}{2} = 0$$

$$R_r = \frac{F}{2}$$

$$\Sigma M_{\text{wall}} = 0$$

$$M_r - \frac{F}{2}\left(\frac{L}{4}\right) = 0$$

$$M_r = \frac{FL}{8}$$

The reaction and moment at the left end of the beam are equal to $R_r$ and $M_r$, respectively. Therefore,

$$R_r = R_l = \frac{F}{2}$$

$$M_r = M_l = \frac{FL}{8}$$

and, as previously calculated, the moment at the center of the span is

$$M = \frac{FL}{8}$$

The shear-force and bending-moment diagrams for the beam with both ends fixed, carrying a concentrated load at midspan, are shown in Fig. 16-13. These diagrams should be compared with Fig. 11-20 for a simply supported beam.

The maximum deflection for the fixed-ends beam is obtained by adding the deflection of the central portion (Fig. 16-11) and the deflection at the free end of the cantilever portion (Fig. 16-12). Applying the

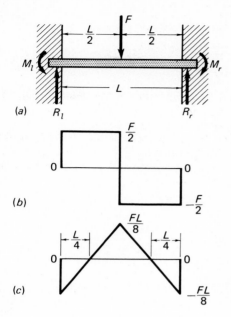

FIGURE 16-13 (a) Beam with both ends fixed and concentrated load at center of span. (b) Shear-force diagram. (c) Bending moment diagram.

deflection formulas, Eqs. (12-11) and (12-16), to the portions of the beam, we obtain

$$y_{max} = \frac{F_1 L_1^3}{48EI} + \frac{F_2 L_2^3}{3EI}$$

but

$$F_1 = F \qquad L_1 = \frac{L}{2} \qquad F_2 = \frac{F}{2} \qquad L_2 = \frac{L}{4}$$

$$y_{max} = \frac{F\left(\dfrac{L^3}{8}\right)}{48EI} + \frac{\dfrac{F}{2}\left(\dfrac{L^3}{64}\right)}{3EI}$$

$$y_{max} = \frac{FL^3}{192EI} \tag{16-1}$$

**Beam with Uniform Load:** A similar procedure may be used to solve fixed-ends beams with uniform loading. The results of such a procedure are summarized in Fig. 16-14.

**Sample Problem 3** An overhead crane rides on a single rail which is rigidly fixed in concrete at the ends to provide a span of 20 ft. An allowable bending stress of 9000 psi is specified for the rail material. The rated capacity of the crane is 20 tons, but an overload of 25 percent is to be provided for. The depth of the rail is limited to 14 in by other

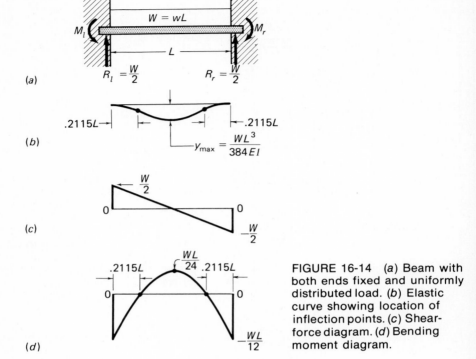

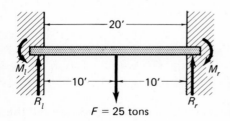

FIGURE 16-14   (a) Beam with both ends fixed and uniformly distributed load. (b) Elastic curve showing location of inflection points. (c) Shear-force diagram. (d) Bending moment diagram.

structural interference. Assuming a static design situation with the crane at midspan, determine the following.

(a)   The minimum centroidal moment of inertia of a 14-in-deep rail if the cross section is symmetrical about the XX axis

(b)   The maximum deflection of the rail, using the moment of inertia from part a, if the material is an alloy steel

**Solution a:**   This physical system is represented by Fig. 16-15, with the

FIGURE 16-15   Beam diagram for Sample Problem 3.

total load of $20 + 20(0.25) = 25$ tons concentrated at midspan. The weight of the rail is assumed negligible for simplification. From Fig. 16-13, p. 399,

$$R_l = R_r = \frac{F}{2} = \frac{25(2000)}{2} = 25\,000 \text{ lb}$$

$$M_l = M_r = M_{\text{midspan}} = \frac{FL}{8} = \frac{50\,000(20)}{8} = 125\,000 \text{ ft·lb}$$

$$s = \frac{Mc}{I}$$

$$I = \frac{Mc}{s}$$

$$c = 7 \text{ in}$$

$$s = 9000 \text{ psi}$$

$$I = \frac{125\,000(12)(7)}{9000} = 1167 \text{ in}^4 \qquad \text{(minimum)}$$

**Solution b:**   From Eq. (16-1),

$$y_{\text{max}} = \frac{FL^3}{192EI}$$

where
$$F = 50\,000 \text{ lb}$$
$$L = 20(12) = 240 \text{ in}$$
$$E = 30 \times 10^6 \text{ psi}$$
$$I = 1167 \text{ in}^4$$

$$y_{\text{max}} = \frac{50\,000(240)^3}{192(30 \times 10^6)(1167)} = 0.103 \text{ in}$$

## PROBLEMS

**16-1.**  A balcony is supported by S6 × 17.25, A36 structural steel beams spaced 10 ft on centers. The balcony projects 6 ft from the wall and carries 200 psf on floor area. The deflection at the end of the beam is prevented by AISI 1020 steel rods fastened from above. Find the size of rods necessary and the maximum stress in the beams. Rod spacing is 10 ft.

**16-2.**  A 10- by 14-in dressed Mountain hemlock beam is fixed at one end and supported at the other. It carries a load of 400 lb per linear foot. The support settles $\frac{1}{2}$ in under the action of the load. Find the reaction of the support and the maximum fiber stress if the span is 17 ft.

*Hint:* If the support were removed, the deflection of the end would be

$$y_1 = +\frac{1}{8}\frac{WL^3}{EI}$$

The support brings the end up:

$$y_2 = \frac{1}{3}\frac{RL^3}{EI}$$

Then
$$y_1 = y_2 + \tfrac{1}{2} \text{ in}$$

Now assume that the support does not settle. Find maximum fiber stress and compare with the first case.

**16-3.** What are the bending and shearing stresses in a beam 9 ft long, fixed at one end and supported at the same level at the other? The rough-sawn cross section of the beam is 6 by 12 in. It carries a uniform load, including its own weight of 820 lb per linear foot. Sketch the shear and moment diagrams. $E = 1\ 200\ 000$ psi.

**16-4.** If the end support in Prob. 16-3 settles 0.4 in, compute the reactions and bending and shearing stresses.

**16-5.** If the end support in Prob. 16-3 is pushed 0.4 in above its previous level, what will be the stresses in the beam?

**16-6.** A run of schedule 40 seamless steel pipe is rigidly fixed into two concrete walls with a clear span of 15 ft between the walls. The outside diameter is 5.563 in and the wall thickness of the pipe is 0.258 in. A load of 1 ton is carried at midspan. Determine the following.
  *a.* The maximum bending stress in the pipe
  *b.* The maximum deflection of the pipe

**16-7.** A platform is supported by two beams, as shown in Fig. Prob. 16-7. Each beam consists of two L4 × 4 × ⅜ sections welded back

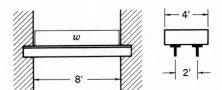

FIGURE PROBLEM 16-7

to back to form a T section. The platform is designed to carry 220 psf. What maximum bending stress is developed in the beams, and how much will they deflect?

# APPENDIX

# A

# Review of Statics

## A-1 FUNDAMENTAL TERMS

The science of mechanics may be subdivided into *statics* and *dynamics*. Statics deals with bodies at rest, while dynamics involves motion. This book is concerned with static systems.

*Strength of materials* is a study of the load-resisting ability of various machine and structural members.

The reader should have a clear understanding of the following technical terms: *length, area, volume, force, pressure, mass, weight, density, load, moment, torque, work,* and *power*. Typical units and definitions of these terms are given on pages 1 and 2.

A *scalar* quantity is specified by magnitude only, such as $40 or 46°F. A *vector* quantity is specified by both magnitude and direction. Force is a vector quantity.

## A-2 FORCE SYSTEMS

Systems of forces acting on members may be classified as either *concurrent* or *nonconcurrent* and as either *coplanar* or *noncoplanar*. This gives four categories of force systems.

1. *Concurrent-coplanar* forces:
   (*a*) The lines of action of all forces pass through a common point.
   (*b*) The forces all lie in the same plane.
2. *Nonconcurrent-coplanar* forces:
   (*a*) The lines of action of all forces *do not* pass through a common point.
   (*b*) The forces all lie in the same plane.
3. *Concurrent-noncoplanar* forces:
   (*a*) The lines of action of all forces pass through a common point.
   (*b*) The forces *do not* all lie in the same plane.
4. *Nonconcurrent-noncoplanar* forces:
   (*a*) The lines of action of all forces *do not* pass through a common point.
   (*b*) The forces *do not* all lie in the same plane.

Two other types of force systems which frequently occur are special cases of those mentioned above. They are as follows.

1'. *Collinear* forces:
    (*a*) All forces act along the same line of action.
    (*b*) This is a special case of *concurrent-coplanar* force systems.
2'. *Parallel* forces:
    (*a*) The lines of action of all forces are parallel.
    (*b*) This is a special case of *nonconcurrent-coplanar* force systems.

## A-3 CONCURRENT-COPLANAR FORCE SYSTEMS

A *resultant* is a single force which can replace two or more concurrent forces and produce the same effect as the original forces. The resultant of a collinear force system lies along the same line of action as the collinear forces. Since collinear forces lie along the same line, *the resultant of collinear forces is their algebraic sum.* However, *the resultant of concurrent forces which are not collinear is their vector sum.* Concurrent force vectors may be added by any of the following methods.

    1. Parallelogram method.   This method is applied to *two* concurrent forces, as in Fig. A-1. Construct or sketch the parallelogram with the vectors as its sides,

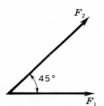

FIGURE A-1  Force diagram.

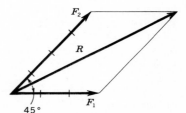

FIGURE A-2  Parallelogram method for determining resultant.

as shown in Fig. A-2. The *diagonal* of the parallelogram is the resultant of the two concurrent forces.

    2. Triangle method.   This method is applied to *two* concurrent forces, as in Fig. A-1. Draw or sketch one vector and connect the tail of the second vector to the head of the first, as shown in Fig. A-3. The line which forms the closing side of the triangle is the resultant of the two concurrent forces. Take note where the arrow on the resultant is placed.

    3. Summation of rectangular components.   This method is applied to *two or more concurrent forces*. All forces are converted to their vertical and horizontal

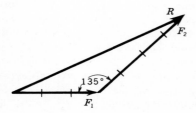

FIGURE A-3  Triangle method for determining resultant.

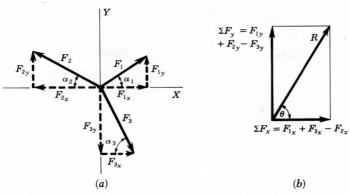

(a)                                    (b)

FIGURE A-4   Component method of finding resultant:
(a) vertical and horizontal components; (b) resultant of
algebraic sum of vertical and horizontal components.

(rectangular) components, as in Fig. A-4a. Both the vertical and horizontal
components are algebraically added, resulting in one vertical vector and one
horizontal vector. The resultant of these two vectors is the resultant of the original
concurrent forces, as in Fig. A-4b.

   a. Determining vertical and horizontal components.   Referring to Fig.
A-4a, the components of force $F_1$ are $F_{1y}$ and $F_{1x}$, which are determined from the
equations

$$F_y = F \sin \alpha$$
$$F_x = F \cos \alpha$$

   b. Determining the resultant.   Referring to Fig. A-4b, the resultant is given
by the equation

$$R = \sqrt{(\Sigma F_x)^2 + (\Sigma F_y)^2}$$

and the angle that the resultant makes with the x axis is given by

$$\tan \theta = \frac{\Sigma F_y}{\Sigma F_x}$$

   4. Polygon method.   This is applied to *more than two* concurrent forces, as
in Fig. A-5a. This method is an extension of the triangle method in which all
concurrent forces are successively joined head to tail. The line which closes the
polygon is the resultant of the concurrent force system; see Fig. A-5b. Note that
the arrowhead on the resultant meets the arrow of the last vector to be drawn.

   Each of the four methods of determining the resultant of concurrent force
systems may be used in either a graphical solution or a trigonometric solution.
The graphical solution requires accurate layout of force vectors, with lengths

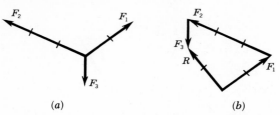

FIGURE A-5   Polygon method of finding resultant:
(a) force diagram; (b) vector diagram.

proportional to force magnitudes and directions correctly aligned. The resultant
is then measured from the drawing. The trigonometric solution for determining
the resultant employs any of several mathematical relationships, the most common
of which are given below.

　　1.  Sine, cosine, tangent (for right triangles only):

$$\sin A = \frac{a}{c} \qquad \sin B = \frac{b}{c}$$

$$\cos A = \frac{b}{c} \qquad \cos B = \frac{a}{c}$$

$$\tan A = \frac{a}{b} \qquad \tan B = \frac{b}{a}$$

　　2.  Pythagorean theorem (for right triangles only):

$$c^2 = a^2 + b^2$$

　　3.  Law of sines (for any triangle):

$$\frac{a}{\sin A} = \frac{b}{\sin B} = \frac{c}{\sin C}$$

　　4.  Law of cosines (for any triangle):

$$c^2 = a^2 + b^2 - 2ab \, \cos C$$

*Note:* If angle $C$ is more than 90°, then substitute for cos $C$ the expression
$[-\cos(180° - C)]$.

# A-4  NONCONCURRENT-COPLANAR FORCE SYSTEMS

The resultant of a nonconcurrent-coplanar force system is a single force which
usually does not pass through the axis of rotation of the body. The magnitude
and direction of the resultant force may be determined by the rectangular
component method using the equations

$$R = \sqrt{(\Sigma F_x)^2 + (\Sigma F_y)^2}$$
$$\tan \theta = \frac{\Sigma F_y}{\Sigma F_x}$$

and the perpendicular distance of the line of action of $R$ from the axis of rotation of the body is given by

$$r = \frac{\Sigma M}{R}$$

where $\Sigma M$ is the net unbalanced moment of the force system about the axis of rotation of the body.

The resultant of a nonconcurrent-coplanar force system is also equivalent to a single force passing through the axis of rotation and a couple.

For a complete discussion of nonconcurrent-coplanar force systems, refer to Chap. 4, Sec. 4-2.

## A-5  EQUILIBRANT

The *equilibrant* of a force system is a single force which would exactly balance the resultant of that system. The equilibrant must be equal in magnitude to the resultant and it must act along the same line of action as the resultant (collinear with the resultant), but in the opposite direction.

## A-6  STATIC EQUILIBRIUM

When a *concurrent-coplanar force system* is in equilibrium, the resultant of the system must be zero. In other words, the vector sum must be zero. In terms of the polygon method of adding force vectors, this means that concurrent force systems in equilibrium form closed polygons. That is, when all force vectors are connected head to tail graphically, the arrowhead of the last vector to be drawn will touch the tail (or starting point) of the first vector drawn. In terms of the component method of adding force vectors, since the force system is in equilibrium there will be no unbalanced net component along any axis, including the $x$ axis and the $y$ axis. Thus, the algebraic sum of the rectangular components for each axis must equal zero, or in equation form,

$$\Sigma F_x = 0 \quad \text{and} \quad \Sigma F_y = 0$$

When a *nonconcurrent-coplanar force system* is in equilibrium, the resultant of the system must be zero. This means that there can be no unbalanced force and no unbalanced moment about any point or axis. These conditions for equilibrium can be expressed by the following equations:

$$\Sigma F_x = 0 \quad \Sigma F_y = 0 \quad \text{and} \quad \Sigma M = 0$$

# APPENDIX
# B
## Tables

**TABLE 1  TYPICAL MECHANICAL PROPERTIES OF METALS***

| Metal | Ultimate Strength | | | | | | Tension Yield Point $s_y$ | | Modulus of Elasticity $E$ | | Modulus of Rigidity $G$ | | Modulus of Rupture $s_r$ | |
|---|---|---|---|---|---|---|---|---|---|---|---|---|---|---|
| | Tension $s_t$ | | Compression $s_c$ | | Shear $s_s$ | | | | | | | | | |
| | $10^3$ psi | $10^6$ Pa | $10^3$ psi | $10^6$ Pa | $10^3$ psi | $10^6$ Pa | $10^3$ psi | $10^6$ Pa | $10^6$ psi | $10^9$ Pa | $10^6$ psi | $10^9$ Pa | $10^3$ psi | $10^6$ Pa |
| Wrought iron | 47 | 325 | 47 | 325 | 38 | 260 | 28 | 190 | 28 | 190 | 10.0 | 70 | | |
| AISI 1020 steel | 65 | 450 | 65 | 450 | 50 | 345 | 45 | 310 | 30 | 200 | 11.5 | 80 | | |
| AISI 1045 steel | 95 | 655 | 95 | 655 | 70 | 480 | 60 | 410 | 30 | 200 | 11.5 | 80 | | |
| AISI 1095 steel | 142 | 980 | 142 | 980 | 105 | 725 | 83 | 570 | 30 | 200 | 11.5 | 80 | | |
| AISI 2340 steel | 135 | 930 | 135 | 930 | 100 | 690 | 116 | 800 | 30 | 200 | 11.5 | 80 | | |
| 302 Stainless steel | 140 | 965 | 140 | 965 | 110 | 760 | 100 | 690 | 28 | 190 | 11.0 | 75 | | |
| Cast-iron Class 20 | 20 | 140 | 80 | 550 | 32 | 220 | | | 11 | 75 | 4.5 | 31 | 53 | 370 |
| Cast-iron Class 40 | 40 | 275 | 125 | 860 | 55 | 380 | | | 16 | 110 | 5.5 | 38 | 78 | 540 |
| Cast-iron Class 60 | 60 | 410 | 170 | 1170 | 65 | 450 | | | 19 | 130 | 8.0 | 55 | 98 | 680 |
| Red brass | 57 | 390 | | | 37 | 255 | 49 | 340 | 15 | 100 | 5.7 | 39 | | |
| Yellow brass | 61 | 420 | | | 40 | 275 | 50 | 345 | 15 | 100 | 5.8 | 40 | | |
| 6061-T6 Aluminum alloy | 45 | 310 | | | 30 | 207 | 40 | 275 | 10.4 | 70 | 3.8 | 26 | | |
| Monel | 80 | 550 | | | 56 | 385 | 40 | 275 | 26 | 180 | 9.5 | 65 | | |
| AZ 31X Magnesium alloy | 40 | 275 | | | 19 | 130 | 30 | 200 | 6.5 | 45 | 2.4 | 16 | | |

* This table is included to provide data for the problems in this book. These values will vary with heat treatment and degree of mechanical working. Allowable stresses for metals specified by the AISC Code for Connections and Welds, and the ASME Code for Bolted Connections, are given in Chap. 9; the AISC allowable stresses for beams are given in Chapter 12.

**TABLE 2**   FACTORS OF SAFETY ($N$)

| Loading Condition | Ductile Metals, Steel, etc. | | Brittle Metals, Cast Iron, etc. |
|---|---|---|---|
| | $N_u$ | $N_y$ | $N_u$ |
| Steady load | 4 | 2 | 7 |
| Varying load | 6 | 3 | 10 |
| Shock load | 10 | 5 | 20 |

**TABLE 3**  UNIFIED AND AMERICAN SCREW THREADS (ABSTRACTED FROM ANSI STANDARD B 1.1-1974)

| Size | Basic Major Diameter, in | Coarse Threads, UNC and NC | | Fine Threads, UNF and NF | |
|------|------|------|------|------|------|
| | | Threads per in | Root (Minor) Diameter, in (External Thread) | Threads per in | Root (Minor) Diameter, in (External Thread) |
| 0 | 0.0600 | . . . . | . . . . . . | 80 | 0.0447 |
| 1 | 0.0730 | 64 | 0.0538 | 72 | 0.0560 |
| 2 | 0.0860 | 56 | 0.0641 | 64 | 0.0668 |
| 3 | 0.0990 | 48 | 0.0734 | 56 | 0.0771 |
| 4 | 0.1120 | 40 | 0.0813 | 48 | 0.0864 |
| 5 | 0.1250 | 40 | 0.0943 | 44 | 0.0971 |
| 6 | 0.1380 | 32 | 0.0997 | 40 | 0.1073 |
| 8 | 0.1640 | 32 | 0.1257 | 36 | 0.1299 |
| 10 | 0.1900 | 24 | 0.1389 | 32 | 0.1517 |
| 12 | 0.2160 | 24 | 0.1649 | 28 | 0.1722 |
| $\frac{1}{4}$ | 0.2500 | 20 | 0.1887 | 28 | 0.2062 |
| $\frac{5}{16}$ | 0.3125 | 18 | 0.2443 | 24 | 0.2614 |
| $\frac{3}{8}$ | 0.3750 | 16 | 0.2983 | 24 | 0.3239 |
| $\frac{7}{16}$ | 0.4375 | 14 | 0.3499 | 20 | 0.3762 |
| $\frac{1}{2}$ | 0.5000 | 13 | 0.4056 | 20 | 0.4387 |
| $\frac{9}{16}$ | 0.5625 | 12 | 0.4603 | 18 | 0.4943 |
| $\frac{5}{8}$ | 0.6250 | 11 | 0.5135 | 18 | 0.5568 |
| $\frac{3}{4}$ | 0.7500 | 10 | 0.6273 | 16 | 0.6733 |
| $\frac{7}{8}$ | 0.8750 | 9 | 0.7387 | 14 | 0.7874 |
| 1 | 1.0000 | 8 | 0.8466 | 12 | 0.8978 |
| $1\frac{1}{8}$ | 1.1250 | 7 | 0.9497 | 12 | 1.0228 |
| $1\frac{1}{4}$ | 1.2500 | 7 | 1.0747 | 12 | 1.1478 |
| $1\frac{3}{8}$ | 1.3750 | 6 | 1.1705 | 12 | 1.2728 |
| $1\frac{1}{2}$ | 1.5000 | 6 | 1.2955 | 12 | 1.3978 |
| $1\frac{3}{4}$ | 1.7500 | 5 | 1.5046 | | |
| 2 | 2.0000 | $4\frac{1}{2}$ | 1.7274 | | |
| $2\frac{1}{4}$ | 2.2500 | $4\frac{1}{2}$ | 1.9774 | | |
| $2\frac{1}{2}$ | 2.5000 | 4 | 2.1933 | | |
| $2\frac{3}{4}$ | 2.7500 | 4 | 2.4433 | | |
| 3 | 3.0000 | 4 | 2.6933 | | |
| $3\frac{1}{4}$ | 3.2500 | 4 | 2.9433 | | |
| $3\frac{1}{2}$ | 3.5000 | 4 | 3.1933 | | |
| $3\frac{3}{4}$ | 3.7500 | 4 | 3.4433 | | |
| 4 | 4.0000 | 4 | 3.6933 | | |

**TABLE 4  W SHAPES: PROPERTIES FOR DESIGNING (SELECTED LISTINGS)**

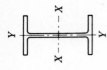

| | | | Flange | | Web | Elastic Properties | | | | | |
| | | | | | | Axis XX | | | Axis YY | | |
| Designation | Area $A$, in² | Depth $d$, in | Width $b_f$, in | Thickness $t_f$, in | Thickness $t_w$, in | $I$, in⁴ | $S$, in³ | $r$, in | $I$, in⁴ | $S$, in³ | $r$, in |
|---|---|---|---|---|---|---|---|---|---|---|---|
| W 36 × 300 | 88.3 | 36.72 | 16.655 | 1.680 | 0.945 | 20 300 | 1 110 | 15.2 | 1 300 | 156 | 3.83 |
| × 260 | 76.5 | 36.24 | 16.551 | 1.440 | 0.841 | 17 300 | 952 | 15.0 | 1 090 | 132 | 3.77 |
| × 230 | 67.7 | 35.88 | 16.471 | 1.260 | 0.761 | 15 000 | 837 | 14.9 | 940 | 114 | 3.73 |
| W 36 × 194 | 57.2 | 36.48 | 12.117 | 1.260 | 0.770 | 12 100 | 665 | 14.6 | 375 | 61.9 | 2.56 |
| × 170 | 50.0 | 36.16 | 12.027 | 1.100 | 0.680 | 10 500 | 580 | 14.5 | 320 | 53.2 | 2.53 |
| × 150 | 44.2 | 35.84 | 11.972 | 0.940 | 0.625 | 9 030 | 504 | 14.3 | 270 | 45.0 | 2.47 |
| W 33 × 220 | 64.8 | 33.25 | 15.810 | 1.275 | 0.775 | 12 300 | 742 | 13.8 | 841 | 106 | 3.60 |
| × 200 | 58.9 | 33.00 | 15.750 | 1.150 | 0.715 | 11 100 | 671 | 13.7 | 750 | 95.2 | 3.57 |
| W 33 × 141 | 41.6 | 33.31 | 11.535 | 0.960 | 0.605 | 7 460 | 448 | 13.4 | 246 | 42.7 | 2.43 |
| × 130 | 38.3 | 33.10 | 11.510 | 0.855 | 0.580 | 6 710 | 406 | 13.2 | 218 | 37.9 | 2.38 |
| W 30 × 124 | 36.5 | 30.16 | 10.521 | 0.930 | 0.585 | 5 360 | 355 | 12.1 | 181 | 34.4 | 2.23 |
| × 116 | 34.2 | 30.00 | 10.500 | 0.850 | 0.564 | 4 930 | 329 | 12.0 | 164 | 31.3 | 2.19 |
| × 108 | 31.8 | 29.82 | 10.484 | 0.760 | 0.548 | 4 470 | 300 | 11.9 | 146 | 27.9 | 2.15 |
| W 27 × 102 | 30.0 | 27.07 | 10.018 | 0.827 | 0.518 | 3 610 | 267 | 11.0 | 139 | 27.7 | 2.15 |
| × 94 | 27.7 | 26.91 | 9.990 | 0.747 | 0.490 | 3 270 | 243 | 10.9 | 124 | 24.9 | 2.12 |

| | 29.5 | 24.00 | 12.000 | 0.775 | 0.468 | 3 000 | 250 | 10.1 | 223 | 37.2 | 2.75 |
|---|---|---|---|---|---|---|---|---|---|---|---|
| W 24 × 100 | 29.5 | 24.00 | 12.000 | 0.775 | 0.468 | 3 000 | 250 | 10.1 | 223 | 37.2 | 2.75 |
| W 24 × 94 | 27.7 | 24.29 | 9.061 | 0.872 | 0.516 | 2 690 | 221 | 9.86 | 108 | 23.9 | 1.98 |
| × 84 | 24.7 | 24.09 | 9.015 | 0.772 | 0.470 | 2 370 | 197 | 9.79 | 94.5 | 21.0 | 1.95 |
| × 76 | 22.4 | 23.91 | 8.985 | 0.682 | 0.440 | 2 100 | 176 | 9.69 | 82.6 | 18.4 | 1.92 |
| W 21 × 73 | 21.5 | 21.24 | 8.295 | 0.740 | 0.455 | 1 600 | 151 | 8.64 | 70.6 | 17.0 | 1.81 |
| × 68 | 20.0 | 21.13 | 8.270 | 0.685 | 0.430 | 1 480 | 140 | 8.60 | 64.7 | 15.7 | 1.80 |
| × 62 | 18.3 | 20.99 | 8.240 | 0.615 | 0.400 | 1 330 | 127 | 8.54 | 57.5 | 13.9 | 1.77 |
| W 18 × 60 | 17.7 | 18.25 | 7.558 | 0.695 | 0.416 | 986 | 108 | 7.47 | 50.1 | 13.3 | 1.68 |
| × 55 | 16.2 | 18.12 | 7.532 | 0.630 | 0.390 | 891 | 98.4 | 7.42 | 45.0 | 11.9 | 1.67 |
| × 50 | 14.7 | 18.00 | 7.500 | 0.570 | 0.358 | 802 | 89.1 | 7.38 | 40.2 | 10.7 | 1.65 |
| W 16 × 58 | 17.1 | 15.86 | 8.464 | 0.645 | 0.407 | 748 | 94.4 | 6.62 | 65.3 | 15.4 | 1.96 |
| × 50 | 14.7 | 16.25 | 7.073 | 0.628 | 0.380 | 657 | 80.8 | 6.68 | 37.1 | 10.5 | 1.59 |
| × 45 | 13.3 | 16.12 | 7.039 | 0.563 | 0.346 | 584 | 72.5 | 6.64 | 32.8 | 9.32 | 1.57 |
| × 40 | 11.8 | 16.00 | 7.000 | 0.503 | 0.307 | 517 | 64.6 | 6.62 | 28.8 | 8.23 | 1.56 |
| × 36 | 10.6 | 15.85 | 6.992 | 0.428 | 0.299 | 447 | 56.5 | 6.50 | 24.4 | 6.99 | 1.52 |
| W 14 × 426 | 125 | 18.69 | 16.695 | 3.033 | 1.875 | 6 610 | 707 | 7.26 | 2 360 | 283 | 4.34 |
| × 342 | 101 | 17.56 | 16.365 | 2.468 | 1.545 | 4 910 | 559 | 6.99 | 1 810 | 221 | 4.24 |
| × 264 | 77.6 | 16.50 | 16.025 | 1.938 | 1.205 | 3 530 | 427 | 6.74 | 1 330 | 166 | 4.14 |
| W 14 × 228 | 67.1 | 16.00 | 15.865 | 1.688 | 1.045 | 2 940 | 368 | 6.62 | 1 120 | 142 | 4.10 |
| × 202 | 59.4 | 15.63 | 15.750 | 1.503 | 0.930 | 2 540 | 325 | 6.54 | 980 | 124 | 4.06 |
| × 176 | 51.7 | 15.25 | 15.640 | 1.313 | 0.820 | 2 150 | 282 | 6.45 | 838 | 107 | 4.02 |
| × 150 | 44.1 | 14.88 | 15.515 | 1.128 | 0.695 | 1 790 | 240 | 6.37 | 703 | 90.6 | 3.99 |

**TABLE 4** W SHAPES: PROPERTIES FOR DESIGNING (SELECTED LISTINGS) (Continued)

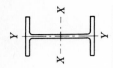

| Designation | Area $A$, in² | Depth $d$, in | Flange Width $b_f$, in | Flange Thickness $t_f$, in | Web Thickness $t_w$, in | Elastic Properties Axis XX $I$, in⁴ | $S$, in³ | $r$, in | Axis YY $I$, in⁴ | $S$, in³ | $r$, in |
|---|---|---|---|---|---|---|---|---|---|---|---|
| W 14 × 136 | 40.0 | 14.75 | 14.740 | 1.063 | 0.660 | 1 590 | 216 | 6.31 | 568 | 77.0 | 3.77 |
| × 111 | 32.7 | 14.37 | 14.620 | 0.873 | 0.540 | 1 270 | 176 | 6.23 | 455 | 62.2 | 3.73 |
| × 87 | 25.6 | 14.00 | 14.500 | 0.688 | 0.420 | 967 | 138 | 6.15 | 350 | 48.2 | 3.70 |
| W 14 × 78 | 22.9 | 14.06 | 12.000 | 0.718 | 0.428 | 851 | 121 | 6.09 | 207 | 34.5 | 3.00 |
| W 14 × 68 | 20.0 | 14.06 | 10.040 | 0.718 | 0.418 | 724 | 103 | 6.02 | 121 | 24.1 | 2.46 |
| W 14 × 34 | 10.0 | 14.00 | 6.750 | 0.453 | 0.287 | 340 | 48.6 | 5.83 | 23.3 | 6.89 | 1.52 |
| × 30 | 8.83 | 13.86 | 6.733 | 0.383 | 0.270 | 290 | 41.9 | 5.74 | 19.5 | 5.80 | 1.49 |
| W 12 × 190 | 55.9 | 14.38 | 12.670 | 1.736 | 1.060 | 1 890 | 263 | 5.82 | 590 | 93.1 | 3.25 |
| × 133 | 39.1 | 13.38 | 12.365 | 1.236 | 0.755 | 1 220 | 183 | 5.59 | 390 | 63.1 | 3.16 |
| × 106 | 31.2 | 12.88 | 12.230 | 0.986 | 0.620 | 931 | 145 | 5.46 | 301 | 49.2 | 3.11 |
| × 92 | 27.1 | 12.62 | 12.155 | 0.856 | 0.545 | 789 | 125 | 5.40 | 256 | 42.2 | 3.08 |
| × 85 | 25.0 | 12.50 | 12.105 | 0.796 | 0.495 | 723 | 116 | 5.38 | 235 | 38.9 | 3.07 |
| × 79 | 23.2 | 12.38 | 12.080 | 0.736 | 0.470 | 663 | 107 | 5.34 | 216 | 35.8 | 3.05 |
| × 72 | 21.2 | 12.25 | 12.040 | 0.671 | 0.430 | 597 | 97.5 | 5.31 | 195 | 32.4 | 3.04 |
| × 65 | 19.1 | 12.12 | 12.000 | 0.606 | 0.390 | 533 | 88.0 | 5.28 | 175 | 29.1 | 3.02 |

| | | | | | | | | | | |
|---|---|---|---|---|---|---|---|---|---|---|
| W 12 × 27 | 7.95 | 11.96 | 6.497 | 0.400 | 0.237 | 204 | 34.2 | 5.07 | 18.3 | 5.63 | 1.52 |
| W 10 × 112 | 32.9 | 11.38 | 10.415 | 1.248 | 0.755 | 719 | 126 | 4.67 | 235 | 45.2 | 2.67 |
| × 89 | 26.2 | 10.88 | 10.275 | 0.998 | 0.615 | 542 | 99.7 | 4.55 | 181 | 35.2 | 2.63 |
| W 10 × 77 | 22.7 | 10.62 | 10.195 | 0.868 | 0.535 | 457 | 86.1 | 4.49 | 153 | 30.1 | 2.60 |
| × 66 | 19.4 | 10.38 | 10.117 | 0.748 | 0.457 | 382 | 73.7 | 4.44 | 129 | 25.5 | 2.58 |
| × 54 | 15.9 | 10.12 | 10.028 | 0.618 | 0.368 | 306 | 60.4 | 4.39 | 104 | 20.7 | 2.56 |
| W 10 × 25 | 7.36 | 10.08 | 5.762 | 0.430 | 0.252 | 133 | 26.5 | 4.26 | 13.7 | 4.76 | 1.37 |
| × 21 | 6.20 | 9.90 | 5.750 | 0.340 | 0.240 | 107 | 21.5 | 4.15 | 10.8 | 3.75 | 1.32 |
| W 8 × 67 | 19.7 | 9.00 | 8.287 | 0.933 | 0.575 | 272 | 60.4 | 3.71 | 88.6 | 21.4 | 2.12 |
| × 48 | 14.1 | 8.50 | 8.117 | 0.683 | 0.405 | 184 | 43.2 | 3.61 | 60.9 | 15.0 | 2.08 |
| × 35 | 10.3 | 8.12 | 8.027 | 0.493 | 0.315 | 126 | 31.1 | 3.50 | 42.5 | 10.6 | 2.03 |
| × 31 | 9.12 | 8.00 | 8.000 | 0.433 | 0.288 | 110 | 27.4 | 3.47 | 37.0 | 9.24 | 2.01 |

Data in Tables 4 through 9 from *Manual of Steel Construction* 7th ed., American Institute of Steel Construction, 1974.

## TABLE 5  S SHAPES: PROPERTIES FOR DESIGNING

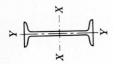

| Designation | Area $A$, in² | Depth $d$, in | Flange Width $b_f$, in | Flange Thickness $t_f$, in | Web Thickness $t_w$, in | Axis XX $I$, in⁴ | Axis XX $S$, in³ | Axis XX $r$, in | Axis YY $I$, in⁴ | Axis YY $S$, in³ | Axis YY $r$, in |
|---|---|---|---|---|---|---|---|---|---|---|---|
| S 24 × 120 | 35.3 | 24.00 | 8.048 | 1.102 | 0.798 | 3 030 | 252 | 9.26 | 84.2 | 20.9 | 1.54 |
| × 105.9 | 31.1 | 24.00 | 7.875 | 1.102 | 0.625 | 2 830 | 236 | 9.53 | 78.2 | 19.8 | 1.58 |
| S 24 × 100 | 29.4 | 24.00 | 7.247 | 0.871 | 0.747 | 2 390 | 199 | 9.01 | 47.8 | 13.2 | 1.27 |
| × 90 | 26.5 | 24.00 | 7.124 | 0.871 | 0.624 | 2 250 | 187 | 9.22 | 44.9 | 12.6 | 1.30 |
| × 79.9 | 23.5 | 24.00 | 7.001 | 0.871 | 0.501 | 2 110 | 175 | 9.47 | 42.3 | 12.1 | 1.34 |
| S 20 × 95 | 27.9 | 20.00 | 7.200 | 0.916 | 0.800 | 1 610 | 161 | 7.60 | 49.7 | 13.8 | 1.33 |
| × 85 | 25.0 | 20.00 | 7.053 | 0.916 | 0.653 | 1 520 | 152 | 7.79 | 46.2 | 13.1 | 1.36 |
| S 20 × 75 | 22.1 | 20.00 | 6.391 | 0.789 | 0.641 | 1 280 | 128 | 7.60 | 29.6 | 9.28 | 1.16 |
| × 65.4 | 19.2 | 20.00 | 6.250 | 0.789 | 0.500 | 1 180 | 118 | 7.84 | 27.4 | 8.77 | 1.19 |
| S 18 × 70 | 20.6 | 18.00 | 6.251 | 0.691 | 0.711 | 926 | 103 | 6.71 | 24.1 | 7.72 | 1.08 |
| × 54.7 | 16.1 | 18.00 | 6.001 | 0.691 | 0.461 | 804 | 89.4 | 7.07 | 20.8 | 6.94 | 1.14 |
| S 15 × 50 | 14.7 | 15.00 | 5.640 | 0.622 | 0.550 | 486 | 64.8 | 5.75 | 15.7 | 5.57 | 1.03 |
| × 42.9 | 12.6 | 15.00 | 5.501 | 0.622 | 0.411 | 447 | 59.6 | 5.95 | 14.4 | 5.23 | 1.07 |

| Designation | | | | | | | | | | | |
|---|---|---|---|---|---|---|---|---|---|---|---|
| S 12 × 50 | 14.7 | 12.00 | 5.477 | 0.659 | 0.687 | 305 | 50.8 | 4.55 | 15.7 | 5.74 | 1.03 |
| × 40.8 | 12.0 | 12.00 | 5.252 | 0.659 | 0.462 | 272 | 45.4 | 4.77 | 13.6 | 5.16 | 1.06 |
| S 12 × 35 | 10.3 | 12.00 | 5.078 | 0.544 | 0.428 | 229 | 38.2 | 4.72 | 9.87 | 3.89 | 0.980 |
| × 31.8 | 9.35 | 12.00 | 5.000 | 0.544 | 0.350 | 218 | 36.4 | 4.83 | 9.36 | 3.74 | 1.00 |
| S 10 × 35 | 10.3 | 10.00 | 4.944 | 0.491 | 0.594 | 147 | 29.4 | 3.78 | 8.36 | 3.38 | 0.901 |
| × 25.4 | 7.46 | 10.00 | 4.661 | 0.491 | 0.311 | 124 | 24.7 | 4.07 | 6.79 | 2.91 | 0.954 |
| S 8 × 23 | 6.77 | 8.00 | 4.171 | 0.425 | 0.441 | 64.9 | 16.2 | 3.10 | 4.31 | 2.07 | 0.798 |
| × 18.4 | 5.41 | 8.00 | 4.001 | 0.425 | 0.271 | 57.6 | 14.4 | 3.26 | 3.73 | 1.86 | 0.831 |
| S 7 × 20 | 5.88 | 7.00 | 3.860 | 0.392 | 0.450 | 42.4 | 12.1 | 2.69 | 3.17 | 1.64 | 0.734 |
| × 15.3 | 4.50 | 7.00 | 3.662 | 0.392 | 0.252 | 36.7 | 10.5 | 2.86 | 2.64 | 1.44 | 0.766 |
| S 6 × 17.25 | 5.07 | 6.00 | 3.565 | 0.359 | 0.465 | 26.3 | 8.77 | 2.28 | 2.31 | 1.30 | 0.675 |
| × 12.5 | 3.67 | 6.00 | 3.332 | 0.359 | 0.232 | 22.1 | 7.37 | 2.45 | 1.82 | 1.09 | 0.705 |
| S 5 × 14.75 | 4.34 | 5.00 | 3.284 | 0.326 | 0.494 | 15.2 | 6.09 | 1.87 | 1.67 | 1.01 | 0.620 |
| × 10 | 2.94 | 5.00 | 3.004 | 0.326 | 0.214 | 12.3 | 4.92 | 2.05 | 1.22 | 0.809 | 0.643 |
| S 4 × 9.5 | 2.79 | 4.00 | 2.796 | 0.293 | 0.326 | 6.79 | 3.39 | 1.56 | 0.903 | 0.646 | 0.569 |
| × 7.7 | 2.26 | 4.00 | 2.663 | 0.293 | 0.193 | 6.08 | 3.04 | 1.64 | 0.764 | 0.574 | 0.581 |
| S 3 × 7.5 | 2.21 | 3.00 | 2.509 | 0.260 | 0.349 | 2.93 | 1.95 | 1.15 | 0.586 | 0.468 | 0.516 |
| × 5.7 | 1.67 | 3.00 | 2.330 | 0.260 | 0.170 | 2.52 | 1.68 | 1.23 | 0.455 | 0.390 | 0.522 |

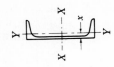

**TABLE 6** AMERICAN STANDARD CHANNELS: PROPERTIES FOR DESIGNING

| Designation | Area $A$, in² | Depth $d$, in | Flange Width $b_f$, in | Flange Thickness $t_f$, in | Web Thickness $t_w$, in | Axis XX $I$, in⁴ | Axis XX $S$, in³ | Axis XX $r$, in | Axis YY $I$, in⁴ | Axis YY $S$, in³ | Axis YY $r$, in | Axis YY $\bar{x}$, in |
|---|---|---|---|---|---|---|---|---|---|---|---|---|
| C 15 × 50 | 14.7 | 15.00 | 3.716 | 0.650 | 0.716 | 404 | 53.8 | 5.24 | 11.0 | 3.78 | 0.867 | 0.799 |
| × 40 | 11.8 | 15.00 | 3.520 | 0.650 | 0.520 | 349 | 46.5 | 5.44 | 9.23 | 3.36 | 0.886 | 0.778 |
| × 33.9 | 9.96 | 15.00 | 3.400 | 0.650 | 0.400 | 315 | 42.0 | 5.62 | 8.13 | 3.11 | 0.904 | 0.787 |
| C 12 × 30 | 8.82 | 12.00 | 3.170 | 0.501 | 0.510 | 162 | 27.0 | 4.29 | 5.14 | 2.06 | 0.763 | 0.674 |
| × 25 | 7.35 | 12.00 | 3.047 | 0.501 | 0.387 | 144 | 24.1 | 4.43 | 4.47 | 1.88 | 0.780 | 0.674 |
| × 20.7 | 6.09 | 12.00 | 2.942 | 0.501 | 0.282 | 129 | 21.5 | 4.61 | 3.88 | 1.73 | 0.799 | 0.698 |
| C 10 × 30 | 8.82 | 10.00 | 3.033 | 0.436 | 0.673 | 103 | 20.7 | 3.42 | 3.94 | 1.65 | 0.669 | 0.649 |
| × 25 | 7.35 | 10.00 | 2.886 | 0.436 | 0.526 | 91.2 | 18.2 | 3.52 | 3.36 | 1.48 | 0.676 | 0.617 |
| × 20 | 5.88 | 10.00 | 2.739 | 0.436 | 0.379 | 78.9 | 15.8 | 3.66 | 2.81 | 1.32 | 0.691 | 0.606 |
| × 15.3 | 4.49 | 10.00 | 2.600 | 0.436 | 0.240 | 67.4 | 13.5 | 3.87 | 2.28 | 1.16 | 0.713 | 0.634 |
| C 9 × 20 | 5.88 | 9.00 | 2.648 | 0.413 | 0.448 | 60.9 | 13.5 | 3.22 | 2.42 | 1.17 | 0.642 | 0.583 |
| × 15 | 4.41 | 9.00 | 2.485 | 0.413 | 0.285 | 51.0 | 11.3 | 3.40 | 1.93 | 1.01 | 0.661 | 0.586 |
| × 13.4 | 3.94 | 9.00 | 2.433 | 0.413 | 0.233 | 47.9 | 10.6 | 3.48 | 1.76 | 0.962 | 0.668 | 0.601 |

| Designation | | | | | | | | | | | | |
|---|---|---|---|---|---|---|---|---|---|---|---|---|
| C 8 × 18.75 | 5.51 | 8.00 | 2.527 | 0.390 | 0.487 | 44.0 | 11.0 | 2.82 | 1.98 | 1.01 | 0.599 | 0.565 |
| × 13.75 | 4.04 | 8.00 | 2.343 | 0.390 | 0.303 | 36.1 | 9.03 | 2.99 | 1.53 | 0.853 | 0.615 | 0.553 |
| × 11.5 | 3.38 | 8.00 | 2.260 | 0.390 | 0.220 | 32.6 | 8.14 | 3.11 | 1.32 | 0.781 | 0.625 | 0.571 |
| C 7 × 14.75 | 4.33 | 7.00 | 2.299 | 0.366 | 0.419 | 27.2 | 7.78 | 2.51 | 1.38 | 0.779 | 0.564 | 0.532 |
| × 12.25 | 3.60 | 7.00 | 2.194 | 0.366 | 0.314 | 24.2 | 6.93 | 2.60 | 1.17 | 0.702 | 0.571 | 0.525 |
| × 9.8 | 2.87 | 7.00 | 2.090 | 0.366 | 0.210 | 21.3 | 6.08 | 2.72 | 0.968 | 0.625 | 0.581 | 0.541 |
| C 6 × 13 | 3.83 | 6.00 | 2.157 | 0.343 | 0.437 | 17.4 | 5.80 | 2.13 | 1.05 | 0.642 | 0.525 | 0.514 |
| × 10.5 | 3.09 | 6.00 | 2.034 | 0.343 | 0.314 | 15.2 | 5.06 | 2.22 | 0.865 | 0.564 | 0.529 | 0.500 |
| × 8.2 | 2.40 | 6.00 | 1.920 | 0.343 | 0.200 | 13.1 | 4.38 | 2.34 | 0.692 | 0.492 | 0.537 | 0.512 |
| C 5 × 9 | 2.64 | 5.00 | 1.885 | 0.320 | 0.325 | 8.90 | 3.56 | 1.83 | 0.632 | 0.449 | 0.489 | 0.478 |
| × 6.7 | 1.97 | 5.00 | 1.750 | 0.320 | 0.190 | 7.49 | 3.00 | 1.95 | 0.478 | 0.378 | 0.493 | 0.484 |
| C 4 × 7.25 | 2.13 | 4.00 | 1.721 | 0.296 | 0.321 | 4.59 | 2.29 | 1.47 | 0.432 | 0.343 | 0.450 | 0.459 |
| × 5.4 | 1.59 | 4.00 | 1.584 | 0.296 | 0.184 | 3.85 | 1.93 | 1.56 | 0.319 | 0.283 | 0.449 | 0.458 |
| C 3 × 6 | 1.76 | 3.00 | 1.596 | 0.273 | 0.356 | 2.07 | 1.38 | 1.08 | 0.305 | 0.268 | 0.416 | 0.455 |
| × 5 | 1.47 | 3.00 | 1.498 | 0.273 | 0.258 | 1.85 | 1.24 | 1.12 | 0.247 | 0.233 | 0.410 | 0.438 |
| × 4.1 | 1.21 | 3.00 | 1.410 | 0.273 | 0.170 | 1.66 | 1.10 | 1.17 | 0.197 | 0.202 | 0.404 | 0.437 |

# TABLE 7 ANGLES, EQUAL LEGS: PROPERTIES FOR DESIGNING (SELECTED LISTING)

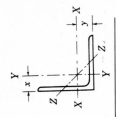

| Size and thickness, in | Weight per foot, lb | Area, in² | Axis XX and Axis YY | | | | Axis ZZ |
|---|---|---|---|---|---|---|---|
| | | | $I$, in⁴ | $S$, in³ | $r$, in | $x$ or $y$, in | $r$, in |
| L 8 × 8 × 1 | 51.0 | 15.0 | 89.0 | 15.8 | 2.44 | 2.37 | 1.56 |
| 3/4 | 38.9 | 11.4 | 69.7 | 12.2 | 2.47 | 2.28 | 1.58 |
| 5/8 | 32.7 | 9.61 | 59.4 | 10.3 | 2.49 | 2.23 | 1.58 |
| 1/2 | 26.4 | 7.75 | 48.6 | 8.36 | 2.50 | 2.19 | 1.59 |
| L 6 × 6 × 1 | 37.4 | 11.0 | 35.5 | 8.57 | 1.80 | 1.86 | 1.17 |
| 3/4 | 28.7 | 8.44 | 28.2 | 6.66 | 1.83 | 1.78 | 1.17 |
| 1/2 | 19.6 | 5.75 | 19.9 | 4.61 | 1.86 | 1.68 | 1.18 |
| 3/8 | 14.9 | 4.36 | 15.4 | 3.53 | 1.88 | 1.64 | 1.19 |
| L 5 × 5 × 3/4 | 23.6 | 6.94 | 15.7 | 4.53 | 1.51 | 1.52 | 0.975 |
| 1/2 | 16.2 | 4.75 | 11.3 | 3.16 | 1.54 | 1.43 | 0.983 |
| 3/8 | 12.3 | 3.61 | 8.74 | 2.42 | 1.56 | 1.39 | 0.990 |
| L 4 × 4 × 3/4 | 18.5 | 5.44 | 7.67 | 2.81 | 1.19 | 1.27 | 0.778 |
| 1/2 | 12.8 | 3.75 | 5.56 | 1.97 | 1.22 | 1.18 | 0.782 |
| 3/8 | 9.8 | 2.85 | 4.36 | 1.52 | 1.23 | 1.14 | 0.788 |
| 1/4 | 6.6 | 1.94 | 3.04 | 1.05 | 1.25 | 1.09 | 0.795 |

| Size | | | | | | | |
|---|---|---|---|---|---|---|---|
| L $3\frac{1}{2} \times 3\frac{1}{2} \times \frac{1}{2}$ | 11.1 | 3.25 | 3.64 | 1.49 | 1.06 | 1.06 | 0.683 |
| $\frac{3}{8}$ | 8.5 | 2.48 | 2.87 | 1.15 | 1.07 | 1.01 | 0.687 |
| $\frac{1}{4}$ | 5.8 | 1.69 | 2.01 | 0.794 | 1.09 | 0.968 | 0.694 |
| L $3 \times 3 \times \frac{1}{2}$ | 9.4 | 2.75 | 2.22 | 1.07 | 0.898 | 0.932 | 0.584 |
| $\frac{3}{8}$ | 7.2 | 2.11 | 1.76 | 0.833 | 0.913 | 0.888 | 0.587 |
| $\frac{1}{4}$ | 4.9 | 1.44 | 1.24 | 0.577 | 0.930 | 0.842 | 0.592 |
| L $2\frac{1}{2} \times 2\frac{1}{2} \times \frac{1}{2}$ | 7.7 | 2.25 | 1.23 | 0.724 | 0.739 | 0.806 | 0.487 |
| $\frac{1}{4}$ | 4.1 | 1.19 | 0.703 | 0.394 | 0.769 | 0.717 | 0.491 |
| L $2 \times 2 \times \frac{3}{8}$ | 4.7 | 1.36 | 0.479 | 0.351 | 0.594 | 0.636 | 0.389 |
| $\frac{1}{4}$ | 3.19 | 0.938 | 0.348 | 0.247 | 0.609 | 0.592 | 0.391 |
| $\frac{1}{8}$ | 1.65 | 0.484 | 0.190 | 0.131 | 0.626 | 0.546 | 0.398 |
| L $1\frac{3}{4} \times 1\frac{3}{4} \times \frac{1}{4}$ | 2.77 | 0.813 | 0.227 | 0.186 | 0.529 | 0.529 | 0.341 |
| $\frac{1}{8}$ | 1.44 | 0.422 | 0.126 | 0.099 | 0.546 | 0.484 | 0.347 |
| L $1\frac{1}{2} \times 1\frac{1}{2} \times \frac{1}{4}$ | 2.34 | 0.688 | 0.139 | 0.134 | 0.449 | 0.466 | 0.292 |
| $\frac{1}{8}$ | 1.23 | 0.359 | 0.078 | 0.072 | 0.465 | 0.421 | 0.296 |
| L $1\frac{1}{4} \times 1\frac{1}{4} \times \frac{1}{4}$ | 1.92 | 0.563 | 0.077 | 0.091 | 0.369 | 0.403 | 0.243 |
| $\frac{1}{8}$ | 1.01 | 0.297 | 0.044 | 0.049 | 0.385 | 0.359 | 0.246 |
| L $1 \times 1 \times \frac{1}{4}$ | 1.49 | 0.438 | 0.037 | 0.056 | 0.290 | 0.339 | 0.196 |
| $\frac{1}{8}$ | 0.80 | 0.234 | 0.022 | 0.031 | 0.304 | 0.296 | 0.196 |

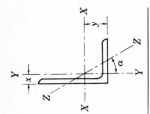

**TABLE 8**  ANGLES, UNEQUAL LEGS: PROPERTIES FOR DESIGNING (SELECTED LISTING)

| Size and thickness, in | Weight per foot, lb | Area, in² | Axis XX | | | | Axis YY | | | | Axis ZZ | |
|---|---|---|---|---|---|---|---|---|---|---|---|---|
| | | | $I$, in⁴ | $S$, in³ | $r$, in | $y$, in | $I$, in⁴ | $S$, in³ | $r$, in | $x$, in | $r$, in | $\tan \alpha$ |
| L 8 × 6 × 1 | 44.2 | 13.0 | 80.8 | 15.1 | 2.49 | 2.65 | 38.8 | 8.92 | 1.73 | 1.65 | 1.28 | 0.543 |
| ¾ | 33.8 | 9.94 | 63.4 | 11.7 | 2.53 | 2.56 | 30.7 | 6.92 | 1.76 | 1.56 | 1.29 | 0.551 |
| ½ | 23.0 | 6.75 | 44.3 | 8.02 | 2.56 | 2.47 | 21.7 | 4.79 | 1.79 | 1.47 | 1.30 | 0.558 |
| L 8 × 4 × 1 | 37.4 | 11.0 | 69.6 | 14.1 | 2.52 | 3.05 | 11.6 | 3.94 | 1.03 | 1.05 | 0.846 | 0.247 |
| ¾ | 28.7 | 8.44 | 54.9 | 10.9 | 2.55 | 2.95 | 9.36 | 3.07 | 1.05 | 0.953 | 0.852 | 0.258 |
| ½ | 19.6 | 5.75 | 38.5 | 7.49 | 2.59 | 2.86 | 6.74 | 2.15 | 1.08 | 0.859 | 0.865 | 0.267 |
| L 7 × 4 × ¾ | 26.2 | 7.69 | 37.8 | 8.42 | 2.22 | 2.51 | 9.05 | 3.03 | 1.09 | 1.01 | 0.860 | 0.324 |
| ½ | 17.9 | 5.25 | 26.7 | 5.81 | 2.25 | 2.42 | 6.53 | 2.12 | 1.11 | 0.917 | 0.872 | 0.335 |
| ⅜ | 13.6 | 3.98 | 20.6 | 4.44 | 2.27 | 2.37 | 5.10 | 1.63 | 1.13 | 0.870 | 0.880 | 0.340 |
| L 6 × 4 × ¾ | 23.6 | 6.94 | 24.5 | 6.25 | 1.88 | 2.08 | 8.68 | 2.97 | 1.12 | 1.08 | 0.860 | 0.428 |
| ½ | 16.2 | 4.75 | 17.4 | 4.33 | 1.91 | 1.99 | 6.27 | 2.08 | 1.15 | 0.987 | 0.870 | 0.440 |
| ⅜ | 12.3 | 3.61 | 13.5 | 3.32 | 1.93 | 1.94 | 4.90 | 1.60 | 1.17 | 0.941 | 0.877 | 0.446 |
| L 6 × 3½ × ½ | 15.3 | 4.50 | 16.6 | 4.24 | 1.92 | 2.08 | 4.25 | 1.59 | 0.972 | 0.833 | 0.759 | 0.344 |
| ¼ | 7.9 | 2.31 | 8.86 | 2.21 | 1.96 | 1.99 | 2.34 | 0.847 | 1.01 | 0.740 | 0.777 | 0.355 |

| Size | | | | | | | | | | | | |
|---|---|---|---|---|---|---|---|---|---|---|---|---|
| ∟ 5 × 3½ × ¾ | 19.8 | 5.81 | 13.9 | 4.28 | 1.55 | 1.75 | 5.55 | 2.22 | 0.977 | 0.996 | 0.748 | 0.464 |
| ∟ 5 × 3½ × ½ | 13.6 | 4.00 | 9.99 | 2.99 | 1.58 | 1.66 | 4.05 | 1.56 | 1.01 | 0.906 | 0.755 | 0.479 |
| ∟ 5 × 3½ × ¼ | 7.0 | 2.06 | 5.39 | 1.57 | 1.62 | 1.56 | 2.23 | 0.830 | 1.04 | 0.814 | 0.770 | 0.492 |
| ∟ 5 × 3 × ½ | 12.8 | 3.75 | 9.45 | 2.91 | 1.59 | 1.75 | 2.58 | 1.15 | 0.829 | 0.750 | 0.648 | 0.357 |
| ∟ 5 × 3 × ¼ | 6.6 | 1.94 | 5.11 | 1.53 | 1.62 | 1.66 | 1.44 | 0.614 | 0.861 | 0.657 | 0.663 | 0.371 |
| ∟ 4 × 3½ × ½ | 11.9 | 3.50 | 5.32 | 1.94 | 1.23 | 1.25 | 3.79 | 1.52 | 1.04 | 1.00 | 0.722 | 0.750 |
| ∟ 4 × 3½ × ¼ | 6.2 | 1.81 | 2.91 | 1.03 | 1.27 | 1.16 | 2.09 | 0.808 | 1.07 | 0.909 | 0.734 | 0.759 |
| ∟ 4 × 3 × ½ | 11.1 | 3.25 | 5.05 | 1.89 | 1.25 | 1.33 | 2.42 | 1.12 | 0.864 | 0.827 | 0.639 | 0.543 |
| ∟ 4 × 3 × ¼ | 5.8 | 1.69 | 2.77 | 1.00 | 1.28 | 1.24 | 1.36 | 0.599 | 0.896 | 0.736 | 0.651 | 0.558 |
| ∟ 3½ × 3 × ½ | 10.2 | 3.00 | 3.45 | 1.45 | 1.07 | 1.13 | 2.33 | 1.10 | 0.881 | 0.875 | 0.621 | 0.714 |
| ∟ 3½ × 3 × ¼ | 5.4 | 1.56 | 1.91 | 0.776 | 1.11 | 1.04 | 1.30 | 0.589 | 0.914 | 0.785 | 0.631 | 0.727 |
| ∟ 3½ × 2½ × ½ | 9.4 | 2.75 | 3.24 | 1.41 | 1.09 | 1.20 | 1.36 | 0.760 | 0.704 | 0.705 | 0.534 | 0.486 |
| ∟ 3½ × 2½ × ¼ | 4.9 | 1.44 | 1.80 | 0.755 | 1.12 | 1.11 | 0.777 | 0.412 | 0.735 | 0.614 | 0.544 | 0.506 |
| ∟ 3 × 2½ × ½ | 8.5 | 2.50 | 2.08 | 1.04 | 0.913 | 1.00 | 1.30 | 0.744 | 0.722 | 0.750 | 0.520 | 0.667 |
| ∟ 3 × 2½ × ¼ | 4.5 | 1.31 | 1.17 | 0.561 | 0.945 | 0.911 | 0.743 | 0.404 | 0.753 | 0.661 | 0.528 | 0.684 |
| ∟ 3 × 2 × ½ | 7.7 | 2.25 | 1.92 | 1.00 | 0.924 | 1.08 | 0.672 | 0.474 | 0.546 | 0.583 | 0.428 | 0.414 |
| ∟ 3 × 2 × ¼ | 4.1 | 1.19 | 1.09 | 0.542 | 0.957 | 0.993 | 0.392 | 0.260 | 0.574 | 0.493 | 0.435 | 0.440 |
| ∟ 2½ × 2 × ¼ | 3.62 | 1.06 | 0.654 | 0.381 | 0.784 | 0.787 | 0.372 | 0.254 | 0.592 | 0.537 | 0.424 | 0.626 |
| ∟ 2½ × 1½ × ¼ | 3.19 | 0.938 | 0.591 | 0.364 | 0.794 | 0.875 | 0.161 | 0.143 | 0.415 | 0.375 | 0.324 | 0.357 |
| ∟ 2 × 1½ × ¼ | 2.77 | 0.813 | 0.316 | 0.236 | 0.623 | 0.663 | 0.151 | 0.139 | 0.432 | 0.413 | 0.320 | 0.543 |
| ∟ 2 × 1½ × ⅛ | 1.44 | 0.422 | 0.173 | 0.125 | 0.641 | 0.618 | 0.085 | 0.075 | 0.448 | 0.368 | 0.326 | 0.558 |
| ∟ 1¾ × 1¼ × ¼ | 2.34 | 0.688 | 0.202 | 0.176 | 0.543 | 0.602 | 0.085 | 0.095 | 0.352 | 0.352 | 0.267 | 0.486 |
| ∟ 1¾ × 1¼ × ⅛ | 1.23 | 0.359 | 0.113 | 0.094 | 0.560 | 0.557 | 0.049 | 0.051 | 0.368 | 0.307 | 0.272 | 0.506 |

**TABLE 9** SECTION MODULUS TABLE FOR SHAPES USED AS BEAMS
(SELECTED LISTING)

| Section Modulus | Shape | Section Modulus | Shape | Section Modulus | Shape |
|---|---|---|---|---|---|
| 1 110 | W 36 × 300 | 140 | W 21 × 68 | 48.6 | W 14 × 34 |
| 952 | W 36 × 260 | 138 | W 14 × 87 | 46.5 | C 15 × 40 |
| | | 128 | S 20 × 75 | 45.4 | S 12 × 40.8 |
| 837 | W 36 × 230 | | | 43.2 | W 8 × 48 |
| 742 | W 33 × 220 | 127 | W 21 × 62 | | |
| 671 | W 33 × 200 | 126 | W 10 × 112 | 42.0 | C 15 × 33.9 |
| 665 | W 36 × 194 | 125 | W 12 × 92 | | |
| 580 | W 36 × 170 | 121 | W 14 × 78 | 41.9 | W 14 × 30 |
| 504 | W 36 × 150 | 118 | S 20 × 65.4 | 38.2 | S 12 × 35 |
| 448 | W 33 × 141 | 116 | W 12 × 85 | 36.4 | S 12 × 31.8 |
| 406 | W 33 × 130 | | | | |
| 355 | W 30 × 124 | 108 | W 18 × 60 | 34.2 | W 12 × 27 |
| 329 | W 30 × 116 | 107 | W 12 × 79 | 31.1 | W 8 × 35 |
| 300 | W 30 × 108 | 103 | W 14 × 68 | 29.4 | S 10 × 35 |
| | | 103 | S 18 × 70 | 27.4 | W 8 × 31 |
| 267 | W 27 × 102 | 99.7 | W 10 × 89 | 27.0 | C 12 × 30 |
| 263 | W 12 × 190 | | | | |
| 252 | S 24 × 120 | 98.4 | W 18 × 55 | 26.5 | W 10 × 25 |
| | | 97.5 | W 12 × 72 | 24.7 | S 10 × 25.4 |
| 250 | W 24 × 100 | 94.4 | W 16 × 58 | 24.1 | C 12 × 25 |
| 243 | W 27 × 94 | 89.4 | S 18 × 54.7 | 21.5 | C 12 × 20.7 |
| 236 | S 24 × 105.9 | | | | |
| | | 89.1 | W 18 × 50 | 21.5 | W 10 × 21 |
| 221 | W 24 × 94 | 88.0 | W 12 × 65 | 20.7 | C 10 × 30 |
| 216 | W 14 × 136 | 86.1 | W 10 × 77 | 18.2 | C 10 × 25 |
| 199 | S 24 × 100 | | | 16.2 | S 8 × 23 |
| | | 80.8 | W 16 × 50 | | |
| 197 | W 24 × 84 | 73.7 | W 10 × 66 | 15.8 | C 10 × 20 |
| 187 | S 24 × 90 | | | | |
| 183 | W 12 × 133 | 72.5 | W 16 × 45 | 14.4 | S 8 × 18.4 |
| 176 | W 14 × 111 | 64.8 | S 15 × 50 | | |
| | | | | 13.5 | C 10 × 15.3 |
| 176 | W 24 × 76 | 64.6 | W 16 × 40 | 13.5 | C 9 × 20 |
| 175 | S 24 × 79.9 | 60.4 | W 10 × 54 | 12.1 | S 7 × 20 |
| 161 | S 20 × 95 | 59.6 | S 15 × 42.9 | | |
| 152 | S 20 × 85 | | | 11.3 | C 9 × 15 |
| | | 56.5 | W 16 × 36 | 11.0 | C 8 × 18.75 |
| 151 | W 21 × 73 | 53.8 | C 15 × 50 | | |
| 145 | W 12 × 106 | 50.8 | S 12 × 50 | 10.6 | C 9 × 13.4 |
| | | | | 10.5 | S 7 × 15.3 |
| | | | | 9.03 | C 8 × 13.75 |

**TABLE 10  TYPICAL PROPERTIES OF LUMBER***

| Species | Compression Parallel to Grain | | Bending | | Horizontal Shear | | Modulus of Elasticity $E$ | |
| | | | Allowable stress | | | | | |
| | psi | $10^6$ Pa | psi | $10^6$ Pa | psi | $10^6$ Pa | $10^6$ psi | $10^9$ Pa |
|---|---|---|---|---|---|---|---|---|
| Douglas fir | 1 050 | 7.24 | 1 400 | 9.65 | 85 | 0.59 | 1.2 | 8.3 |
| Eastern white pine | 725 | 5.00 | 975 | 6.72 | 65 | 0.45 | 1.1 | 7.6 |
| Mountain hemlock | 925 | 6.38 | 1 250 | 8.62 | 90 | 0.62 | 1.1 | 7.6 |
| Ponderosa pine | 800 | 5.52 | 1 000 | 6.90 | 65 | 0.45 | 1.1 | 7.6 |
| Sitka spruce | 875 | 6.03 | 1 150 | 7.93 | 70 | 0.48 | 1.3 | 9.0 |

* This table is included to provide data for the problems in this book. These values will vary with moisture content, cross-sectional size, and grade. For further information see *National Design Specification for Stress-Grade Lumber*, National Forest Products Association, 1974.

**TABLE 11** LUMBER AND TIMBER, AMERICAN STANDARD SIZES*: PROPERTIES FOR DESIGNING, NATIONAL FOREST PRODUCTS ASSOCIATION

| Nominal Size, in | American Standard Dressed Size, in | Area of Section, in² | Weight per Foot, lb | Moment of Inertia, in⁴ | Section Modulus, in³ |
|---|---|---|---|---|---|
| 2 × 4 | 1½ × 3½ | 5.25 | 1.46 | 5.36 | 3.06 |
| 6 | 5½ | 8.25 | 2.29 | 20.8 | 7.56 |
| 8 | 7¼ | 10.9 | 3.03 | 47.6 | 13.1 |
| 10 | 9¼ | 13.9 | 3.86 | 98.9 | 21.4 |
| 12 | 11¼ | 16.9 | 4.69 | 178 | 31.6 |
| 14 | 13¼ | 19.9 | 5.53 | 291 | 43.9 |
| 16 | 15¼ | 22.9 | 6.36 | 443 | 58.1 |
| 3 × 4 | 2½ × 3½ | 8.75 | 2.43 | 8.93 | 5.10 |
| 6 | 5½ | 13.8 | 3.83 | 34.7 | 12.6 |
| 8 | 7¼ | 18.1 | 5.03 | 79.4 | 21.9 |
| 10 | 9¼ | 23.1 | 6.42 | 165 | 35.7 |
| 12 | 11¼ | 28.1 | 7.81 | 297 | 52.7 |
| 14 | 13¼ | 33.1 | 9.19 | 485 | 73.1 |
| 16 | 15¼ | 38.1 | 10.6 | 739 | 96.9 |
| 4 × 4 | 3½ × 3½ | 12.2 | 3.39 | 12.5 | 7.15 |
| 6 | 5½ | 19.2 | 5.33 | 48.5 | 17.6 |
| 8 | 7¼ | 25.4 | 7.06 | 111 | 30.7 |
| 10 | 9¼ | 32.4 | 9.00 | 231 | 49.9 |
| 12 | 11¼ | 39.4 | 10.9 | 415 | 73.8 |
| 14 | 13¼ | 46.4 | 12.9 | 678 | 102 |
| 16 | 15¼ | 53.4 | 14.8 | 1 034 | 136 |

| Nominal Size, in | American Standard Dressed Size, in | Area of Section, in² | Weight per Foot, lb | Moment of Inertia, in⁴ | Section Modulus, in³ |
|---|---|---|---|---|---|
| 10 × 10 | 9½ × 9½ | 90.3 | 25.0 | 679 | 143 |
| 12 | 11½ | 109 | 30.3 | 1 204 | 209 |
| 14 | 13½ | 128 | 35.6 | 1 948 | 289 |
| 16 | 15½ | 147 | 40.9 | 2 948 | 380 |
| 18 | 17½ | 166 | 46.1 | 4 243 | 485 |
| 20 | 19½ | 185 | 51.4 | 5 870 | 602 |
| 22 | 21½ | 204 | 56.7 | 7 868 | 732 |
| 24 | 23½ | 223 | 62.0 | 10 274 | 874 |
| 12 × 12 | 11½ × 11½ | 132 | 36.7 | 1 458 | 253 |
| 14 | 13½ | 155 | 43.1 | 2 358 | 349 |
| 16 | 15½ | 178 | 49.5 | 3 569 | 460 |
| 18 | 17½ | 201 | 55.9 | 5 136 | 587 |
| 20 | 19½ | 224 | 62.3 | 7 106 | 729 |
| 22 | 21½ | 247 | 68.7 | 9 524 | 886 |
| 24 | 23½ | 270 | 75.0 | 12 437 | 1 058 |
| 14 × 14 | 13½ × 13½ | 182 | 50.6 | 2 768 | 410 |
| 16 | 15½ | 209 | 58.1 | 4 189 | 541 |
| 18 | 17½ | 236 | 65.6 | 6 029 | 689 |
| 20 | 19½ | 263 | 73.1 | 8 342 | 856 |
| 22 | 21½ | 290 | 80.6 | 11 181 | 1 040 |
| 24 | 23½ | 317 | 88.1 | 14 600 | 1 243 |

| Nominal | Dressed | | | | |
|---|---|---|---|---|---|
| 6 × 6 | 5½ × 5½ | 30.3 | 8.40 | 76.3 | 27.7 |
| 8 | 7½ | 41.3 | 11.4 | 193 | 51.6 |
| 10 | 9½ | 52.3 | 14.5 | 393 | 82.7 |
| 12 | 11½ | 63.3 | 17.5 | 697 | 121 |
| 14 | 13½ | 74.3 | 20.6 | 1 128 | 167 |
| 16 | 15½ | 85.3 | 23.6 | 1 707 | 220 |
| 18 | 17½ | 96.3 | 26.7 | 2 456 | 281 |
| 20 | 19½ | 107.3 | 29.8 | 3 398 | 349 |
| 8 × 8 | 7½ × 7½ | 56.3 | 15.6 | 264 | 70.3 |
| 10 | 9½ | 71.3 | 19.8 | 536 | 113 |
| 12 | 11½ | 86.3 | 23.9 | 951 | 165 |
| 14 | 13½ | 101.3 | 28.0 | 1 538 | 228 |
| 16 | 15½ | 116.3 | 32.0 | 2 327 | 300 |
| 18 | 17½ | 131.3 | 36.4 | 3 350 | 383 |
| 20 | 19½ | 146.3 | 40.6 | 4 634 | 475 |
| 22 | 21½ | 161.3 | 44.8 | 6 211 | 578 |

| Nominal | Dressed | | | | |
|---|---|---|---|---|---|
| 16 × 16 | 15½ × 15½ | 240 | 66.7 | 4 810 | 621 |
| 18 | 17½ | 271 | 75.3 | 6 923 | 791 |
| 20 | 19½ | 302 | 83.9 | 9 578 | 982 |
| 22 | 21½ | 333 | 92.5 | 12 837 | 1 194 |
| 24 | 23½ | 364 | 101 | 16 763 | 1 427 |
| 18 × 18 | 17½ × 17½ | 306 | 85.0 | 7 816 | 893 |
| 20 | 19½ | 341 | 94.8 | 10 813 | 1 109 |
| 22 | 21½ | 376 | 105 | 14 493 | 1 348 |
| 24 | 23½ | 411 | 114 | 18 926 | 1 611 |
| 26 | 25½ | 446 | 124 | 24 181 | 1 897 |
| 20 × 20 | 19½ × 19½ | 380 | 106 | 12 049 | 1 236 |
| 22 | 21½ | 419 | 116 | 16 150 | 1 502 |
| 24 | 23½ | 458 | 127 | 21 089 | 1 795 |
| 26 | 25½ | 497 | 138 | 26 945 | 2 113 |
| 28 | 27½ | 536 | 149 | 33 795 | 2 458 |
| 24 × 24 | 23½ × 23½ | 552 | 153 | 25 415 | 2 163 |
| 26 | 25½ | 599 | 166 | 32 472 | 2 547 |
| 28 | 27½ | 646 | 180 | 40 727 | 2 962 |
| 30 | 29½ | 693 | 193 | 50 275 | 3 408 |

* All properties and weights are given in dressed sizes only. The weights given are based on an assumed average weight of 40 lb/ft³.
Thickness up to 5 in is classified as lumber. Thickness of 5 in and above is classified as timber.
Lumber indicated is dry (19 percent moisture content or less). Timber indicated is green (more than 19 percent moisture content).

**TABLE 12** RECTANGULAR AND POLAR MOMENT OF INERTIA, SECTION MODULUS, AND RADIUS OF GYRATION OF SIMPLE AREAS

| | Rectangular | Polar |
|---|---|---|
| | $I_x = \dfrac{bh^3}{12}$ <br> $S_x = \dfrac{bh^2}{6}$ <br> $r_x = 0.289h$ | $J = \dfrac{bh(b^2 + h^2)}{12}$ |
| | $I_x = \dfrac{bh^3}{36}$ <br> $S_x = \dfrac{bh^2}{24}$ <br> $r_x = 0.236h$ | |
| | $I_x = \dfrac{\pi d^4}{64}$ <br> $S_x = \dfrac{\pi d^3}{32}$ <br> $r_x = \dfrac{d}{4}$ | $J = \dfrac{\pi d^4}{32}$ <br> $S' = \dfrac{\pi d^3}{16}$ |
| | $I_x = \dfrac{\pi(d_o{}^4 - d_i{}^4)}{64}$ <br> $S_x = \dfrac{\pi(d_o{}^4 - d_i{}^4)}{32d_o}$ <br> $r_x = \dfrac{\sqrt{d_o{}^2 + d_i{}^2}}{4}$ | $J = \dfrac{\pi(d_o{}^4 - d_i{}^4)}{32}$ <br> $S' = \dfrac{\pi(d_o{}^4 - d_i{}^4)}{16d_o}$ |
| | $I_x = 0.11r^4$ <br> $S_x = 0.191r^3$ <br> $r_x = 0.264r$ <br> $I_y = 0.393r^4$ <br> $S_y = 0.393r^3$ <br> $r_y = \dfrac{r}{2}$ | $J = 0.503r^4$ |

**TABLE 13**   LIST OF SELECTED SI METRIC UNITS

| Quantity | Unit Name | Unit Symbol |
|---|---|---|
| Length | meter | m |
| Mass | kilogram | kg |
|  | metric ton | t (= 1000 kg) |
| Time | second | s |
|  | hour | h |
| Temperature | kelvin* | K ($\simeq$ 273 + °C) |
| Plane angle | radian | rad (= $180/\pi \simeq 57.3°$) |
| Force | newton | N (= 1 kg·m/s$^2$) |
| Pressure, Stress | pascal | Pa (= 1 N/m$^2$) |
| Work, Energy | joule | J (= 1 N·m) |
| Power | watt | W (= 1 J/s) |

* A difference of 1 K is equal to a difference of 1°C.

**TABLE 14** CONVERSION FACTORS: U.S. CUSTOMARY UNITS TO SI UNITS

| Quantity | Multiply | By | To Get |
|---|---|---|---|
| Length | inch (in) | $2.54 \times 10^{-2}$ | meter (m) |
| | feet (ft) | $3.048 \times 10^{-1}$ | meter (m) |
| | mile (mi) | $1.609 \times 10^{3}$ | meter (m) |
| | yard (yd) | $9.144 \times 10^{-1}$ | meter (m) |
| Area | inch$^2$ (in$^2$) | $6.452 \times 10^{-4}$ | meter$^2$ (m$^2$) |
| | foot$^2$ (ft$^2$) | $9.290 \times 10^{-2}$ | meter$^2$ (m$^2$) |
| | yard$^2$ (yd$^2$) | $8.361 \times 10^{-1}$ | meter$^2$ (m$^2$) |
| Volume | inch$^3$ (in$^3$) | $1.639 \times 10^{-5}$ | meter$^3$ (m$^3$) |
| | foot$^3$ (ft$^3$) | $2.832 \times 10^{-2}$ | meter$^3$ (m$^3$) |
| | U.S. gallon (gal) | $3.785 \times 10^{-3}$ | meter$^3$ (m$^3$) |
| | U.S. gallon (gal) | 3.785 | liter (L or $l$) |
| Angle | degree (deg. or °) | $1.745 \times 10^{-2}$ | radian (rad) |
| Velocity | foot/minute (ft/min) | $5.080 \times 10^{-3}$ | meter/second (m/s) |
| | foot/second (ft/s) | $3.048 \times 10^{-1}$ | meter/second (m/s) |
| | mile/hour (mi/hr) | $4.470 \times 10^{-1}$ | meter/second (m/s) |
| | mile/hour (mi/hr) | 1.609 | kilometer/hour (km/h) |
| Acceleration | foot/second$^2$ (ft/s$^2$) | $3.048 \times 10^{-1}$ | meter/second$^2$ (m/s$^2$) |
| Density | pound-mass/foot$^3$ (lb/ft$^3$) | $1.602 \times 10^{1}$ | kilogram/meter$^3$ (kg/m$^3$) |
| Force | pound (lb) | 4.448 | newton (N) |

| | | | |
|---|---|---|---|
| Pressure, Stress | pound/inch² (lb/in²) | $6.895 \times 10^3$ | pascal (Pa) |
| | pound/foot² (lb/ft²) | $4.788 \times 10^1$ | pascal (Pa) |
| Moment, Torque | foot-pound (ft·lb) | $1.356$ | newton-meter (N·m) |
| | inch-pound (in·lb) | $1.130 \times 10^{-1}$ | newton-meter (N·m) |
| Energy, Work | foot-pound (ft·lb) | $1.356$ | joule (J) |
| Power | foot-pound/minute (ft·lb/min) | $2.260 \times 10^{-2}$ | watt (W) |
| | horsepower (hp) | $7.457 \times 10^2$ | watt (W) |
| Mass | slug (lb·s²/ft) | $1.459 \times 10^1$ | kilogram (kg) |
| | pound-mass (lbm) | $4.536 \times 10^{-1}$ | kilogram (kg) |

**TABLE 15** LIST OF FORMULAS

| | | |
|---|---|---|
| Pythagorean theorem | $c^2 = a^2 + b^2$ | (2-1) |
| Law of sines | $\dfrac{a}{\sin A} = \dfrac{b}{\sin B} = \dfrac{c}{\sin C}$ | (2-2) |
| Law of cosines | $c^2 = a^2 + b^2 - 2ab \cos C$ | (2-3) |
| Components of a force | $F_x = F \cos \alpha;\ F_y = F \sin \alpha$ | (2-4), (2-5) |
| Resultant of force components | $R = \sqrt{(\Sigma F_x)^2 + (\Sigma F_y)^2}$ | (2-6) |
| Slope of resultant | $\tan \theta = \dfrac{\Sigma F_y}{\Sigma F_x}$ | (2-7) |
| Equilibrium of concurrent forces | $\Sigma F_x = 0;\ \Sigma F_y = 0$ | (2-8) |
| Equilibrium of a force system | $\Sigma F = 0;\ \Sigma M = 0$ | (3-1), (3-2) |
| Resultant of noncoplanar force components | $R = \sqrt{(\Sigma F_x)^2 + (\Sigma F_y)^2 + (\Sigma F_z)^2}$ | (5-2) |
| Coefficient of friction | $f = \dfrac{F_m}{N}$ | (6-1) |
| Simple stress $\quad s = \dfrac{F}{A}\quad$ Strain | $\epsilon = \dfrac{\delta}{l}$ | (7-1), (8-1) |
| Modulus of elasticity | $E = \tan \theta = \dfrac{s}{\epsilon};\quad E = \dfrac{Fl}{A\delta}$ | (8-2), (8-3) |
| Percent reduction in area $= \dfrac{\text{original area} - \text{final area}}{\text{original area}}(100)$ | | (8-5) |
| Percent elongation $= \dfrac{\text{change in gage length}}{\text{original gage length}}(100)$ | | (8-6) |
| Factor of safety $\quad N_u = \dfrac{\text{ultimate stress}}{\text{allowable stress}};\ N_y = \dfrac{\text{yield stress}}{\text{allowable stress}}$ | | (8-7), (8-8) |
| Modulus of rigidity | $G = \dfrac{E}{2(1 + \mu)}$ | (8-9) |
| Thermal expansion | $\delta = \alpha l\ \Delta t$ | (8-10) |
| Thermal stress | $s = E\alpha\ \Delta t$ | (8-11) |
| Load on bolted and riveted joint | $F = \left[ n\left(\dfrac{\pi d^2}{4}\right) \right] s_s$ | (9-2a) |
| Load on bolted and riveted joint | $F = [(b - nD)t]s_t$ | (9-3) |
| Load on bolted and riveted joint | $F = [ntd]s_c$ | (9-4) |
| Efficiency of bolted joint $\quad \eta = \dfrac{\text{strength of the joint}}{\text{tensile strength of the gross area}} \times 100$ | | (9-5) |
| Allowable fillet-weld force | $F = F'(L)$ | (9-6) |
| Longitudinal section of pressure vessel | $s_t = \dfrac{pD_c}{2t};\ F = \dfrac{pD_c l}{2}$ | (9-7), (9-8) |
| Transverse section of pressure vessel | $s_t = \dfrac{pD_c}{4t};\ F = \dfrac{pD_c l}{4}$ | (9-9), (9-10) |
| Centroid location $\quad \bar{x} = \dfrac{\Sigma Ax}{\Sigma A};\ \bar{y} = \dfrac{\Sigma Ay}{\Sigma A}$ | | (10-2), (10-2) |

**TABLE 15**   LIST OF FORMULAS (Continued)

Formulas for centroid locations and moments of intertia of
  simple areas                                                    Tables 10-1, 10-2

Transfer formula      $I_{a-a} = I_x + Ad^2$                       (10-7)

Conditions for static equilibrium      $\Sigma F_x = 0; \ \Sigma F_y = 0; \ \Sigma M = 0$      (11-1)

Stress due to bending      $s = \dfrac{Mc}{I} ; \ s = \dfrac{M}{S}$      (12-1), (12-2)

Modulus of rupture      $s_r = \dfrac{M_r c}{I}$                    (12-3)

Horizontal shear stress      $s_h = \dfrac{Va\bar{y}}{Ib}$          (12-4)

Vertical web shear stress      $s_v = \dfrac{V}{td}$               (12-7)

Formulas for beam deflections                                     Table 12-2

Allowable stress, laterally unsupported beam      $s = \dfrac{12\,000\,000C_b}{ld/A_f}$      (12-20)

Torsional shear stress      $s_s = \dfrac{Tc}{J} ; \ s_s = \dfrac{T}{S'}$      (13-1), (13-2)

Angle of twist      $\theta = \dfrac{s_s l}{Gc} ; \ \theta = \dfrac{Tl}{JG}$      (13-5), (13-6)

Horsepower      $\text{hp} = \dfrac{Tn}{63\,000}$                  (13-7a)

Metric Power      $W = \dfrac{Tn}{9.55}$                          (13-7b)

Combined axial and bending stress      $s = \dfrac{F}{A} \pm \dfrac{Mc}{I}$      (14-1)

Combined stress, eccentrically loaded members      $s = \dfrac{F}{A} \pm \dfrac{Fec}{I}$      (14-2)

Maximum resultant tensile stress      $(s_t)_{max} = \dfrac{s_t}{2} + \sqrt{s_s^2 + \left(\dfrac{s_t}{2}\right)^2}$      (14-3)

Maximum resultant shear stress      $(s_s)_{max} = \sqrt{s_s^2 + \left(\dfrac{s_t}{2}\right)^2}$      (14-5)

Equivalent bending moment      $M_e = (s_t)_{max} S = \dfrac{M}{2} + \dfrac{1}{2}\sqrt{T^2 + M^2}$      (14-6)

Equivalent torque      $T_e = (s_s)_{max} S' = \sqrt{T^2 + M^2}$      (14-7)

Radius of gyration      $r = \sqrt{\dfrac{I}{A}}$      Euler's equation      $\dfrac{F}{A} = \dfrac{\pi^2 E}{(l/r)^2}$      (15-2), (15-4)

Design equations for metal columns                               Table 15-2

Design equation for wood columns (NFPA)

$\dfrac{F}{A} = \dfrac{0.30E}{\left(\dfrac{l}{b}\right)^2}$        $\dfrac{F}{A} \le s_{allowable}$        $\dfrac{l}{b} \le 50$      (15-8)

# ANSWERS TO SELECTED PROBLEMS

## CHAPTER 2

**2-1.** $R$ = 23.3 N; $\theta$ = 30.95° or 0.54 rad.    **2-2.** $R$ = 100 lb; $\theta$ = 60° with either force.    **2-3.** $R$ = 1.73$P$ lb; $\theta$ = 30° with either force.    **2-4.** $R$ = 87 lb; $\theta$ = 16.7° or 16°42' with the 40-lb force, or 13.3° (13°18') with the 50-lb force. **2-5.** $F$ = 42.7 lb; $\theta$ = 39° with the 60-lb force.    **2-7.** $m$ = 8.52 kg.    **2.9. AB** = 60 kN (tension); **BC** = 98 kN (compression).

**2-11.**

|     | $\theta°$ | $F_x$ (lb) | $F_y$ (lb) |
|-----|-----------|------------|------------|
| (a) | 10        | 197        | 35         |
| (b) | 35        | 164        | 115        |
| (c) | 55        | 115        | 164        |
| (d) | ,80       | 35         | 197        |
| (e) | 90        | 0          | 200        |

**2-13.** Any load greater than $W$ = 97 lb.    **2-15.** $F$ = 1050 lb.    **2-17.** Cable tensions = 1.37 tons at 30° and 1.065 tons at 40°.    **2-19. BC** = 288 lb (tension); **AB** = 577 lb (compression).    **2-21.** $F$ = 200 + 0.0995$W$ N.    **2-23.** Force in each arm = 100 lb (tension); $R_s = R_n$ = 173 lb.    **2-25.** $F$ = 2.67 kN.

## CHAPTER 3

**3-1.** $R_l$ = 350 lb; $R_r$ = 250 lb.    **3-2.** $R$ = 775 N; $F$ = 475 N.    **3-3.** $R_1$ = 8500 lb; $R_2$ = 13 500 lb.    **3-4.** $R_1$ = 2280 lb; $R_2$ = 2160 lb.    **3-5.** 58.3 in from rear axle.    **3-7.** $R_1$ = 5520 lb; $R_2$ = 7080 lb.    **3-9.** $R_1$ = 500 lb; $R_2$ = 100 lb; right support 5 ft from right end.    **3-11.** $F$ = 2795 lb.    **3-13.** $P$ = 4.69 kN. **3-15.** $F$ = 44 kN; $p$ = 318 kPa.    **3-17.** $M$ = 180 in·lb; $M$ = 180 in·lb.

## CHAPTER 4

**4-1.** $F$ = 17.33 lb between upper and lower cylinders; $F$ = 8.66 lb between lower cylinders and sides of box; $F$ = 45 lb between lower cylinders and bottom of the box.    **4-2.** $R_a$ = 1667 lb at 36.9° with horizontal; $R_b$ = 1333 lb horizontal to left. **4-3. BD** = 1.11 kN (compression); $R_c$ = 1.93 kN at 55° with horizontal. **4-4.** $R_{ax}$ = 190 lb; $R_{ay}$ = 289 lb; $R_{bx}$ = 190 lb; $R_{by}$ = 11 lb; $R_{cx}$ = −190 lb; $R_{cy}$ = 211 lb.    **4-5. BC** = 6.5 kN; $R_{ax}$ = 4.86 kN; $R_{ay}$ = 0.18 kN.    **4-7.** Upper hinge $R$ = 3.08 kN horizontal to left; lower hinge $R$ = 3.55 kN at 29.75° with horizontal.    **4-9.** $R_b$ = 2600 lb; $R_c$ = 2152 lb; $R_d$ = 2810 lb.    **4-11. CG** = 1794

lb; **DE** = 1115 lb; $R_{ax}$ = 1078 lb; $R_{ay}$ = 1889 lb; $R_{bx}$ = 1794 lb; $R_{by}$ = 1600 lb.
**4-13. AB** = 34 k (compression; **AH** = 16 k (tension); **BH** = 10 k (tension);
**BC** = 24 k (compression); **HG** = 16 k (tension); **BG** = 17 k (tension); **CG** = 10
k (compression); $R_1 = R_2$ = 30 k.     **4-15.** $P$ = 5550 lb (tension); $F_{bx}$ = 3330 lb;
$F_{by}$ = 8440 lb; **EF** = 4710 lb (tension); **BF** = 7770 lb (compression); **BE** = 950
lb (compression); **BC** = 2660 lb (compression); **CE** = 1330 lb (tension); **CD** =
2660 lb (compression); **DE** = 3760 lb (tension).     **4-17.** $R$ = 676 lb acting at
102.8° from horizontal.     **4-19.** Upper horizontal members (left to right): 26 kip
(compression); 46.22 kip (compression); 54.89 kip (compression); 46.22 kip
(compression); 26 kip (compression). Lower horizontal members (left to right):
13 kip (tension); 36.11 kip (tension); 50.55 kip (tension); 50.55 kip (tension);
36.11 kip (tension); 13 kip (tension). Slanted members (left to right): 26 kip
(compression); 26 kip (tension); 20.22 kip (compression); 20.22 kip (tension);
8.67 kip (compression); 8.67 kip (tension); 8.67 kip (tension); 8.67 kip (compression); 20.22 kip (tension); 20.22 kip (compression); 26 kip (tension); 26 kip
(compression).

## CHAPTER 5

**5-1 AB** = 2.0 tons (compression); **BC** = **BD** = 1.14 tons (compression).     **5-2.**
**DE** = 50.6 kN (tension); **AD** = 14.6 kN (compression); **CD** = 19.9 kN
(compression).     **5-3. BD** = 7500 lb (tension).     **5-4. BD** = 10 600 lb (tension).
**5-5.** $P$ = 12.5 tons (tension) in hoisting cable and in each pulley cable; $F$ = 48
tons (compression) in each leg of crane; $F$ = 37.1 tons (tension) in stay cable.
**5-7. CD** = 2415 lb.     **5-8. DF** = 648 lb (compression).     **5-9.** (a) **DE** = 15.88
tons (tension) (b) **DF** = 21.12 tons (compression); (c) **AD** = 11.68 tons (tension);
**BD** = 15 tons (tension).     **5-11. BC** = 16 450 lb (tension).

## CHAPTER 6

**6-1.** $P$ = 8.55 lb.     **6-2.** Body will slide without additional force.     **6-7.** $f$ =
0.118.     **6-9.** No. Total $F_m$ = 123.6 lb; component of 60-lb body parallel to
plane = 42.42 lb; tension in cord = 20.82 lb; $F$ = 166 lb to start 200-lb body to
the right.     **6-11.** $F$ = 1064 lb; $P$ = 1684 lb.     **6-13.** No. At base of ladder, $F_m$
= 35.4 N, but horizontal force = 66 N.     **6-15.** Yes. Minimum $F$ required = 63.2
lb.

## CHAPTER 7

**7-1.** $s_t$ = 4000 psi.     **7-2.** $s_c$ = 4.6 MPa.     **7-3.** $F$ = 3890 lb.     **7-4.** $s_c$ = 4.17
MPa.     **7-5.** $d$ = 112 mm.     **7-7.** $h$ = 0.358 in (using root area in tension);
$h$ = 0.50 in (using shank area in tension).     **7-9.** $F$ = 666 kN.     **7-11.** Use 12
posts.     **7-13.** $s_t$ = 47 300 psi (root area).     **7-15.** (a) $d_o$ = 66 mm, $d_i$ = 44 mm;
(b) $d_o$ = 107 mm; $d_i$ = 95 mm.

## CHAPTER 8

**8-1.** $s_t$ = 15 300 psi; $\delta$ = 0.0734 in.     **8-2.** $F$ = 238 kN.     **8-3.** $\epsilon$ = 0.001088
m/m; final length = 993.08 mm.     **8-4.** Original length = 84 in.     **8-5.** $s_t$ = 280
MPa, $\epsilon$ = 1.40 × 10$^{-3}$ m/m; $s_t$ = 297 MPa, $\epsilon$ = 1.48 × 10$^{-3}$ m/m; $s_t$ = 382 MPa

(above yield point), $\epsilon = 1.91 \times 10^{-3}$ m/m (not true strain).   **8-9.** $\delta = 0.00496$ in.   **8-11.** $E = 134$ GPa.   **8-13.** $N_u = 6.83$.   **8-15.** (a) $d = 1.09$ in; (b) $d = 1.09$ in.   **8-17.** Use 10 in by 10 in.   **8-19.** 3.24 mm; 65.5 MPa (compression). **8-21.** 0.0825 in, 330 psi.   **8-23.** (a) $\delta = 1.94$ mm; $s = 0$; (b) $\delta = 0$; $s = 108$ MPa (compression); (c) $F = 136$ kN.   **8-25.** $F = 158\,000$ lb.   **8-27.** (a) $s_c = 11\,800$ psi; (b) $s_c = 13\,300$ psi; (c) $s_c = 15\,200$ psi.

# CHAPTER 9

**9-1.** $F = 26\,500$ lb (shear); ($F = 34\,400$ lb, tension; $F = 36\,400$ lb, bearing). **9-2.** $F = 14\,200$ lb (shear); ($F = 19\,500$ lb, tension; $F = 18\,800$ lb, bearing). **9-3.** Use 3 bolts; ($n = 2.2$, shear; $n = 1.98$, bearing).   **9-4.** $d = 0.56$ in; use two $\frac{9}{16}$ in bolts.   **9-5.** $F = 106\,000$ lb (shear); ($F = 180\,000$ lb tension).   **9-7.** Member $A$: The bolting is safe; ($s_s = 8300$ psi, $s_c = 16\,300$ psi, $s_t = 18\,700$ psi in angles at hole). Member $B$: $F = 67\,500$ lb (double shear); ($F = 198\,000$ lb, compression angles; $F = 166\,700$ lb, bearing).   **9-9.** Member $A$: Use 2 bolts; ($n = 2.79$, double shear; $n = 1.58$, bearing in gusset plate). Member $B$: Use 2 bolts; ($n = 2.17$, double shear; $n = 1.23$, bearing in gusset plate; tensile stress in angles at hole = 9300 psi).   **9-11.** Joint strength $F = 60\,600$ lb (bearing); efficiency = 64%; ($F = 115\,600$ lb, double shear; $F = 85\,000$ lb, row 1 tension; $F = 88\,200$ lb, row 2 tension).   **9-13.** Joint strength $F = 66\,300$ lb (bearing); efficiency = 70%; ($F = 95\,200$ lb, double shear; $F = 72\,800$ lb (row 1 tension); $F = 78\,100$ lb, row 2 tension; $F = 96\,500$ lb, row 3 tension).   **9-15.** (a) Each side weld = 1.43 in + $\frac{7}{8}$ in corner returns; (b) Yes, required $L = 2.86$ in + $\frac{7}{8}$ in corner returns. **9-17.** (a) $F = 24\,000$ lb (weld); ($F = 27\,500$ lb, plate); (b) $F = 27\,500$ lb (using plate allowable stress for butt weld).   **9-19.** (a) Overlap = 4.3 in.; (b) $F = 48\,100$ lb (plate, AISC); ($F = 141\,000$ lb, weld).   **9-21.** (a) $s_t = 7840$ psi (tension, longitudinal section); (b) No, allowable $s_t = 5720$ psi for steady load.   **9-23.** $p = 1833$ psi (maximum).   **9-25.** $t = 0.273$ in (minimum).   **9-27.** Spacing $l = 5.3$ in (maximum).   **9-29.** $p = 189$ psi (maximum); (joint strength $F = 5680$ lb, bearing, for $l = 2\frac{1}{2}$ in).   **9-31.** $D = 41.8$ ft (maximum); (joint strength $F = 18\,800$ lb, tension, for $l = 3$ in).   **9-33.** $h = 34.5$ m (maximum).   **9-35.** (a) $t = \frac{3}{8}$ in (minimum); (single bead of weld); (b) $t = 0.205$ in (minimum).   **9-37.** (a) $h = 115$ ft; (joint strength $F = 16\,200$ lb, shear, for $l = 3$ in); (b) $h = 154$ ft; (c) $h = 168$ ft.

# CHAPTER 10

**10-1.** $\bar{y} = 1.16$ in, $\bar{x} = 2.5$ in (by inspection) from point $O$.   **10-2.** $\bar{y} = 7.21$ in, $\bar{x} = 2.82$ in from lower right corner.   **10-3.** $\bar{y} = 131$ mm, $\bar{x} = 76.2$ mm from lower left corner.   **10-4.** $\bar{y} = 5.29$ in from base.   **10-5.** $\bar{y} = 7.12$ ft, $\bar{x} = 7.33$ ft from point $D$.   **10-7.** $\bar{y} = 0.58$ in, $\bar{x} = 4$ in (by inspection) from lower left corner.   **10-9.** $\bar{y} = 66$ mm, $\bar{x} = 207$ mm from lower left corner.   **10-11.** $\bar{y} = 10.58$ in from base on the vertical centerline.   **10-13.** Maximum $b = 3.72$ m, ($a = 3.04$ m).   **10-15.** Volume = 589 ($10^3$) mm³. **10-17.** Volume = 6140 $\pi$ in³ or 19 300 in³.   **10-19.** (a) $I_x = 47.6$ in⁴, $I_y = 24.3$ in⁴; (b) $I_x = I_y = 7.36$ in⁴; (c) $I_x = 1.03$ in⁴, $I_y = 3.68$ in⁴; (d) $I_x = I_y = 0.515$ in⁴.   **10-21.** $I_x = 8.23$ in⁴, $I_y = 7.92$ in⁴.   **10-23.** $I_x = 1126$ in⁴, $I_y = 1308$ in⁴.   **10-25.** $I_x = 494$ in⁴, $I_y = 186$ in⁴.   **10-27.** $I_x = 1864$ in⁴.   **10-29.** $I_x = 6010$ in⁴ (bolt holes deducted); $I_x = 6690$ in⁴, (bolt holes not deducted).

## CHAPTER 11

**11-1.** $R_l = 1200$ lb, $R_r = 800$ lb; $V_l = 1200$ lb $= V_4$, $V'_4 = -800$ lb $= V_r$.    **11-2.** $R_1 = 12$ kN, $R_2 = 10$ kN; $V_l = 12$ kN $= V_1$, $V_1' = -1$ kN $= V_3$, $V_3' = -10$ kN $= V_r$.    **11-3.** $R_l = 8.25$ kN, $R_r = 14.25$ kN; $V_l = 8.25$ kN, $V_{3.5} = 3$ kN, $V_{3.5}' = -12$ kN, $V_r = -14.25$ kN.    **11-4.** $R_1 = 2040$ lb, $R_2 = 3760$ lb; $V_l = 2040$ lb, $V_{12} = 240$ lb $= V_{18}$; $V_{18}' = -3760$ lb $= V_r$.    **11-5.** $R = 18$ kN; $V_l = -8$ kN $= V_2$, $V_2' = -18$ kN $= V_r$.    **11-7.** $R_1 = 1735$ lb, $R_2 = 1640$ lb; $V_l = 1735$ lb, $V_5 = 610$ lb, $V_9 = 110$ lb, $V_{9.4} = 0$, $V_{14} = -1265$ lb, $V_r = -1640$ lb.    **11-9.** $R_1 = 10.2$ kN, $R_2 = 16.8$ kN; $V_l = 0$, $V_1 = -1.4$ kN, $V_1' = 8.8$ kN, $V_2 = 7.4$ kN $= V_3$, $V_3' = -8.6$ kN $= V_{3.5}$, $V_{4.5} = -9.8$ kN $= V_5$, $V_5' = 7$ kN $= V_r$.    **11-11.** $M_{max} = 4800$ ft·lb 4 ft from left support.    **11-13.** $M_{max} = 19.69$ kN·m, 1.5 m from right support. **11-15.** $M_{max} = -43$ kN·m at wall; ($M_2 = -16$ kN·m).    **11-17.** $M_{max} = 7320$ ft·lb 9.4 ft from left support; ($M_5 = 5870$ ft·lb, $M_9 = 7290$ ft·lb, $M_{14} = 4360$ ft·lb).    **11-19.** $M_{max} = 14.85$ kN·m, 3 m from left end of beam; ($M_1 = -0.7$ kN·m, $M_2 = 7.42$ kN·m, $M_{3.5} = 10.56$ kN·m, $M_{4.5} = 1.39$ kN·m, $M_5 = -3.5$ kN·m).    **11-21.** (a) $R_1 = (P/L)(L - a)$, $R_2 = Pa/L$; (b) $V_{max} = (P/L)(L - a)$ at all sections from $R_1$ to $P$, if $a < L/2$; (c) $M_{max} = Pa(L - a)/L$ located at $P$; (d) $M_x = Px(L - a)/L$.    **11-23.** $R = 14\ 700$ lb, $M = -138\ 850$ ft·lb (at wall). **11-25.** $M_{max} = 735$ kN·m, occurs under rear axle (153 kN load) when it is 6.4 m from left support and front axle is 1.4 m from right support. ($M_{max} = 654$ kN·m for 81 kN at 8.5 m from left support; $M_{max} = 451$ kN·m for 36-kN load at 10 m from left support.)

## CHAPTER 12

**12-1.** $s_{0.9} = 16.2$ MPa; $s_{1.5} = 27$ MPa; $s_{2.4} = 10.8$ MPa.    **12-2.** $W = 3910$ lb (maximum).    **12-3.** $L = 10.8$ ft (maximum).    **12-4.** $d_o = 140.8$ mm, $d_i = 70.4$ mm (minimum).    **12-5.** $N_u = 144$ (top fiber compression), $N_u = 75.4$ (bottom fiber tension); ($\bar{y} = 3.22$ in from base, $I_x = 155$ in$^4$, $M_{max} = 3192$ ft·lb 3 ft from $R_1$).    **12-7.** $L = 31.1$ ft (maximum, neglecting beam weight); ($L = 29.3$ ft, including beam weight).    **12-9.** S8 × 18.4 (required $S = 14.4$ in$^3$, neglecting beam weight; S8 × 23 (required $S = 14.8$ in$^3$, including beam weight).    **12-11.** $w = 267$ psf of floor area (maximum floor load in addition to beam weight); ($w = 271$ psf with beam weight neglected).    **12-13.** $d = 3.57$ in (minimum; shaft weight included); ($d = 3.45$ in, neglecting shaft weight).    **12-15.** $d = 2.47$ in (minimum, based on ultimate stress with $N_u = 6$); ($d = 2.22$ in, based on yield stress with $N_y = 3$).    **12-17.** Yes, Eastern white pine is satisfactory. ($s_b = 0.95$ MPa with beam weight).    **12-19.** Use dressed 6 in × 14 in (required $S = 150$ in$^3$ for bending, required $A = 68$ in$^2$ for shear; beam weight neglected).    **12-21.** $F = 5760$ lb (maximum, for nominal size); $F = 4850$ lb (for dressed size). **12-23.** (a) $W = 16\ 800$ lb (maximum, for nominal size); $W = 15\ 300$ lb (for dressed size); (b) $s_b = 616$ psi (nominal size); $s_b = 637$ psi, (dressed size). **12-25.** Bolt spacing $= 11.3$ in center to center, two rows (maximum; beam weight neglected).    **12-27.** $s_v = 3300$ psi (maximum web shear at supports).    **12-29.** $s_h = 11\ 800$ psi at base of flange at supports for $W = 130\ 000$ lb; ($s_v = 12\ 300$ psi maximum web shear).    **12-31.** $W = 8360$ lb (maximum, for dressed size; shear controls); ($W = 9600$ lb for nominal size).    **12-33.** $s_b = 16\ 150$ psi at midspan, $s_v = 4130$ psi at supports, $y = 0.751$ in at midspan.    **12-35.** (a) No, $y = 0.539$ in (beam weight neglected); ($y = 0.561$ in with beam weight); (b) $s_b = 14\ 500$ psi,

$s_v$ = 1715 psi (beam weight neglected); ($s_b$ = 15 100 psi, $s_v$ = 1765 psi with beam weight).    **12-37.** (a) y = 0.0923 in (including shaft weight); (b) $s_h$ = 59.8 psi at supports, $s_b$ = 4360 psi at midspan. **12-39.** S15 × 42.9 (required S = 50 in³; beam weight neglected).

# CHAPTER 13

**13-1.** $T$ = 7.95 kN·m (maximum).    **13-2.** No; allowable $s_s$ = 18 350 psi, actual $s_s$ = 29 400 psi.    **13-3.** (a) d = 48.9 mm (minimum); (b) $d_o$ = 50 mm, $d_i$ = 25 mm.    **13-4.** L = 411 in or 34.2 ft (maximum). **13-5.** θ = 0.043 rad.    **13-7.** d = 41.3 mm (minimum, based on ultimate with $N_u$ = 10).    **13-9.** n = 223 rpm. **13-11.** L = 53.3 mm (bearing); (L = 32 mm, shear).    **13-13.** $N_s$ = 1.79 (shear), $N_c$ = 1.19 (bearing).    **13-15.** $s_s$ = 38 MPa (shear), $s_c$ = 24 MPa (bearing). **13-17.** d = 38.4 mm (based on angle of twist); d = 23.1 mm (based on shear stress).

# CHAPTER 14

**14-1.** Bottom fiber maximum tension = 21.9 MPa, top fiber maximum compression = 15.9 MPa (both at midspan).    **14-2.** h = 10.24 in (minimum depth). **14-3.** Right-side maximum tension = 293 MPa (at midspan).    **14-4.** F = 97 500 lb.    **14-5.** Right-side maximum compression = 4680 psi; left-side maximum tension = 2100 psi.

**14-7.**

BASE PRESSURES (psf)

| | Footing weight | 120-k load | 85-k load | Total |
|---|---|---|---|---|
| Left-front corner | 300 | 16 670 | −14 180 | 2 790 |
| Left-rear corner | 300 | 16 670 | 4 720 | 21 690 |
| Right-rear corner | 300 | −3 330 | 23 620 | 20 590 |
| Right-front corner | 300 | −3 330 | 4 720 | 1 690 |

(Negative values are tension)

**14-9.** F = 1940 lb (maximum, based on shock loading; tension controls).

**14-11.**

STRESSES (MPa)

| | At A | At B |
|---|---|---|
| Due to 4.5-kN load | 30 | 30 |
| Due to 450-N load | 36 | −36 |
| Due to 600-N load | 21.8 | −13.8 |
| Total | 87.8 | −19.8 |

(Negative values are compression)

**14-13.** $N_u = 1.43$.    **14-15.** $s_s = 13\ 580$ psi (no eccentricity).    **14-17.** $d = 1.03$ in, use $1\frac{1}{8}$-in. bolts.    **14-19.** Maximum $s_s = 2980$ psi on lower right-hand bolt; (lower left $= 2760$ psi; center right $= 2280$ psi; center left $= 1990$ psi; upper right $= 1680$ psi; upper left $= 1260$ psi).    **14-21.** (a) $F = 7350$ lb (bolt $A$ controls); (b) $F = 8830$ lb (no eccentricity).    **14-23.** hp $= 68.5$    **14-25.** $d = 2.97$ in (minimum; from maximum principal stress); ($d = 2.96$ in, minimum; from maximum shear stress).    **14-27.** (a) $d = 75.8/\sqrt[3]{(s_s)_{\max}}$; (b) $d = 90.6/\sqrt[3]{(s_t)_{\max}}$.    **14-29.** $d_o = 90$ mm (minimum), $d_i = 45$ mm (from maximum principal stress).

## CHAPTER 15

**15-1.** $F = 103\ 600$ lb.    **15-2.** $F = 203\ 000$ lb.    **15-3.** $F = 69\ 000$ lb.    **15-4.** Maximum safe load $F = 15.26$ kN (intermediate column; machine-design formula). The member is satisfactory. **15-5.** $L = 13.95$ ft or $167.5$ in.    **15-7.** $F = 695\ 000$ lb ($I_y = 481.5$ in$^4$).    **15-9.** $F = 51\ 000$ lb.    **15-11.** Two $3\frac{1}{2}'' \times 3\frac{1}{2}'' \times \frac{1}{4}''$ angles (can carry 34 200 lb).    **15-13.** S $10 \times 25.4$ (can carry 49 000 lb). **15-15.** $0.515 \times 1.03$ in (slender column; machine-design formula). **15-17.** The column is safe. Maximum safe load $F = 24\ 900$ lb (dressed size).    **15-19.** 11.2 in $\times$ 11.2 in. Use $12 \times 12$ (can carry 109,000 lb).    **15-21.** For left column, use 6 in $\times$ 6 in, (or 6 in diam.); for right column, use 6 in $\times$ 6 in, or 4 in $\times$ 8 in (or 4.7 in diam.).

## CHAPTER 16

**16-1.** Force in rod $F = 4500$ lb; rod diameter $d = 0.594$ in for AISI 1020 with steady load (use $\frac{5}{8}$-in-diameter rod). Maximum bending stress $s_b = 12\ 300$ psi at wall.    **16-2.** Support reaction $R_1 = 2170$ lb; maximum bending stress $s_b = 868$ psi at wall (for dressed size); ($R_1 = 2100$ lb; $s_b = 810$ psi for nominal size). If support does not settle, $R_1 = 2550$ lb; $s_b = 600$ psi at wall (for dressed size). **16-3.** $s_b = 692$ psi at wall; $s_h = 96$ psi at wall. ($M_{\max} = 8300$ ft·lb; $V_{\max} = 4610$ lb).    **16-4.** $s_b = 1435$ psi at wall; $s_h = 117$ psi at wall. ($M_{\max} = 17\ 200$ ft·lb; $V_{\max} = 5600$ lb).    **16-5.** $s_b = 720$ psi at section 4.58 ft from support; $s_h = 78.3$ psi at support. ($M_{\max} = 8640$ ft·lb; $V_{\max} = 3760$ lb).    **16-7.** $s_b = 9150$ psi (compression) at bottom fiber at wall; $y_{\max} = 0.0307$ in at midspan.

# Index

Allowable stress, 155
Alloy, definition, 152
American Institute of Steel Construction, allowable stresses for beams, 287
   code for rivets, bolts and threaded parts, 175, 176
   code for welds, 188
   column formulas, 380
   steel shapes, tables, 412–424
   stress in laterally unsupported beams, 308
American Society of Mechanical Engineers, code for bolted connections, 177
American Society for Testing Materials, tension test, 146–151
American standard, channels, table, 418, 419
   lumber and timber sizes, table, 426, 427
   S shapes, table, 416, 417
American Welding Society, fillet and butt welds, 187
Angle of twist in shafts, 322, 323
Angles, equal legs, table, 420, 421
   unequal legs, table, 422, 423
Area, definition, 1
   determination by centroid methods, 217
   percent reduction of, 153

Beams, AISC allowable stresses, 287
   assumptions and limitations, 232
   bending-moment diagram, 244–249
Beams, bending stress, 269–271
   applied couple, 255–259
   bending theory, 233–236
   cantilever, 232
   concentrated load, 232–235
   conditions for static equilibrium, 232, 407

Beams (continued):
   deflection, 288–306
     tables, of formulas, 304
   design of, 306
   diagram relationships, 249, 250
   horizontal and vertical shear stresses, 278–287
   lateral buckling, 308, 309
   moments from shear diagram area, 252–254
   moving loads, 259–265
   neutral axis, 233, 234
   overhanging, 232, 233
   reactions, 235, 236
     by lever rule, 237, 238
   section modulus table, 424, 425
   selection for economy, 278
   shear-force diagram, 239–244
   simply supported, 232
   statically determinate, 232
   statically indeterminate, 232, 389
     both ends fixed, 396–401
     one end fixed and one end supported, 389–396
   types, 232, 233
   uniformly distributed load, 232, 235, 236
   vertical web shear, 287
Bearing stress in bolted joint, 173, 174
Bending, combined with axial stress, 340–345
   combined with torsion, 359–363
Bending moment, diagram, 244–249
   equivalent, combined bending and torsion, 360
   shear-diagram area method, 252–254
Bolted joints, AISC code, 175, 176
   ASME code, 177
   design for maximum efficiency, 182–185
   eccentrically loaded, 352–357
   edge distance, 175
   efficiency, 179–182